Forests, Carbon Cycle and Climate Change

Edited by
Denis Loustau

Éditions Quæ
c/o Inra, RD 10, F-78026 Versailles Cedex

Collection *Update Sciences & Technologies*

La démarche qualité dans la recherche publique et l'enseignement supérieur
Claude Granier, Léandre-Yves Mas, Luc Finot, Bernard Arnoux, Nathalie Pasqualini,
Vincent Dollé, coordinateurs
2009, 376 p.

Homme et animal, la question des frontières
Valérie Camos, Frank Cézilly, Pierre Guenancia et Jean-Pierre Sylvestre, coordinateurs
2009, 216 p.

Le golfe du Lion. Un observatoire de l'environnement en Méditerranée
André Monaco, Wolfgang Ludwig, Mireille Provansal, Bernard Picon, coordinateurs
2009, 344 p.

Politiques agricoles et territoires
Francis Aubert, Vincent Piveteau, Bertrand Schmitt, coordinateurs
2009, 224 p.

La mise à l'épreuve. Le transfert des connaissances scientifiques en questions
Christophe Albaladejo, Philippe Geslin, Danièle Magda, Pascal Salembier, coordinateurs
2009, 280 p.

Contaminations métalliques des agrosystèmes et écosystèmes péri-industriels
Philippe Cambier, Christian Schvartz, Folkert van Oort, coordinateurs
2009, 308 p.

Conceptual basis, formalisations and parameterization of the STICS crop model
Nadine Brisson, Marie Launay, Bruno Mary, Nicolas Beaudoin, editors
2008, 304 p.

Landscape: from knowledge to action
Martine Berlan-Darqué, Yves Luginbühl, Daniel Terrasson
2008, 312 p.

Multifractal analysis in hydrology. Application to time series
Pietro Bernardara, Michel Lang, Éric Sauquet, Daniel Schertzer, Ioulia Tchiriguyskaia
2008, 58 p.

© Éditions Quæ, 2010   ISBN : 978-2-7592-0384-0   ISSN : 1773-7923

Foreword

Forest carbon cycle, greenhouse effect and climate change

Forests feature commonly in fairy tales where they serve as a metaphor for wilderness and for that reason they are attractive but also a little frightening, at least for children. They also serve human needs for food and shelter, providing wood for housing and heating, berries, mushrooms and game. Forests cover a little less than 30% of the world ice-free continental area, and contain many species of trees. Tree size varies between 2 and 120 m, and trees may live from 20 to 5000 years. Such a long life makes us feel that forests are permanent features of our landscape. Yet they change with harvests, fires, pest attacks and climatic stresses. The present book deals with the carbon balance of forests, a subject that has recently become popular for several reasons. Firstly, tropical forests are disappearing at an alarming rate, although they are believed to contain over half of the living species on Earth. Secondly, forests regulate the water cycle: during drought periods, deep roots allow trees to access deep soil water and to maintain transpiration long after grasses have dried out. Thirdly, trees have a long life during which they store carbon in wood and soil. The total amount of carbon stored in tree wood is about half of that stored in the atmosphere as CO_2. Thus small variations in tree growth or in forest area have important consequences on the global carbon cycle, a major driver of climate change.

Atmospheric CO_2 has increased from about 280 ppmv in 1750 to 385 ppmv in 2008. CO_2 is a greenhouse gas: it has absorption bands in the infrared spectrum emitted by the Earth's surface, and these bands have widened with the increase in CO_2 concentration. The has resulted in an increase in the amount of infrared radiation absorbed and emitted by the atmosphere, leading to a warming of the terrestrial surface by what is called the greenhouse effect. Global warming was 0.6°C during the 20th century. It is predicted to range between 1.8°C and 4°C (with extremes from 1.1 to 6.4°C) in the 21st century, depending on emission scenarios and on climate models (IPCC, 2007). Other greenhouse gases participate in the warming but CO_2 presently accounts for over half of the total. Warming is more pronounced than average at high northern latitudes. Higher temperatures induce higher evaporation rates over the oceans and thus an acceleration of the water cycle. However, the increase in precipitation is unevenly distributed: some dry regions will become drier, while wet ones will experience more frequent flooding. Moreover, warming will lead to an increase in sea level of 0.2 to 0.6 m in 2100 (i.e. thermal expansion of seawater and decrease in continental ice), and likely much more in the following centuries threatening coastal areas.

Consequences of such a strong climate change may be favourable for northern countries such as Canada or Russia, but may be devastating for regions that either lack water (i.e. Mediterranean countries) or are submitted to flooding (i.e. Bangladesh and Indonesia). Thus a consensus has emerged suggesting that it would be wise to keep global warming below a limit of 2°C, and thus keep greenhouse gases concentration below 550 ppmv of CO_2 equivalent (Stern, 2006). Since this concentration was 455 ppmv in 2005, this objective implies a strong reduction in present world emissions by 2050 (roughly by 2, and to be fair, by 4 for the emissions of industrialized countries). This will be very difficult to fulfil considering the rapid rise in CO_2 emissions in developing countries such as China and India, and the slowness in reducing emissions in developed countries. The Kyoto Protocol asked for a modest but significant decrease in greenhouse gas emissions of industrialized countries (5.2% in average between 1990 and 2010). It was proposed in 1997, but was only ratified in 2005 (without the USA) and even this modest effort may not be achieved. Clearly stronger changes are needed, the stakes are high, and the cost of preventive measurements is significantly lower than that of repairing damage caused by climatic change after it has occurred (Stern, 2006).

Yet the increase in atmospheric CO_2 only represents less than half of the emissions from fossil fuels and deforestation, the remainder being absorbed by oceans and the terrestrial biosphere. Forests play a special role in this respect. On one hand, tropical deforestation releases about 20% of the total CO_2 emissions, on the other hand, the remaining forests seem to grow faster and thus accumulate carbon. The net effect is a global terrestrial carbon sink whose intensity is quite variable from year to year, depending on climate. This terrestrial sink is presently known only by the difference between the emissions and the atmospheric and oceanic sinks. Why is there such a terrestrial sink? How is it distributed in space and time? What is the role of forests in this sink? How does the carbon sink vary with local climate, tree species and soil type? Models of terrestrial biosphere have attempted to answer these questions, but they still require careful validation, especially for long-life species such as trees that may suffer from climatic stress several years after it has occurred.

Thus a regional study on French forests is welcome for several reasons. Firstly, France has a temperate climate with an Atlantic influence in the west and a Mediterranean one in the south; warming causes a northern movement of the Mediterranean influence, which is already discernable, making the territory a sensible location. Secondly, forest inventory is well developed in France with over 100,000 plots sampled every decade, and this has been used to study species distribution of forest trees, carbon stocks and variations. Thirdly, there are several intensive flux measurement sites for both deciduous and coniferous species, some being operational for over a decade. Most sites also support process studies on trees and soil, suitable for model parameterization.

What do we want to know about the carbon cycle of temperate forests? We wish: (i) to quantify their role as a carbon sink; (ii) to understand the factors acting on this sink; and (iii) to incorporate our knowledge into models predicting the evolution of this sink in the coming decades. We also wish to know the impact of climate change on wood production for the main forest trees in order to adapt forest management to these changes. Global warming is likely to increase the frequency of climatic stresses such as heat waves or long dry periods. What will be the effect of such stresses on trees in terms of forest fires or pest attacks? These questions are addressed here.

The present book is organized into three parts: the first one presents the actual forest carbon cycle in temperate and Mediterranean climates, including the dynamics of soil carbon and the total carbon stock of French forests based on forest inventories. The second part uses models to simulate the effects of climate change on tree phenology and forest carbon balance. The third part deals with the impact of climate change on forest vulnerability: change in geographical distribution of forest tree species and pathogenic fungi, and the consequences on forest fires and pest attacks.

Professor Bernard Saugier
Member of Académie d'Agriculture de France

References

IPCC, 2007. Summary for Policymakers. *In: Climate Change 2007: The Physical Science Basis. Contribution of Working Group I to the Fourth Assessment Report of the Intergovernmental Panel on Climate Change* (S. Solomon, D. Qin, M. Manning, Z. Chen, M. Marquis, K.B. Averyt, M. Tignor, H.L. Miller, eds). Cambridge University Press, Cambridge. [http://www.ipcc.ch/pdf/assessment-report/ar4/syr/ar4_syr_spm.pdf]

Stern N., 2006. *The Economics of Climate Change: The Stern Review.* [http://www.hm-treasury.gov.uk/stern_review_report.htm]

Acknowledgements

The results presented in this book originate from the project CARBOFOR (2002–05) funded by the French Ministère de l'Écologie et du Développement Durable (MEDD) and Ministère de l'Agriculture et de la Pêche (MAP) within the framework of the governmental programme called "Management and Impacts of Climate Change" (GICC). The project involved 14 participants belonging to INRA, CNRS, IFN, CEA, CIRAD, CNRM/Météo-France and the Universities of Paris-XI-Orsay and Orléans. We are grateful to M.M. Maurice Muller (MEDD, GICC), Claude Millier (Agro ParisTech, GICC), Guy Landmann (MAP, GIP Ecofor) and Professor Bernard Saugier (University of Paris XI - Orsay) who supervised the implementation of the project. The authors would like to thank all partners and collaborators who contributed to the implementation and performance of this research work. I am pleased to acknowledge Professor G.M.J. Mohren (University of Wageningen) and Professor J. Grace (University of Edinburgh) who coordinated the pioneer FERN, LTEEF-1 and LTEEF-2 European projects from which this research was inspired.

Beyond the project CARBOFOR itself, the edition and publication of its results constituted an additional challenge that could not be met without the support of many organizations and people: the Ministry of Ecology, ECOFOR (Jean-Luc Peyron and Guy Landmann) and INRA (mission MICCES, Bernard Seguin) who maintained an interest in the project and provided funds for this publication. Marie-Pierre Reviron, Anne-Marie Bouchon, Stéphanie Hayes and QUAE editors, Anne-Lise Prodel, Valerie Howe and graphic assistant, Joëlle Delbrayère played an important role in the shaping of this book. All authors, reviewers and assistant editors must also be thanked for the efforts they have made to write this book in English and present their results as clearly as possible.

Denis Loustau

Contents

Part II
Forest under a changing climate

Part III
Climate change and forest vulnerability

Executive summary

The French metropolitan area occupies a unique ecological place in Europe in the context of climate change being at the confluence of four climate zones – oceanic, continental, Mediterranean and mountain. Around 28% is covered by forests managed by public and private landowners. French forests offer, therefore, a great diversity of silvicultural types. The French metropolitan area includes the northern, southern or altitudinal margins of a large number of tree and pathogen species. This range makes it challenging to observe, analyse and simulate the effects of climate change on forests at regional and sub-regional scales. The aims of the CARBOFOR project were thus:

– to assess the present status of French forest ecosystems from the point of view of the carbon cycle (Part I);

– to describe the potential effects of climate change on its phenology, productivity and carbon cycle at a range of spatial scales (Part II);

– to provide some insights into the possible future of tree species biogeography, pathogens and fire risks (Part III).

An important characteristic of climate change is its spatial variability. Recent modelling studies at the sub-regional level show that changes in precipitation and temperature regimes vary at the sub-regional scale, for example, between the south and north of Europe. Since the geographical limits for many tree species are shaped by climate constraints such as temperature and drought, a change in climate can have a dramatic and asymmetric effect at the margins of natural areas, removing low temperature limitations towards the poles whilst increasing water deficits and high temperatures at lower latitudinal limits.

Climate change can potentially affect tree species directly as well as indirectly through local site characteristics that control the availability of resources. Assessing these effects on tree and stand functioning, therefore, requires a quantitative description of changes in the variables of interest at the local level. In this report, we summarized the results of predictions based on 50 × 50 km grid climate scenarios, we analysed the interactive effects between management scenarios and site fertility on forest growth and carbon balance, potential habitat areas for tree species, areas at increasing risk of pathogen infection and fires and its potential impacts on flora in the Mediterranean zone.

Present status of the forest carbon cycle and productivity

The comparative analysis of carbon flux and carbon balance of forest sites led to the conclusion that the main source of variability in the carbon cycle of French and tropical forests is their physical structure. Differences in carbon, water and energy exchanges attributed to variations in tree age and species composition are mainly driven by variations in leaf area index, canopy height and relative importance of the canopy layers (Chapter 2). Accordingly, the carbon dynamics in humus organic layers exhibit a strong temporal pattern during the ecosystem life cycle, building up the OH and OF layers within a period of 30 to 50 years in *Fagus* and *Pinus* forest of the western part of France (Chapter 3).

Several options used for estimating the amount of carbon stored in forest tree biomass are reviewed and illustrated (Chapter 4). The estimate of the total amount of carbon in French metropolitan forest and its temporal dynamics are presented in Chapter 5. Interestingly, this analysis confirms the increase of carbon stock in managed temperate forests and shows how errors and uncertainties in parameter estimates propagate in the calculation of carbon stock value.

Climate scenario

The French national meteorological office Météo-France atmospheric model ARPEGE/ Climate (3.0) has been used to simulate present climate and 21st century climate through a 140-year numerical experiment (Chapter 6). The greenhouse gas and aerosol concentrations were prescribed by the so-called IPCC B2 scenario. Ocean surface temperatures are provided by a model with a coarser resolution coupled to an oceanic water circulation scheme. The radiative forcing scheme includes four greenhouse gases (CO_2, CH_4, N_2O and CFC) in addition to water vapour and ozone, and five aerosol classes – land, marine, urban, desert and sulphate, respectively. The model predicts an increase in temperature reaching +4°C in summer over south-western Europe and a shift in seasonal precipitation from summer to winter by 50 mm together with significant sub-regional variations.

Phenology

The phenology of vegetation is a major component of forest ecosystem productivity and influences the annual balance of energy and gas exchange by forest ecosystems. A review of the variability of leaf unfolding dates in major forest trees shows that on average leaf unfolding has been advancing at a mean rate of 2.9 days per decade since 1950 in tree species from the temperate zone with some species variation (Chapter 7). Although current patterns can be estimated from satellites, we still lack the ability to predict accurately and widely future trends in the response of phenology to climate change because leaf phenology and its intra- and inter-population variability are difficult to parameterize. The observed changes in tree phenophases during the last decades across the French ICP forest network show that leaf unfolding, growing season duration and leaf colouring have shifted in the last few decades to earlier dates than those previously observed (Chapter 8).

If changes in phenology remain linear with warming, using present trends we can estimate that leaf unfolding should advance on average at a rate of 5.4 to 10.8 days per decade over the period 2000–50. Thus, by 2050, leaf unfolding of forest trees could occur on average 27 to 54 days earlier than at present. Such a change should have major consequences on forest productivity and on the specific species composition of forests over a given area. A few species, with chilling requirements, would be delayed by a warming climate. However, the dual action of temperature on phenology (i.e. the action of cool temperature to break dormancy followed by the action of warmer temperatures promoting cell growth during quiescence) should lead to a non-linear response of phenological change to warming.

Forest growth and biogeochemical cycles

Biophysical models such as CASTANEA, GRAECO and the large-scale ORCHIDEE model were used to simulate the annual carbon and water balances and wood production of forests averaged over complete forest rotations. These models were first evaluated against data collected across a network of flux sites of the CARBOEUROPE project (Chapter 9) and then used in predictive mode with a climate scenario based upon the IPCC B2 economic scenario of CO_2 accumulation in the atmosphere (Chapter 10). A range of management scenarios and site conditions were considered.

The models predict a slight increase in potential forest production until 2030–50 followed by a plateau or a declining phase in 2070–2100 sensitive to geographical variation, with the northern part of the temperate zone being more favourable for wood production than the southern temperate and Mediterranean zones. In the southern temperate and Mediterranean forests where the largest increase in the growing season water deficit occurs, the CO_2 enhancement of gross primary production was overshadowed by drought impacts. The changes in forest production, as predicted for different forest management options and site conditions, are explained by the counterbalancing effects of rising CO_2, water deficit and the fact that ecosystem sensitivity to climate decreases with age. This interaction between climate, CO_2, nitrogen and water availability and management regime is an important outcome of the modelling analysis.

In terms of the geographical variation of climate change impacts, our analysis refines the conclusions published on the global impacts of climate change on European forests so far and confirms the hypothesis for a strong regional pattern in the 1990–2050 predictions for age-independent net ecosystem productivity, with larger increases in net ecosystem productivity for the boreal zone and a decline across Mediterranean forests.

Biogeography, pathogens and risks

Using botanical inventory data collected by the National Forest Inventory, empirical models relating the frequency of a given species to climate parameters such as minimum and maximum temperature, monthly precipitation and Penman's potential evapotranspiration were established and used for predicting the change in potential habitat areas under the climate scenario considered (Chapter 11). This approach predicts a dramatic change in the geographical distribution of potential areas for tree species with an extension of the southern temperate and Mediterranean species by 150 to 250 km northwards by 2100, together with a similar recession of most oak species, silver fir and beech at their southern edges. These predictions are consistent with observations of the decline of some beech and Scots pine forests at their southern margins, for example, in the plains of southern France and the southern Alps.

Several types of models were used to simulate the effect of the climate change scenario on pathogens and diseases: statistical biogeographical models based on distribution data from specific surveys, a specific epidemiological model and the generic model CLIMEX (Chapter 12). Unsurprisingly, poleward extension of thermophilic pathogen and insect species and associated damage risk is predicted. However, the effect of warming would be counterbalanced by the negative effect of decreased summer rainfall for some species. Due to the high dispersal potential of many fungi, the colonization of new regions becoming climatically favourable could put them into contact with naive host populations (i.e. with no co-evolution or co-adaptation history), with the same potentially dramatic consequences as those observed with introduced parasites.

In Mediterranean and southern temperate forests, the duration of the annual period when fire risk is high will be extended by climate change. Together with ongoing land abandonment and the increase in urban areas and peri-urban forest areas, where ignition frequency is highest, the risk of fire in southern, mostly unmanaged, forest ecosystems will increase as already observed since the 1970s. Under a changing climate, the fire return interval might decrease from 72 to 62 years for Mediterranean forests and from 20 to 16 years for shrublands. In turn, fire frequency curbs the extension of forests into southern Europe as increased fire frequencies leads to a domination of fast growing shrubs or resprouting species (Chapter 13).

Lessons for managing forest in an uncertain future

Where climate change effects are beneficial to forest functions, in northern temperate, continental and boreal forests, our results suggest that optimizing forest management should aim at reducing the effects of limiting factors, for instance, through fertilization. Conversely, where detrimental effects of future climate are expected through increased water deficit, for example, in southern temperate and Mediterranean forests, enhancing the ecosystems resistance to drought and fire using species substitution, understorey control, site preparation and reductions in the maximal value of leaf area index could be appropriate strategies to adopt.

For future research, the complex interaction between climate and atmospheric composition calls for: (i) an assessment of the entire environment rather than single factor response; and (ii) regional studies to account for local features of the forest environment and management. The forest function must also be considered.

Since climate change is provoking a continuous, but not monotonous, change in site productivity, the management of a given forest must be revised dynamically along its life course. At the southern margin of geographical areas, management aiming at an optimal adaptation of forests should be considered, favouring, for example, multi-age and mixed forest stands including pre-existing species and their southern variants and maximizing intra-specific diversity.

Disease management in forest ecosystems has to rely on an anticipatory and preventive approach based on risk analysis. Since simulated geographical ranges are only potential envelopes in which parasites may establish, depending on their dispersal ability, the application of strict hygiene measures, based on the most probable dissemination pathways of organisms (in seeds, wood and plants) is necessary in order to delay the establishment of parasites in climatically favourable zones.

Part I

The carbon cycle in temperate and Mediterranean forests

Chapter 1
The forest carbon cycle:
generalities, definitions and scales

Denis Loustau, Serge Rambal

Introduction

Forests are open dynamic systems which exchange mass and energy with their environment. Both forest vegetation and soil organic matter are composed of carbohydrate molecules. Therefore, a practical way of describing and analysing forest functioning is to describe and measure the carbon exchanges between system components, i.e. the atmosphere, biomass and soil. Carbon is also the major component of the anthropogenic greenhouse gases CO_2, CH_4 and CFCs. The fact that forests represent the main terrestrial reservoir of carbon has led several international programmes to study the role of forests in the carbon cycle since the pioneer International Biosphere Program (Lieth, 1975). The storage of carbon in forest ecosystems has been identified by international agreements (e.g. United Nations Framework Convention on Climate and the Kyoto Protocol) as a possible way of counterbalancing the anthropogenic build-up of carbon in the atmosphere. As a result, over the last few decades, the assessment of the amount of carbon stored in forest ecosystems, and their past and future changes, have became important issues in ecological research.

Carbon flow exchanged from forest vegetation can be assessed using two different approaches: the "flux" and "stock change" approaches. Both are commonly applied, independently or in combination, and some confusion may have arisen due to the sharing and comparison of results by users from different communities, such as foresters, atmosphere physicists and ecologists (Janssens *et al.*, 2003); a "net" flux for the atmosphere is not necessarily "net" from the ecosystem point of view. In this chapter, we attempt to provide a consistent framework of the main concepts and methods used for analysing the carbon cycle in forests, as later described in this book.

Spatial and temporal domains

This chapter and subsequent chapters refer mainly to forest ecosystems defined vertically as the top of the tallest tree to the depth of the deepest roots, and horizontally as one square kilometre. This horizontal limitation is of course arbitrary.

Forest ecosystems are composed of long-living trees and their physiology, size and structure are strongly dependent on their age and size. Few forests have a steady age distribution and are, therefore, close to a stationary state. Conversely, a majority of boreal and temperate forests are subject to more or less periodic disturbances, fires, storms, clear-cuts, etc. Disturbances themselves are important events in the forest carbon cycle. Most of them cause an immediate and abrupt change in the carbon flux and stocks, but they also have delayed effects, which affect the ecosystem dynamics for years, decades and even centuries (Dupouey *et al.*, 2002). The shortest time scale for assessing the forest carbon cycle must, therefore, encompass the duration of a whole rotation between two disturbances, typically a few decades for plantation forests or boreal forests and up to one to two centuries for natural reserves under temperate climate.

The forest ecosystem

The forest carbon cycle involves four major pools: the atmosphere, soil, biomass and harvested products. Their turnover-rate decreases from the atmosphere to the biomass, soil, fossil carbon and sediments. The residence time of carbon is typically one to two orders of magnitude less in the atmosphere (5–7 years) than in biomass (1–250 years) and soil (5000–10,000 years). Figure 1.1 depicts the carbon fluxes exchanged between these pools. Carbon is withdrawn by vegetation from the atmosphere through photosynthesis, i.e. gross primary production (GPP), which fixes carbon atoms in carbohydrate molecules such as sugars. The release of CO_2 associated with photosynthesis, photorespiration, cannot be separated from photosynthesis and is included in GPP. When the energy incorporated by photosynthesis in covalent bonds is used by living organisms in the presence of oxygen, carbon is oxidized to form CO_2, which returns to the atmosphere. This mineralization occurs as a metabolic process called respiration and commonly qualified as "autotrophic" respiration in plants, R_a, and "heterotrophic" respiration in animals and decomposers, R_h. Carbon is also released from organic matter by physical processes such as the aerobic combustion of organic material by fires (not shown in Figure 1.1). In addition, a significant amount of carbon is exchanged between the vegetation and atmosphere as non-CO_2 gaseous molecules, C_x, such as methane (CH_4) and isoprene. This flux contributes marginally to the exchange between the ecosystem and the atmosphere in well-drained forest ecosystems under normal conditions. Methane is mainly produced under anaerobic conditions during the decomposition of soil organic matter, and its emission may be taken into account for wetlands and peatlands. Methane deposition from the atmosphere into the ecosystem (not shown in Figure 1.1), and its subsequent oxidation by methanotrophic bacteria, is also significant, corresponding typically to a net flux of 0.5 $gC·m^{-2}·year^{-1}$, in well-drained forest ecosystems (Zhuang *et al.*, 2007). Methane has also been demonstrated to be produced by green leaves, the cause and the relevance of this emission within the carbon cycle is however, under debate (Keppler *et al.*, 2006). The soil carbon pool is characterized by a wide range of turnover rates, from 1 to 10^{-4} $year^{-1}$ (see

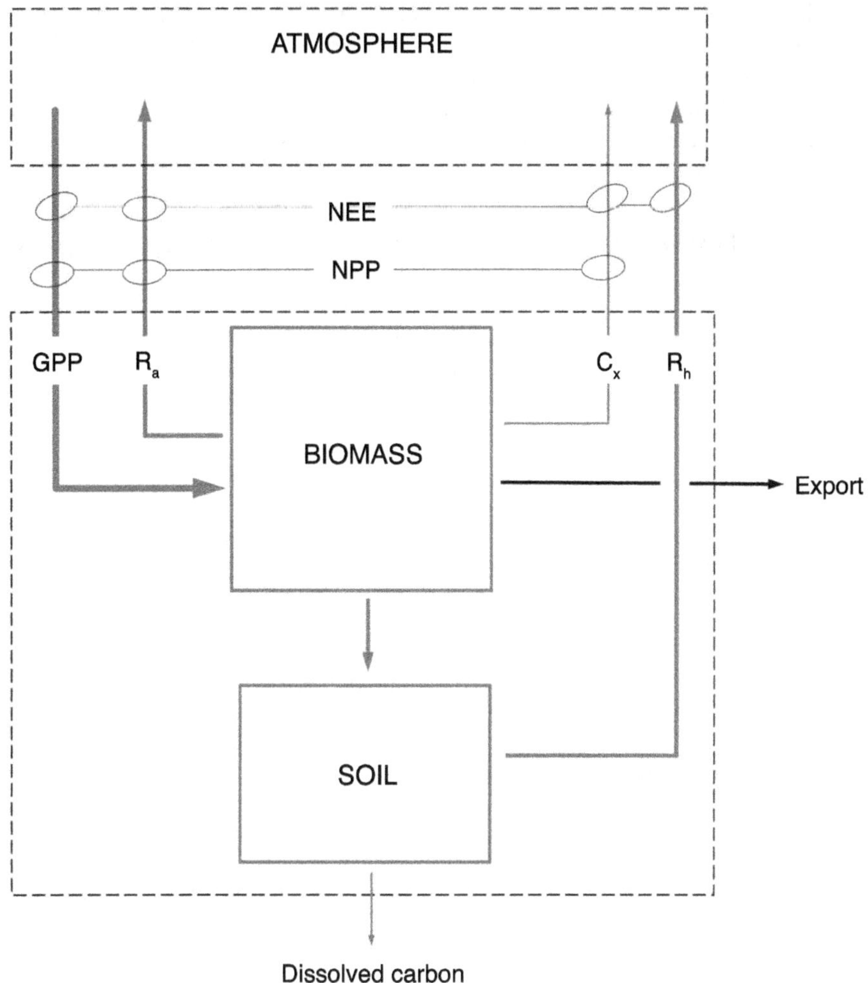

Figure 1.1. The forest carbon cycle showing definitions of the net ecosystem exchange (NEE) and net primary production (NPP).

Chapter 3), depending on the chemical quality of the organic matter and the biophysical environment (e.g. temperature, oxygen, water, pH, etc.).

Solid carbon is withdrawn and subsequently exported out of the forest ecosystem (the "export" flow in Figure 1.1) by biological entities such as herbivores and humans during animal migration and time of harvest, respectively, and to a lesser extent by wind, flooding (erosion) or fire (soot). The export of carbon via fires, flooding, herbivory or harvesting may represent a very large fraction of the carbon budget of forest ecosystems (Ciais *et al.*, 2008). The processes involved in carbon export are essentially discontinuous and difficult to measure accurately. Herbivory is largely neglected in forest carbon budget estimates.

The dissolved carbon flow includes both organic and inorganic compounds. Its relative importance in the forest carbon cycle is two to three orders of magnitude less than the main flux, photosynthesis or respiration, but it may account for several percent of the ecosystem carbon balance (Qualls *et al.*, 2000; Jonsson *et al.*, 2007). The flow of dissolved carbon in groundwater is largest in forest ecosystems subject to seasonal flooding (Richey *et al.*, 2002).

Carbon balance, production and carbon stock changes

The balance of the carbon that is exchanged between a given forest ecosystem and the atmosphere is the net ecosystem exchange (NEE). NEE can be considered as the net flux "seen" from a tower or aircraft above the ecosystem. In the definition, the adjective "net" refers to the atmosphere not to the ecosystem, since the carbon export and dissolved carbon flow are not accounted for. Theoretically NEE should be better referred to as a net atmospheric exchange limited to the volume of atmosphere exchanging carbon with the forest ecosystem. In this book, we shall not however, add this new term to the already large set of terms used by the carbon science community.

$$NEE = GPP + R_a + R_h + Cx \tag{1}$$

The second concept, the net primary production (NPP) is centred on the vegetation. It may be viewed as the net carbon uptake by the vegetation. NPP may be evaluated at tree or stand scale and is equal to:

$$NPP = GPP + R_a + Cx \qquad \left(= NEE - R_h\right) \tag{2}$$

The concept of net ecosystem production (NEP) has been introduced by different authors as the sum of all carbon flux excluding disturbances, and the net biome production (NBP), as the overall balance of carbon including harvest, fire and other disturbances (Schulze and Heimann, 1998; Hyvönen *et al.*, 2007). Although the NEP definition is still a matter of debate, a common understanding of NEP as the net exchange of carbon in the ecosystem is emerging:

$$NEP = GPP + R_a + R_h + Cx + \text{export} + \text{dissolved carbon} \tag{3}$$

We prefer not to use NBP because a clear consensus has not been reached on its temporal and spatial definitions among the international community (Roy and Saugier, 2001; Randerson *et al.*, 2002; Chapin *et al.*, 2006; Hyvönen *et al.*, 2007).

Measurements

A wide range of measurement methods is available for assessing the carbon fluxes involved in the forest carbon cycle based on chambers, turbulent transport and stock changes. It is not within the scope of this introduction to detail these methods (see Chapters 2–4), but rather to review briefly their applicability and accuracy. Figure 1.2 summarizes the temporal and spatial domains of different techniques used to measure the carbon balance in terrestrial forest ecosystems. A description of chambers and eddy-covariance methods is provided in Chapter 2 and biomass stock measurement in Chapter 4.

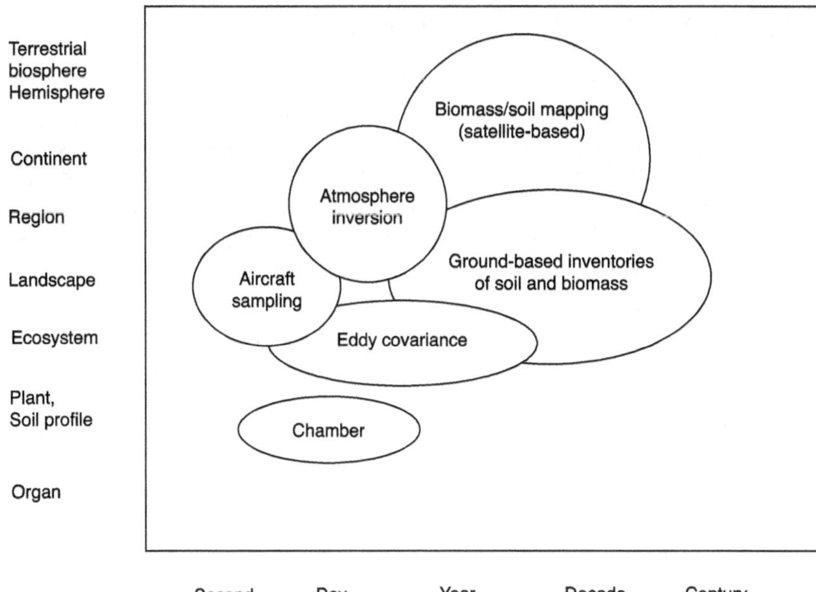

Figure 1.2. Temporal and spatial scales of the main technical methods used in estimating the ecosystem carbon fluxes and carbon balance (redrawn from Randerson *et al.*, 2002).

All methods are characterized by time and space limitations, and none of them is applicable in a continuous mode due to either technical or theoretical limitations. Methods based on open or closed chambers are discontinuous; the eddy-covariance method requires minimal turbulence and, therefore, does not function under stable conditions; most inversion and remote sensing approaches rely on non-stationary air- or spacecrafts. Most methods are also limited in space, and spatial interpolation methods must be used for assessing the flux and carbon balance over a specific spatial domain, thus adding a second major source of uncertainty. Therefore, in order to calculate carbon flux and carbon balance over defined time periods, it is necessary to interpolate measured data using empirical or mechanistic models, which is a major source of uncertainty in estimating forest carbon budget.

Flux can also be assessed via repeated measurements of biomass, soil or atmospheric stocks, and the net exchange from a given carbon pool can be deduced from temporal changes in its total carbon content. For a long time, repeated measurements of biomass, combined with the use of litter traps, were the only available methods for assessing above-ground NPP such as during the pioneer International Biosphere Program. For decades now, foresters have also been carrying out inventories of the commercial stem volume standing in forests. In combination with conversion and expansion factors or allometric equations, the resulting data have been widely used to provide robust estimates of the biomass carbon stock in forests and its changes (see Chapter 4 of this volume; UNECE/FAO, 2000; Nabuurs *et al.*, 2004; De Vries *et al.*, 2006).

When analysing the carbon balance of entire ecosystems, combining data provided by these various measurement methods is challenging (Curtis *et al.*, 2002; Janssens *et al.*, 2003; see Chapter 2 of this volume).

Conclusion and future challenges

The rapid development of our technical ability to measure and quantify carbon fluxes and stocks in forest ecosystems has considerably increased our understanding of the carbon cycle at different organizational levels. The successful series of European projects EUROFLUX, CARBOEUROFLUX and CARBOEUROPE helped considerably to design the methodological and scientific contour lines of the flux tower network covering the continents with more than 500 locations in 2008. Nevertheless, we have not reached the same level of understanding for each of the physical and biological processes at work in the forest carbon cycle. Despite the considerable amount of literature accumulated over the last 15 years and data based on eddy-covariance technique (Baldocchi, 2008), the integration of these processes at different organizational levels remains a research challenge.

A complete assessment of ecosystem functions is also a matter of research. Most process models of forest ecosystem tend implicitly to consider carbon as the main or even unique limiting factor of ecosystem primary production against the observational evidence of other nutrient limitations. Biological beings assemble molecules made not only from carbon but also from oxygen, hydrogen, nitrogen, phosphorus, etc., under fixed stoechiometric constraints. As far as we know, no existing model can pretend to provide a satisfactory description of the main biophysical and biogeochemical processes which determine the behaviour of the ecosystem in the short and long term.

Undeniably, we are still far from a clear mechanistic description of ecosystem biogeochemistry in non-stationary environments. This hampers a clear understanding and prognosis of the future of forests. The question raised here is the behaviour of complex biological systems, forests under dynamic regimes of limiting factors, that is, atmospheric concentrations of CO_2 and O_3, enhanced nitrogen deposition, water limitation, temperature and radiation regimes, changing practices, etc. The answer might rely on the biogeochemistry and biophysics of natural systems, which should be considered and understood in their entirety.

Finally, the different spatial and temporal levels imbricated in the carbon cycle have to be considered because long-term and short-term predictions must account for the dynamic coupling and various feedbacks between the biosphere, hydrosphere and atmosphere. In this respect, there is a large gap in our knowledge and efforts to integrate further research projects and disciplines to achieve such an integrated view are required. It is the main condition for mechanistically up-scaling carbon cycle and ecosystem functioning over landscapes, regions and continents.

References

Baldocchi D.D., 2008. Breathing of the terrestrial biosphere: lessons learned from a global network of carbon dioxide flux measurement systems. *Australian Journal of Botany*, 56 (1), 1–26.

Chapin F.S., Woodwell G.M., Randerson J.T., Rastetter E.B., Lovett G.M., Baldocchi D.D., Clark D.A., Harmon M.E., Schimel D.S., Valentini R., Wirth C., Aber J.D., Cole J.J., Goulden M.L., Harden J.W., Heimann M., Howarth R.W., Matson P.A., McGuire A.D., Melillo J.M., Mooney H.A., Neff J.C., Houghton R.A., Pace M.L., Ryan M.G., Running S.W.,

Sala O.E., Schlesinger W.H., Schulze E.D., 2006. Reconciling carbon-cycle concepts, terminology, and methods. *Ecosystems*, 9 (7), 1041–1050.

Ciais P., Borges A.V., Abril G., Meybeck M., Folberth G., Hauglustaine D., Janssens I.A., 2008. The impact of lateral carbon fluxes on the European carbon balance. *Biogeosciences*, 5 (5), 1259–1271.

Curtis P.S., Hanson P.J., Bolstad P., Barford C., Randolph J.C., Schmid H.P., Wilson K.B., 2002. Biometric and eddy-covariance based estimates of annual carbon storage in five eastern North American deciduous forests. *Agricultural and Forest Meteorology*, 113 (1–4), 3–19.

De Vries W., Reinds G.J., Gundersen P., Sterba H., 2006. The impact of nitrogen deposition on carbon sequestration in European forests and forest soils. *Global Change Biology*, 12 (7), 1151–1173.

Dupouey J.-L., Dambrine E., Laffite J.D., Moares C., 2002. Irreversible impact of past land use on forest soils and biodiversity. *Ecology*, 83 (11), 2978–2984.

Hyvönen R., Agren G.I., Linder S., Persson T., Cotrufo M.F., Ekblad A., Freeman M., Grelle A., Janssens I.A., Jarvis P.G., Kellomaki S., Lindroth A., Loustau D., Lundmark T., Norby R.J., Oren R., Pilegaard K., Ryan M.G., Sigurdsson B.D., Stromgren M., van Oijen M., Wallin G., 2007. The likely impact of elevated [CO_2], nitrogen deposition, increased temperature and management on carbon sequestration in temperate and boreal forest ecosystems: a literature review. *New Phytologist*, 173 (3), 463–480.

Janssens I.A., Freibauer A., Ciais P., Smith P., Nabuurs G.J., Folberth G., Schlamadinger B., Hutjes R.W.A., Ceulemans R, Schulze E.D., Valentini R., Dolman A.J., 2003. Europe's terrestrial biosphere absorbs 7 to 12% of European anthropogenic CO_2 emissions. *Science*, 300 (5625), 1538–1542.

Jonsson A., Algesten G., Bergstrom A.K., Bishop K., Sobek S., Tranvik L.J., Jansson M., 2007. Integrating aquatic carbon fluxes in a boreal catchment carbon budget. *Journal of Hydrology*, 334 (1–2), 141–150.

Keppler F., Hamilton J.T.G., Brass M., Rockmann T., 2006. Methane emissions from terrestrial plants under aerobic conditions. *Nature*, 439 (7073), 187–191.

Lieth H.F.H., 1975. Primary production of the major vegetation units of the world. *In: Primary Productivity of the Biosphere* (H. Lieth, R.H. Whittaker, eds). *Ecological Studies* 14. Springer-Verlag, New York, pp. 203–215.

Nabuurs G.J., Ravindranth N.H., Paustina K., Freibauer A., Hohenstein W., Makundi W., 2004. Land use change and forestry sector good practice guidance. Part 3.2 Forest land. *In: Good Practice Guidance for Land Use, Land Use Change and Forestry* (J. Penman, M. Gytarstky, T. Hiraishi, T. Krug, D. Kruger, R. Pipatti, L. Buendia, T. Ngara, K. Tanabe, F. Wagner, eds). IPCC (GIEC)/Cambridge University Press, Cambridge, pp. 3.23–3.68.

Qualls R.G., Haines B.L., Swank W.T., Tyler S.W., 2000. Soluble organic and inorganic nutrient fluxes in clearcut and mature deciduous forests. *Soil Science Society of America Journal*, 64, 1068–1077.

Randerson J.T., Chapin III F.S., Harden J.W., Neff J.C., Harmon M.E., 2002. Net ecosystem production: a comprehensive measure of net carbon accumulation by ecosystems. *Ecological Applications*, 12, 937–947.

Richey J.E., Melack J.M., Aufdenkampe A.K., Ballester V.M., Hess L.L., 2002. Outgassing from Amazonian rivers and wetlands as a large tropical source of atmospheric CO_2. *Nature*, 416 (6881), 617–620.

Roy J., Saugier B., 2001. Terrestrial primary productivity: definitions and milestones. *In: Terrestrial Global Productivity* (J. Roy, B. Saugier, H.A. Mooney, eds). Academic Press, London, pp. 1–9.

Schulze E.D., Heimann M., 1998. Carbon and water exchange in terrestrial systems, *In: Asian Change in the Context of Global Change* (J.N. Galloway, J. Melillo, eds). *International Biosphere – Geosphere Publications* Series 3. Cambridge University Press, Cambridge, pp. 145–162.

UNECE/FAO, 2000. *Forest Resources of Europe, CIS, North America, Australia, Japan and New Zealand (TBFRA 2000). Main Report. UNECE/FAO Contribution to the Global Forest Resources Assessment.* United Nations, New York.

Zhuang Q., Melillo J.M., McGuire A.D., Kicklighter D.W., Prinn R.G., Steudler P.A., Felzer B.S., Hu S., 2007. Net emissions of CH_4 and CO_2 in Alaska: implications for the region's greenhouse gas budget. *Ecological Applications*, 17 (1), 203–212.

Chapter 2

Environmental control of carbon fluxes in forest ecosystems in France: a comparison of temperate, Mediterranean and tropical forests

BERNARD LONGDOZ, ANDRÉ GRANIER, DENIS LOUSTAU, MARK BAKKER, SYLVAIN DELZON, ANDREW S. KOWALSKI, SERGE RAMBAL, ÉRIC DUFRÊNE, DAMIEN BONAL, YANN NOUVELLON

Introduction

Since 1990, the French national forest-wood industry sector has been a fast growing carbon sink, thus contributing to the fulfilment of national commitments to the Kyoto Protocol. The French forests' carbon sink is mainly due to the imbalance between the harvest and production rates, which in turn, are partly a consequence of a continuous increase in stand productivity (see Chapter 5). This enhancement in productivity is attributed in part to the global change effects, for example, increase in nitrogen deposition rate, temperature and CO_2 concentration, and in part to the age structure of French forests because, like most European forests, they are largely in a growing phase (Spiecker, 1999; Nellemann and Thomsen, 2001; Vries *et al.*, 2006). Hence, the national carbon balance of French forest, typical of European temperate forests, is a result of an interaction between environment and management, for example, the age structure and effects. Previous studies aimed at comparing the response of European forests to climate led to the conclusion that they all behave in a comparable way in terms of water and carbon fluxes. The differences between them have been ascribed to latitude (Valentini *et al.*, 2000), air temperature and tree age (Magnani *et al.*, 2007), which are mainly mediated by the difference in leaf area index (LAI) (Watson, 1947; Granier *et al.*, 2002, 2003). These comparative analyses of carbon fluxes measured across a network of forest ecosystem flux sites revealed a powerful tool for understanding and quantifying the management-environment interaction.

In this part, we analyse and synthesize the main features in the carbon and water fluxes variation with climate and stand age among a range of French and remote forest ecosystems investigated by the national project CARBOFOR. The sites examined in this section cover a range of species, site and climate conditions, which were chosen in order to satisfy the fetch and homogeneity requirements of the eddy-covariance methodology. In the main, they are mature sites belonging to the high productivity site class. The sites investigated here are equipped with eddy-covariance towers measuring energy and mass fluxes exchanged between forest and atmosphere and the main methods used will be briefly described below. The sites are located in forests growing under contrasted climates and soil types (see Table 2.1). Four sites are located in France and one in the Republic of Congo. Some additional data, from a recently installed site in a tropical rainforest of French Guiana, are also included in this study.

This sample does not cover the life cycle of a forest. The age of trees is, however, a major factor in the forest carbon cycle in temperate and Mediterranean forests (Gower *et al.*, 1996; Bond-Lamberty *et al.*, 2004; Magnani *et al.*, 2007). In the final part of this chapter, we complete the comparative analysis of French forest "flux" sites with an analysis of the causes and impacts of age on the primary productivity of maritime pine forest, the first species for softwood production in France. The age effect on the soil carbon balance is also covered in Chapter 3.

Methods of measuring CO_2 fluxes and incertitude

Net ecosystem exchange, ecosystem gross primary productivity and ecosystem respiration

In normal windy conditions and when the assumption of horizontal homogeneity of the canopy can be accepted, the CO_2 mass balance equation applied to the terrestrial ecosystem leads to correspondence between net ecosystem exchange (NEE) and CO_2 turbulent vertical flux above the canopy, F_c combined with CO_2 storage in the canopy air (Aubinet *et al.*, 2000; Finnigan *et al.*, 2003). Following the definitions proposed in Chapter 1, this holds when non-CO_2 atmospheric exchange is neglected. The CO_2 turbulent vertical flux, F_c, is equal to the temporal average of the product of the CO_2 concentration and vertical wind velocity temporal variances. The time average is performed in a period during which the micrometeorological conditions can be considered as constant (typically one half-hour). The temporal variances correspond to the difference between the instantaneous value and its average over a period ranging from a few minutes to a complete half-hour. The determination of the net CO_2 flux requires the measurement of CO_2 concentration (C) and vertical wind velocity (w) at a reference level, which may be optimally immediately above the roughness layer and at a frequency of at least 10 Hz.

When wind turbulence is low, essentially during quiet nights, the CO_2 stored in the canopy air and the CO_2 transported by processes other than vertical turbulence (e.g. horizontal advection) represent a non-negligible quantity in comparison with the vertical turbulent flux (Aubinet *et al.*, 2005). The storage flux in the air layer below the measurement level can be easily obtained by the difference between two successive estimations of CO_2 content as deduced from repeated CO_2 concentration measurements along a vertical profile between the soil and the measurement levels (Yang *et al.*, 2007). The measurement of the other fluxes is more difficult and the determination of a correct methodology is still in progress. Presently, to overcome these difficulties, the eddy-covariance data, corrected by the air storage, are selected to represent the net CO_2 ecosystem exchange when the friction velocity ($u*$, variable representing the turbulence level) is above a threshold that has to be determined for each site and year (Gu *et al.*, 2005; Papale *et al.*, 2006). The $u*$ threshold corresponds to the value below which the night flux decreases without any change of micrometeorological drivers (e.g. w).

The eddy-covariance system is composed of a three-dimensional sonic anemometer measuring the three wind velocity components and an infrared gas analyser measuring the CO_2 and water vapour concentrations. These two instruments must operate at a frequency of 10 Hz or more. The anemometer is installed above the canopy. If the analyser includes an internal cell for the infrared absorption (closed path analyser), the air sample has to be sucked from the top of the tower (close to the anemometer) by a pump at a high rate (a few litres per minute). In the case of an open path analyser (air is free to circulate between the infrared source and the receptor), then this analyser is simply placed near the anemometer with care not to perturb the wind flow measured. The air CO_2 vertical profile can be measured with a system combining a pump, an infrared gas analyser and automated vanes that switch the air taken up to the infrared analyser between different levels.

Some corrections and post processing have to be performed on the raw data from the eddy-covariance system before calculating the CO_2 flux. If a closed path is used, there is a time lag between the measurements of w and C, respectively, due to the time of air transfer from the sampling point to the analyser. The transfer time value corresponds to the time for which the maximum of w and C covariance is found. The wind data must be delayed from this time lag before computing the flux with the w and C covariance (Aubinet *et al.*, 2000). The second correction introduced by a closed path analyser comes from the reduction of the C high frequency oscillations due to the air mixing during its transfer to the analyser. The factor correcting this effect can be obtained by comparison of the co-spectra of C with a theoretical co-spectra (Kaimal *et al.*, 1972) or with the sensible heat flux, which is not affected by air transfer as it is calculated only from the anemometer data. When an open path analyser is used, the air density variations introduce some errors in the determination of CO_2 concentration from the analyser measurement that gives the number of CO_2 mole by volume unit (Webb *et al.*, 1980). The correction factor can be calculated from values of air temperature and atmospheric pressure.

Finally, the corrected fluxes have to be selected to keep only values recorded when the eddy-covariance system works correctly. A first step in data screening can be operated on high frequency data (10 Hz) by statistical analysis to avoid electronic and other technical failings from the database (Vickers and Mahrt, 1997; Longdoz *et al.*, 2008). The fluxes obtained when $u*$ is below the threshold (see above) should also be excluded and the measurements performed during rain or fog events should be taken with care. Moreover, the presence of water on the sensor surfaces (open path analyser and anemometer) is able to perturb the sound or infrared transmission and induce some abnormal values. The data rejected by these different tests have to be replaced to estimate the annual carbon sequestration by the ecosystem. Different data gap filling methods could be applied. For gaps smaller than 2 hours, linear regression is recommended. Otherwise, the reference table, neuronal network and multiple imputations could be used (Falge *et al.*, 2001; Papale and Valentini, 2003; Hui *et al.*, 2004, Reichstein *et al.*, 2005, Moffat *et al.*, 2007).

When the non-CO_2 flux from the ecosystem can be neglected, F_c can be equated to NEE. Under such conditions, it is possible to infer from F_c estimations of the ecosystem respiration (Reco) and gross primary productivity (GPP). The net fluxes obtained during nights and leafless periods (for deciduous forest) correspond to Reco. These values can be used to calibrate some Reco responses to temperature and soil water content. With the assumption of identical respiration behaviour from the forest components during daylight and night, the response function driven by daylight micrometeorological data provides a surrogate for the daylight Reco. During daylight, the difference between NEE measured and Reco gives GPP (Falge *et al.*, 2002a; Reichstein *et al.*, 2005).

Soil CO_2 efflux

The most widely used method for monitoring CO_2 efflux from soil is based on different kinds of chambers that intercept the flux coming out across the soil surface. The chambers are directly placed on the soil or laid on collars previously inserted in the soil. Soil respiration chamber systems can be grouped into three categories based on their working principle. In a closed chamber system, the CO_2 accumulates (closed system) and flux is determined from the concentration increase during a known period of time. Closed chamber systems can be further divided into two major categories: closed dynamic and closed static.

In a closed static chamber system (also known as non-steady-state non-flow-through chambers), CO_2 is usually trapped with chemicals such as sodium hydroxide or soda lime (Raich *et al.*, 1990). Thus, normally, the CO_2 concentration within the chamber remains relatively stable. The CO_2 efflux can be calculated from the amount of CO_2 bound in the trapping solution, exposure time and the cross-sectional area of the soil intercepted by the chamber. In some static chambers, successive air samples are extracted with a syringe and analysed separately with a CO_2 analyser (either an infrared analyser or gas chromatograph).

Another major category of closed chambers is the closed dynamic chamber system (also known as non-steady-state flow-through chambers). In these systems, the air is pumped from the chamber through a CO_2 analyser before it returns to the chamber and so on (Norman *et al.*, 1992). At each passage in the chamber, the soil efflux adds some CO_2 to the air and the CO_2 concentration increase is measured regularly. The CO_2 efflux is calculated from the slope of the concentration change within the chamber's headspace versus time, the system volume and the soil surface intercepted by the chamber. As with the static system, this procedure is valid while CO_2 concentration increases linearly in the system. When CO_2 concentration in the chamber's headspace becomes high enough to reduce significantly the vertical gradient, the saturation effect reduces the flux. The measurement must account for this saturation, for example, by using a non-linear equation in the flux calculation. The value obtained with this kind of system is very sensitive to the difference in pressure between the inside of the chamber and the atmosphere. An overpressure/depression of a few tenths of a Pascal in the chamber is able to block/suck important quantities of CO_2 from the soil and then divide/multiply the soil CO_2 efflux by a factor higher than two.

In an open chamber system (also known as steady-state flow-through chambers), airflow passes through the chamber (owing to a pump system). The CO_2 efflux is determined from the difference between CO_2 concentration at the inlet and outlet of the chamber, the airflow rate and the soil surface intercepted by the chamber (Rayment and Jarvis, 1997). The determination of the CO_2 efflux is very sensitive to accurate concentration of compensation air (the air fed into the chamber). If not measured accurately and synchronized in time, the results can be misleading. Before performing the measurement, the CO_2 concentration within the chamber must reach steady state (equilibrium) after a time that depends on airflow rate and the CO_2 efflux from the soil.

CO_2 flux from the stem and leafy branches

Similarly, the CO_2 flux from stem sections and branches can be measured using either closed or open chambers (Dufrêne *et al.*, 1993; Ceschia *et al.*, 2002; Bosc *et al.*, 2003; Damesin, 2003). These techniques were automated and used in the field during international programmes, such as BOREAS and CARBOEUROPE (Rayment *et al.*, 2002).

Main characteristics of the studied sites

The methods described have been applied since 1996 in a range of sites located in the French metropolitan area, French Guiana (Paracou site) (Bonal *et al.*, 2008) and Congo (Hinda site) (Tables 2.1 and 2.2). Mean annual temperature ranges between 8.5°C and 15°C for the French sites; it reaches 24°C at Hinda in Congo (Figure 2.1) and 25.7°C at Paracou in French Guiana. A large variation in annual radiation, photosynthetic active radiation (PAR), is observed between the sites, between 8000 and 10,000 $mol·m^{-2}·year^{-1}$ in France and up to 13,000 $mol·m^{-2}·year^{-1}$ at Paracou. In France, the lower annual values were recorded at Hesse, the highest at Puéchabon, linked to latitude and cloudiness.

Table 2.1. Sites and years selected for the analysis (black cells). Years with significant missing data (grey cells) could not be used to establish annual balances but were used to fit the response functions.

	Location	1996	1997	1998	1999	2000	2001	2002	2004
Puéchabon	South France				▨	▨	■		
Le Bray	South-west France	▨	■	■	■		■	■	
Bilos	South-west France								■
Hesse	East France	▨	■	■	■	▨	■		
Paracou	French Guiana								■
Hinda	Congo						■		

Table 2.2. Main characteristics of the six studied stands.

Site	Species	Age in 2002	Mean temperature (°C)	Rain (mm·year⁻¹)	LAI ($m^2·m^{-2}$)	Density (ha⁻¹)	Height (m)
Puéchabon	Mediterranean oak	60	13.5	883	2.9	8500	6
Le Bray	Maritime pine	32	12.5	950	4.2	425	20
Bilos	Maritime pine	0	12.5	950	1.5	–	1.5
Hesse	Beech	36	9.2	820	7.3	3800	13
Hinda	Eucalypt	4	25.5	1200	1.5	700	12
Paracou	Mixed tropical	unknown	25.7	2875	7.0	603	35

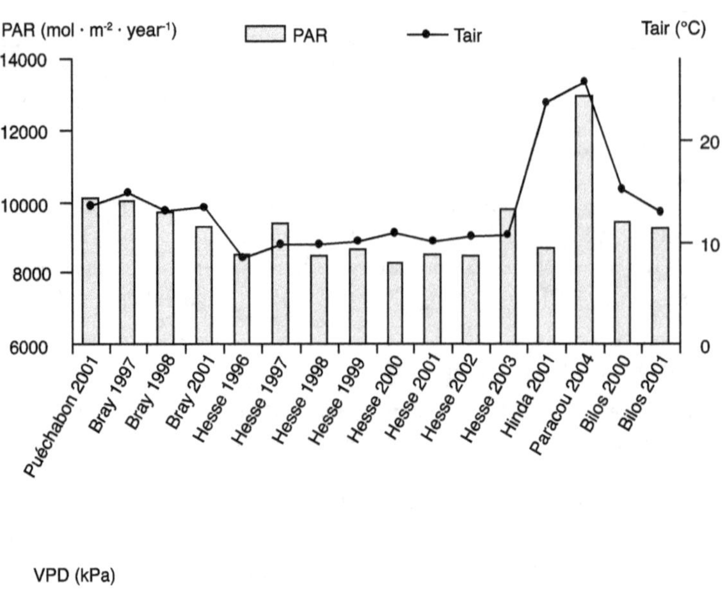

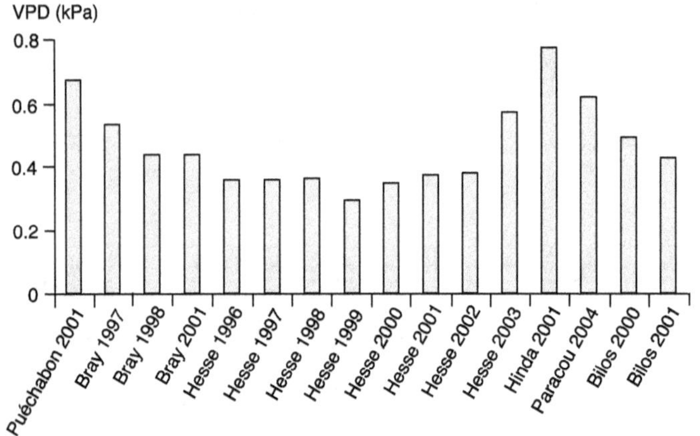

Figure 2.1. Annual mean of air temperature and cumulated photosynthetic active radiation (PAR) in the investigated sites (upper). Annual mean of water vapour pressure saturation deficit (VPD) (lower).

Comparison of the carbon and water annual balances among ecosystems

Water and carbon balances

Evapotranspiration versus available energy

The annual sums of net radiation were comparable from site to site, except at Bilos where regrowing vegetation after clear-cut was still low and sparse (Kowalski *et al.*, 2003) and at Paracou. As an example, time courses of latent heat flux (λE) and net

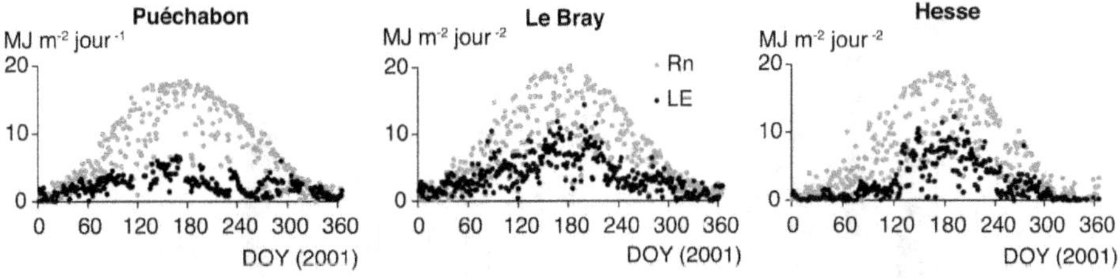

Figure 2.2. Annual courses of net radiation (R_n) and of latent heat flux (λE) in Puéchabon, Le Bray and Hesse (daily data). DOY : day of the year.

radiation (R_n) are presented in Figure 2.2 in three sites. At Puéchabon, we can observe that λE remained low during the year compared with R_n. Furthermore, the latent heat flux showed a large decrease at the end of summer during periods when precipitation was very low (DOY 200-230 and DOY 237-262). At Le Bray, the ratio between λE and R_n was constant over the year except, as in Puéchabon, for a decrease during a drought period (around DOY 240). At Hesse, λE increased sharply in spring after budburst took place (DOY 120).

The ratio $\lambda E/R_n$, that is, the proportion of available energy transformed into evapotranspiration (ETR) flux, equalled on average 0.5. Nevertheless, this ratio was variable according to sites and years, from 0.31 (Puéchabon, 2001) to 0.83 (Bilos, 2000 and 2001). Very close values were calculated at Le Bray and Hesse during the leafy period from May to October (see Figure 2.3). The low $\lambda E/R_n$ ratio at Puéchabon is probably linked to both the low LAI of this Mediterranean stand (2.9) and to the severe water stress and hence stomatal closure. The impact of water stress on the ratio $\lambda E/R_n$ is also shown in Figure 2.3(a) at Hesse for 7 successive years of measurements: the more the water deficit is important, the less available energy is transformed into latent heat flux (thus, the sensible heat flux increases). The eucalypt stand, despite its low LAI (1.5) exhibits high transpiration rates. This clone (EPF1 1-41) shows particularly low water use efficiency (see below).

Fluxes of carbon: net ecosystem exchange, gross assimilation and ecosystem respiration

Neglecting the non-CO_2 flux, the annual net carbon balance (i.e. NEE) was calculated in all the investigated sites as the sum of half-hourly CO_2 flux measurements corrected for CO_2 variation in the air between the soil surface and eddy-covariance systems. Gaps due to missing data were filled on a daily time step, using simple statistical models.

Annual NEE variation between sites and years is presented in Figure 2.4. Negative values indicate fluxes entering the ecosystem, while positive values correspond to carbon release into the atmosphere. A huge variability is observed: NEE varied between − 582 gC·m⁻²·year⁻¹ and + 229 gC·m⁻²·year⁻¹. The Bilos site, a recent clear-cut, is a large carbon source. However, there was a tendency for NEE at Bilos to decrease during the second year of measurement (i.e. 2 years after clear-cut), due to vegetation regrowth.

At Le Bray, annual NEE showed a low inter-annual variability (in the range of − 450 gC·m⁻²·year⁻¹ to − 550 gC·m⁻²·year⁻¹) (Berbigier *et al.*, 2001), while at Hesse the

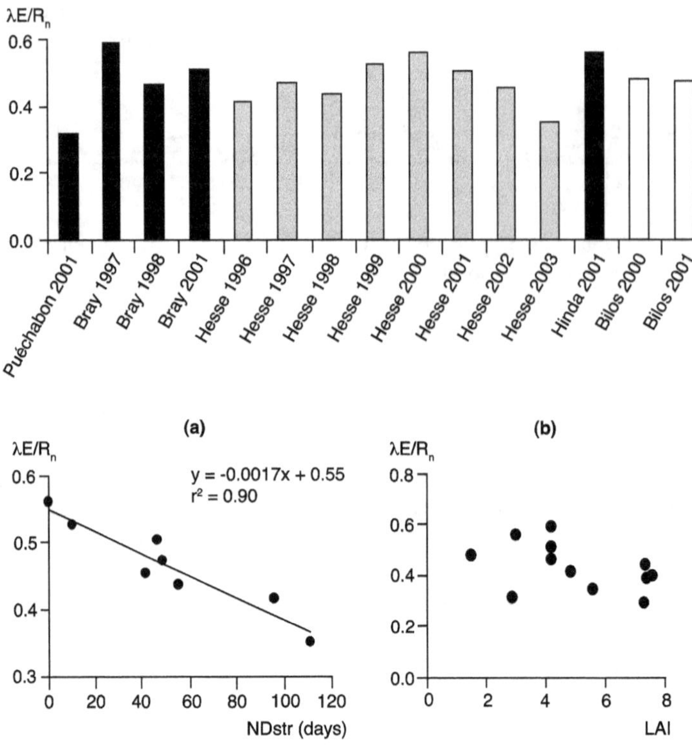

Figure 2.3. Inter-site and inter-annual variation in the ratio of latent heat flux to net radiation ($\lambda E/R_n$). Calculation was performed for complete years, except at Hesse where it was restricted to the leafy period (May to October). (a) Effect of water stress on the ratio $\lambda E/R_n$ at Hesse, expressed as the number of days (NDstr) during which the relative extractable water in the soil is below 0.4; (b) relationship between $\lambda E/R_n$ and the leaf area index (LAI) for the same sites and years.

variability was much higher (from -68 gC·m^{-2}·year^{-1} to -582 gC·m^{-2}·year^{-1}) (Granier *et al.*, 2000b, 2008).

The GPP is calculated as NEE minus Reco. Reco is estimated during the periods when photosynthesis does not occur, during the night and, in deciduous stands, leafless periods. Nevertheless, even when corrected for CO_2 storage variation in the air, a Reco estimate still remains problematic, especially under low turbulence conditions (i.e. low wind speed). Therefore, the important uncertainty in Reco estimates feeds directly into the values of GPP.

Except for Bilos, where photosynthetic rates were very low due to reduced plant cover, GPP varied over a wide range, between -1000 gC·m^{-2}·year^{-1} and -2100 gC·m^{-2}·year^{-1} (Figure 2.5). At Hesse, the between-year variation was much higher. Moreover, a regular increase in GPP was observed from 1996 to 2001 stabilizing in 2002. A multivariate analysis showed that inter-annual variation in GPP at Hesse was explained (at *ca.* 70%) by both LAI and the duration and intensity of water stress during the leafy period.

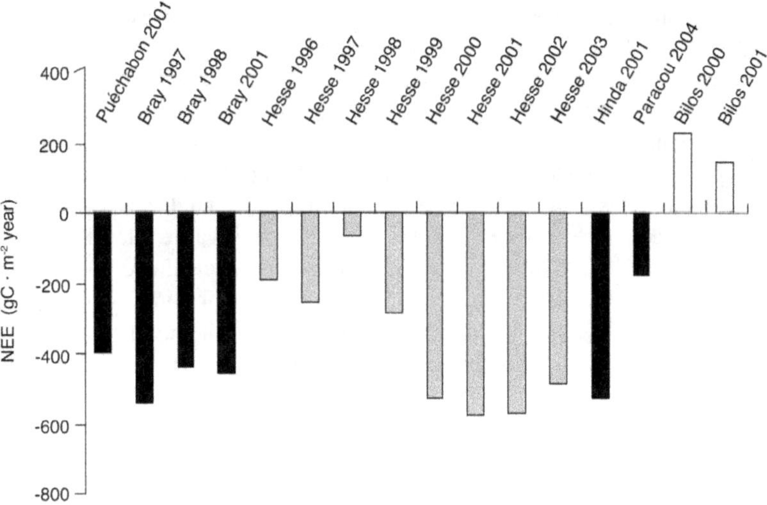

Figure 2.4. Variation of the annual net carbon balance (i.e. NEE) of six forest ecosystems. At Le Bray and Hesse, 3 and 7 years, respectively, of measurement are presented. Negative values indicate a carbon uptake, and positive values a carbon release from the ecosystem to the atmosphere.

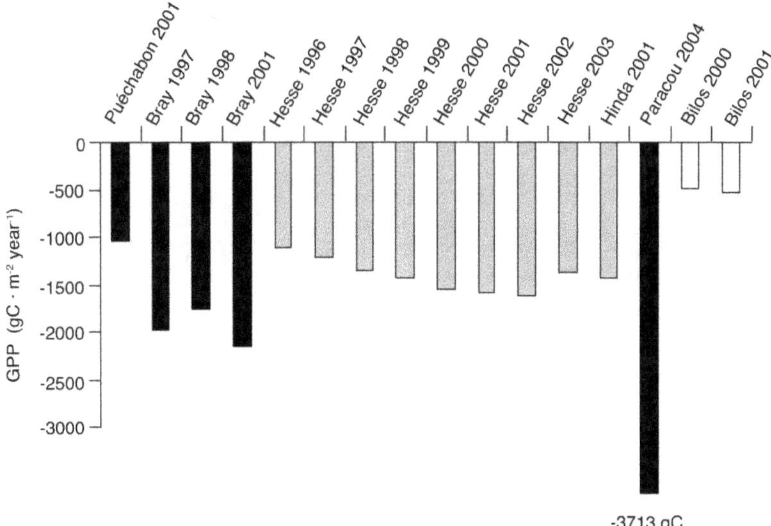

Figure 2.5. Annual gross assimilation (GPP) between sites and years.

The variation in GPP among investigated ecosystems is also explained by the LAI values. Another source of variation is the annual incident PAR. Therefore, the GPP/PAR ratio is correlated to LAI (see Figure 2.6, $r^2 = 0.73$). At Hinda and Le Bray, the ratio GPP/PAR (absolute value) is relatively higher than in other sites, indicating higher light-use efficiency. A probable explanation lies in the canopy structure of both stands: the eucalypt trees at Hinda have narrow and elongated canopies, while at Le Bray and, to a lower extent, Hesse in 1999 (just after thinning), the stands were open allowing a high light interception. In any case, photosynthetic capacity at leaf level probably plays also an important role in light-use efficiency. At Paracou, the value of GPP/PAR is very different from all the other sites, reaching $- 23.9$ mmol·mol^{-1}, indicating a contrasting photosynthetic response to the intercepted radiation of the tropical rainforest. The relationship in Figure 2.6 is slightly curvilinear: light-use efficiency does not saturate at the highest LAI values.

When considering all sites and years, there is a strong positive relationship between absolute values of GPP and Reco (Figure 2.7). Indeed, on an annual time basis, more carbon is fixed by photosynthesis, more growth rates are high and more living tissues respire, leading, therefore, to large autotrophic respiration fluxes. Nevertheless, the heterotrophic component included in Reco (originating from leaf and root litter, woody debris and soil organic matter decomposition) can reduce this relationship and introduce some lag effects into respiration. This was probably what happened in 1998 at Hesse where the important activity of decomposer organisms with abnormally high respiration rates was observed in August and September on woody debris remaining on the ground from a previous thinning.

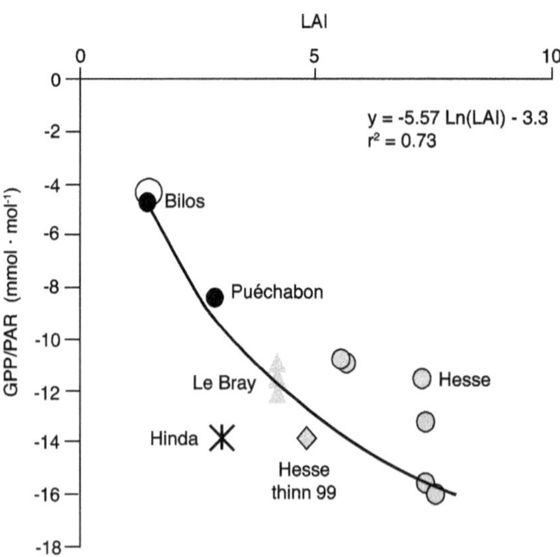

Figure 2.6. Relationship between the gross primary production to photosynthetically active radiation ratio (GPP/PAR annual values) and the leaf area index (LAI). Data from Paracou are not included here.

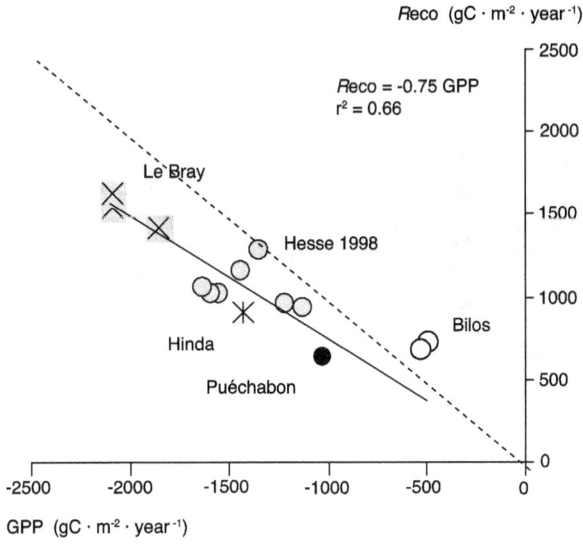

Figure 2.7. Relationship between ecosystem respiration (Reco) and gross primary production (GPP) (annual sums). The regression line does not take into account data from Bilos.

Water and carbon coupling: water use efficiency

Water use efficiency (WUE) is calculated as the ratio between annual biomass production and water consumption by vegetation. Unfortunately, wood increment was not measured at all the investigated sites. We found values of WUE ranging between 0.8 mmol C·mol H_2O^{-1} and 2.4 mmol C·mol H_2O^{-1} (see Figure 2.8). WUE is higher at Hesse, but at Le Bray, the water consumption of the vegetation includes ETR from the abundant herbaceous vegetation, while the annual biomass increment was measured only for the pines. Former measurements at Le Bray showed that understorey vegetation transpiration was on average a quarter to a third of the total ETR. When calculating WUE for pines only, we obtained on average (years 1997, 1998, 2000 and 2001) a value of 2.08 mmol C·mol H_2O^{-1}, very close to that of Hesse (2.05 mmol C·mol H_2O^{-1}). At Hinda biomass increment is rather low while stand ETR is high, leading to a low WUE compared with the two other stands.

Canopy water use efficiency (WUEc, see Figure 2.9) is calculated as the ratio of GPP to evapotranspiration (GPP/ETR). Contrasting WUEc are observed among sites: a high WUEc at Hesse, between – 5 mmol C·mol H_2O^{-1} and – 7 mmol C·mol H_2O^{-1}, with a significant inter-annual variation, the higher value being observed for the dry year 2003. At Puéchabon, Le Bray and Hinda, WUEc ranges between – 3 mmol C·mol H_2O^{-1} and – 4 mmol C·mol H_2O^{-1}. At Le Bray, when calculating WUEc for pines alone (see above), higher values are obtained, close to Hesse WUEc (Figure 2.9). At Bilos, WUEc is very low, around – 1 mmol C·mol H_2O^{-1}.

Intrinsic water use (WUEi) represents the ratio between GPP and canopy conductance for water vapour (gc). We calculated gc from the Penman-Monteith equation, assuming the big-leaf hypothesis. Similar ranking of the sites for WUEi than for WUEc was obtained (data not shown).

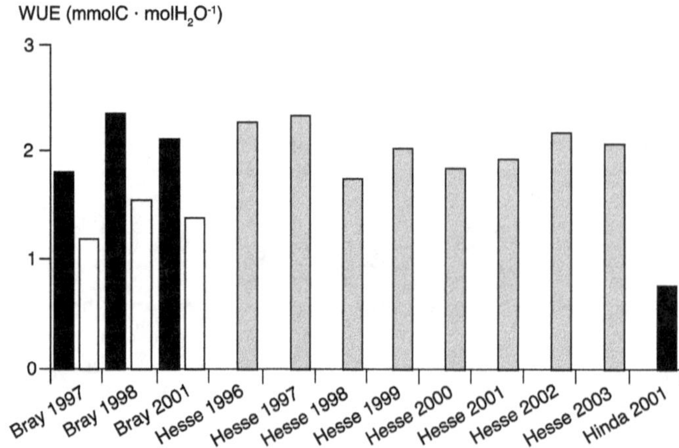

Figure 2.8. Water use efficiency (WUE) (annual values) in three CARBOFOR sites over several years of measurement. At Le Bray, WUE is calculated either: (i) with the total evapotranspiration (see text) (right bars); or (ii) with an estimate of pine transpiration only (left bars).

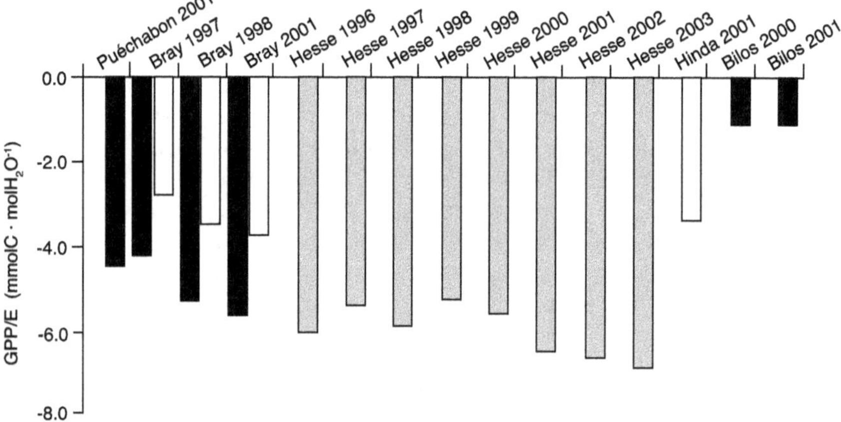

Figure 2.9. Canopy water use efficiency WUEc, calculated as GPP/ETR in five CARBOFOR sites. At Le Bray, WUEc was calculated in two ways (see Figure 2.8).

Carbon balances and carbon stocks

When comparing annual NEE and biomass increment, some discrepancies are found (see Figure 2.10). The annual biomass increment can be higher or lower than the annual carbon balance. At Le Bray, a good linear relationship, close to the 1:1 line, was observed for the 1996–2000 period. However, in 2002 (a very dry year in this site), this relationship no longer matched: while NEE was close to 0, the biomass increment was *ca.* 300 gC·m⁻². At least two different reasons could explain the discrepancies between both annual carbon balance estimates. First, the carbon flux measurements were performed in

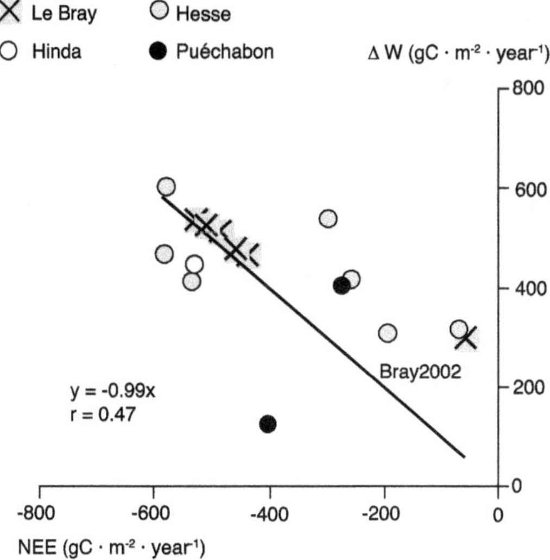

Figure 2.10. Relationship between the annual net carbon balance (NEE) and biomass increment of the stands.

the footprint area whose extent and direction vary according to aerodynamic conditions. Ground measurements at tree level were never performed in the same area, the sampled trees being at the same locations in the stands. Second, other compartments (aerial and below-ground) than the measured standing trees can experience significant variation in carbon stocks: soil organic matter and non-living biomass are still difficult to estimate in an annual time step. Finally, the Bilos data, not shown in this graph, highlight the difference between NEE and biomass increment. For an ecosystem that respires more carbon than it fixes, the positive values of NEE do not correspond to a null stand increment but rather to changes in soil carbon stocks. Such changes are neglected in the biomass increment estimates at other sites, and in any event cannot be detected over time scales of a few years.

In order to try to match more closely the two estimates of carbon balance, we calculated the sum net ecosystem exchange $(NEP) = \Delta W + D - R_h$, where ΔW is the annual stand increment in biomass and D is the leaf plus fine root production. The relationship obtained for the three sites where the data were available is shown in Figure 2.11. A better agreement between NEE and NEP than between NEE and ΔW (Figure 2.10) is observed. However, the most important problem is the fine root production that is poorly known in most forest ecosystems.

Response functions of the canopies

Canopy conductance for water vapour

The canopy conductance was calculated half-hourly from water flux measurements above stands and from climate measurements by inverting the Penman-Monteith equation

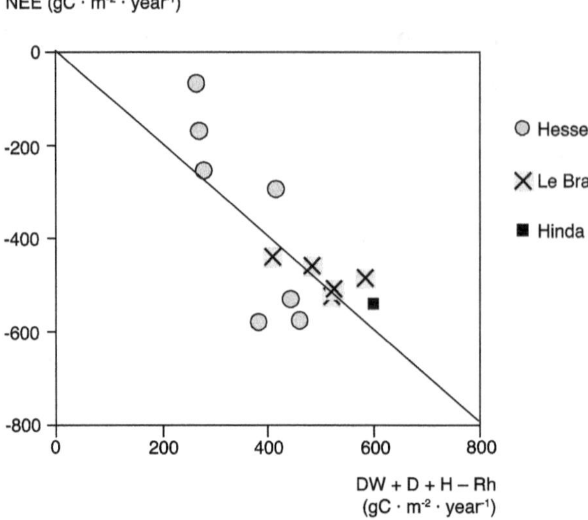

Figure 2.11. Relationship between the sum of biomass increment plus fine root and leaf production less the heterotrophic respiration $(\Delta W + D - R_h)$ and annual carbon balance (NEE).

(Granier and Loustau, 1994; Granier and Bréda, 1996). Non-linear fits were performed for each investigated site using the following model:

$$gc = gc_{max}\left[Rg\ (Rg + Rg_0)\right]^*\left[1\ (1 + k\ VPD)\right] \tag{1}$$

where Rg is the global radiation, VPD is the water vapour pressure saturation deficit and gc_{max}, Rg_0, k are the fitted parameters.

Under high VPD conditions (> 1 kPa), the canopy conductance is close for Hesse, Le Bray and Hinda, while it is much lower at Puéchabon (Figure 2.12). At Hinda, the response of gc to radiation differs from that of the other sites: saturation is reached at lower radiation level, maybe due to the particular architecture of the eucalypts. The models of gc variation can be included in stand transpiration models to estimate day-to-day variation in soil water content (Granier *et al.*, 2000c).

Diurnal net carbon flux

The relationship between net carbon flux and radiation (PAR) was fitted according to the following hyperbolic model:

$$NEE = Reco + a\ PAR\ \left[1 - (PAR/2000) + (a\ PAR/Fc\_opt)\right] \tag{2}$$

where Reco, a and Fc_opt are the fitted parameters; Reco is the ecosystem respiration, a is the slope of NEE at the intercept and Fc_opt is the value of NEE when PAR = 2000 $\mu mol \cdot m^{-2} \cdot s^{-1}$. The fitted parameters are given in Table 2.3. Figure 2.13 shows the temporal variation of this relationship obtained at Hesse; it is stable for most of the growing season. A decrease was only observed from September, when leaf yellowing took place.

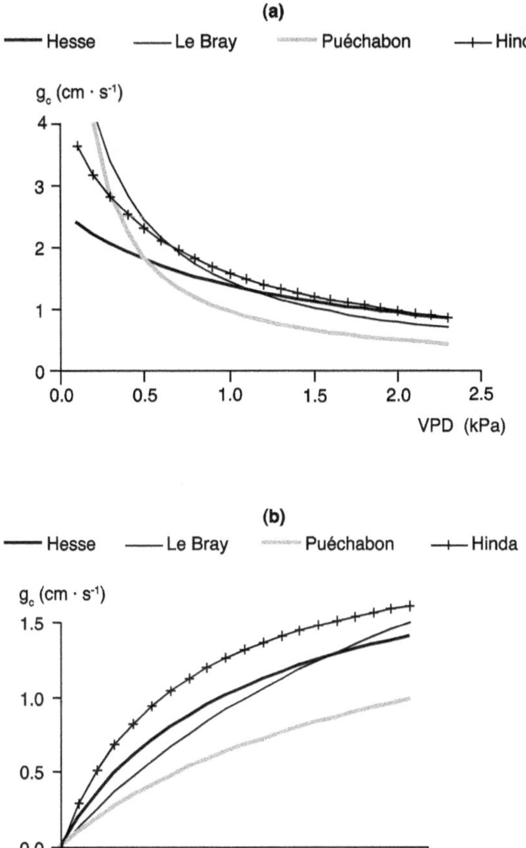

Figure 2.12. Variation of canopy conductance (g_c) to water vapour for four CARBOFOR sites, fitted according to the Equation (1). Effects of air vapour pressure deficit at Rg = 900 W·m^{-2} (a) and global radiation (b) at VPD = 1 kPa.

Table 2.3. Fitted parameters of Equation (2) for different sites and years.

Site	Date	a	Fc_opt	Reco	r^2
	June 1997	− 0.064	− 26.9	6.90	0.74
	June 1998	− 0.063	− 26.1	7.32	0.59
	June 1999	− 0.073	− 27.2	8.29	0.62
	June 2000	− 0.127	− 30.1	9.09	0.61
Hesse	June 2001	− 0.096	− 29.0	6.85	0.68
	June 2002	− 0.069	− 28.7	5.31	0.57
	July 2002	− 0.124	− 32.0	8.88	0.66
	Aug. 2002	− 0.100	− 30.6	7.89	0.65
	Sept. 2002	− 0.060	− 23.1	3.61	0.59

Table 2.3.   (continued)

Site	Date	a	Fc_opt	Reco	r^2
	June 1997	− 0.043	− 23.6	5.89	0.73
	June 1998	− 0.039	− 20.4	5.72	0.60
Le Bray	June 1999	− 0.061	− 21.2	7.78	0.72
	June 2000	− 0.065	− 27.6	8.87	0.67
	June 2001	− 0.051	− 23.0	6.00	0.53
Puéchabon	June 1999	− 0.031	− 11.6	4.63	0.59
	June 2000	− 0.060	− 14.7	5.23	0.50
Hinda	June 2001	− 0.026	− 22.8	2.44	0.69
Bilos	June 2001	− 0.036	− 10.4	4.41	0.46

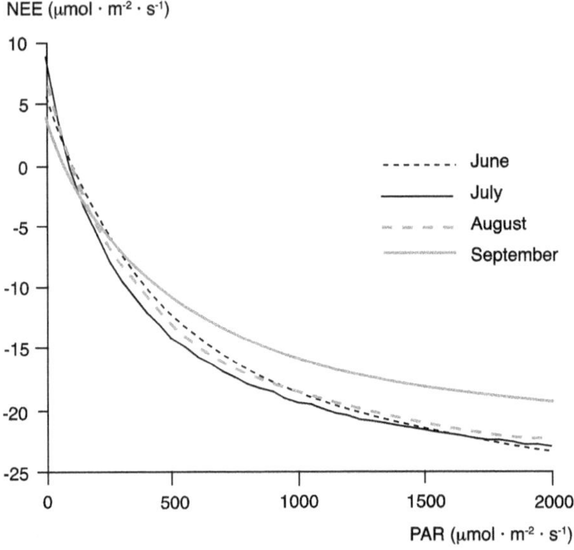

Figure 2.13. Variation of net ecosystem exchange (NEE) as a function of photosynthetic active radiation (PAR) at Hesse for the June to September period in 2002 (wet year) from model 2. The coefficients of regression range between 0.6 and 0.7.

There is a large variation of NEE (PAR) relationships between the sites (Figure 2.14). The lower values of NEE (absolute values) are observed at Puéchabon and Bilos, below 10 μmol·m^{-2}·s^{-1}. At Le Bray, a large inter-annual variation was found. The largest NEE values were observed at Hesse, greater than − 20 μmol·m^{-2}·s^{-1} under saturating PAR.

Night carbon fluxes

This analysis took into account only temperature effects on ecosystem respiration. We fitted the same exponential model using air temperature because soil temperature was not measured at the same depth in all the sites.

The fitted relationships are shown in Figure 2.15, and the parameters are given in Table 2.4. Although the relationships are very variable from site to site, the maximum

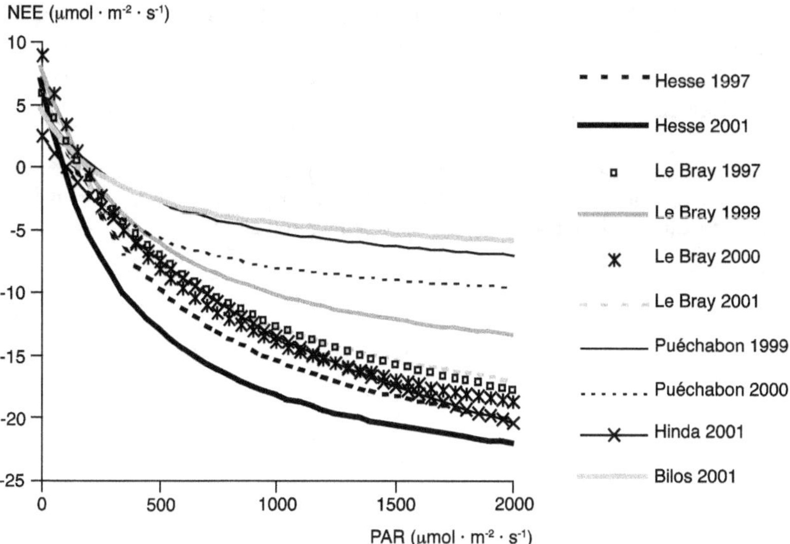

Figure 2.14. Variation of net ecosystem exchange (NEE) as a function of photosynthetic active radiation (PAR) during June in several sites and years. Curves are the best fits of Equation (2), obtained for the June measurements. The coefficients of regression range between 0.6 and 0.7.

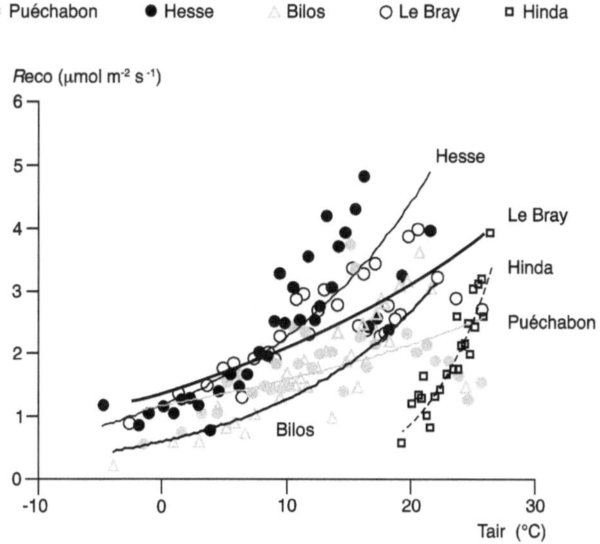

Figure 2.15. Relationship between ecosystem respiration (Reco, night measurements) and air temperature in five CARBOFOR sites during 2001. Data points are bin averages of 10 measurements.

values are similar. Therefore, the temperature does not explain the variation of Reco among the sites. At Puéchabon, the negative effect of soil drought can be observed at the highest temperatures. At Hinda, there is a strong increase in Reco with temperature that varies in a limited range (less than 10°C).

Table 2.4. Estimated parameters of the relationship Reco = a exp (b Tair), and coefficient of determination (r^2) values in five sites (year 2001). The column "maxi" corresponds to the estimated ecosystem respiration at the highest observed temperature in each site.

	a	b	r^2	maxi
Puéchabon	1.13	0.031	0.31	2.52
Le Bray	1.39	0.040	0.64	3.88
Hesse	1.13	0.068	0.75	4.86
Hinda	0.02	0.205	0.82	3.07
Bilos	0.59	0.076	0.70	3.75

The comparative response to climate of carbon and water fluxes among French forest sites: discussion

This work consisted of a cross-analysis of the flux variation to environmental factors among a contrasted range of forest ecosystems (e.g. climate, species and stand structure).

We showed that energy, water and carbon variation in fluxes mainly depended on climate and canopy structure and, to a lesser extent, on the tree species. For instance, the transformation of radiation to latent heat flux related well to: (i) LAI as clearly seen in Figure 2.2 for seasonal variation (Granier *et al.*, 2000a), and in Figure 2.3(b) for inter-annual variation; and (ii) water stress (Figure 2.3a) that reduces the ratio $\lambda E/Rn$, especially at Puéchabon under a Mediterranean climate.

There is still some uncertainty in the estimation of gross assimilation and ecosystem respiration from the net carbon fluxes as measured using eddy covariance. The low turbulence at night in most sites is the most important methodological problem. Future algorithms would probably change the data presented above.

Nevertheless, gross assimilation among the investigated sites seems to be in the range of 1000–2000 gC·m^{-2}·year^{-1}, that is, 10–20 tC·ha^{-1}·year^{-1}. In the French Guiana site (Paracou), the annual value is much larger, close to 35 tC·ha^{-1}·year^{-1}. The determinants of the inter-annual variation in GPP and Reco fluxes, and therefore in NEE, are still poorly known. At Hesse, the progressive increase of GPP from 1996 to 2001 seems to be mostly driven by interaction between drought during the growing season and the intercepted radiation. The eucalypt plantation at Hinda showed higher light conversion efficiency than the other stands that could be related to both a higher physiological performance and a more efficient structure in terms of radiation interception.

Understorey vegetation is an important component of energy, water and carbon cycles in some forest ecosystems, especially in open stands like those at Le Bray, where a large LAI of grasses (*Molinia*) develops as a consequence of the high proportion of radiation it receives (Misson *et al.*, 2007; Jarosz *et al.*, 2008). Such understorey is characterized by very active water extraction from the soil and high photosynthetic rates, especially during the dry periods and, therefore, plays a determinant role in water and carbon balances.

Inter-annual variation in carbon balance is closely dependent on the variation in ecosystem respiration (Valentini *et al.*, 2000; Janssens *et al.*, 2001): low NEE values are observed when Reco is high. Its determinants are much less well known than those for GPP, being more complex, involving numerous above- and below-ground compartments and different processes, each responding to a specific set of environmental variables (see Falge *et al.*, 2002a, 2002b). Reco results from: (i) respiratory fluxes linked to the growth and maintenance of all living tissues in the ecosystem; and (ii) carbon fluxes resulting from the decomposition of organic matter, such as litter, roots and coarse debris. In forest ecosystems, soil respiration represents more than 50% of Reco (Epron *et al.*, 2001). The woody debris can result from the natural mortality of management practices (e.g. thinning and clear-cut). At Hesse, the coarse debris resulting from thinning amounts to 400–500 gC·m⁻², every 5 to 6 years. In the same site, the natural coarse litter (mainly branches) is between 50 gC·m⁻² and 70 gC·m⁻².

Forest management also drives inter-annual variability in carbon balance. Reaction of forests to thinning can be rapid, especially in young fast-growing stands, where light-use efficiency (i.e. ratio GPP/PAR) temporarily increases in the year after thinning.

Carbon balance course during the life cycle of forests

The main source of variability in carbon cycling among managed forest stands in Europe is the age of the stand. Immediately after disturbance, the carbon balance of a forest ecosystem is negative, GPP and net primary production (NPP) are reduced to zero and increase subsequently as the canopy develops. They reach a maximum near canopy closure, and finally decrease in older stands (Ryan *et al.*, 1997). This decline in above-ground wood production after canopy closure, long known by foresters, is a common pattern in even-aged forest stands around the world (Gower *et al.*, 1996; Ryan *et al.*, 1997, 2004). One compelling need is to explain why NPP is reduced in ageing forest stands: the growth patterns are pronounced and predictable but the underlying mechanisms remain unclear. Moreover, recent investigations have confirmed that natural (fire and storm) or anthropogenic (harvest and thinning) disturbances convert mature forest sinks into carbon sources requiring years to decades to recover their sink status depending on their growth rate (Bond-Lamberty *et al.*, 2004; Kowalski *et al.*, 2004). The importance of these disturbances on the maritime pine forest carbon balance was demonstrated in the Landes de Gascogne forest using eddy-covariance measurements simultaneously over a clear-cut site and a neighbouring mature forest site in 2001 (Kowalski *et al.*, 2003). Data in Table 2.5 show that both components of NEE, GPP and Reco, were lower in the clear-cut site in comparison to the mature stand.

The clear-cut site, covered with remaining understorey woody and herbaceous species, maintained about one-third of the GPP of the mature stand. Due to past accumulated carbon stocks in the soil and canopy (representing harvest residues), total respiration by the clear-cut site reached about the two-thirds that of the mature stand, suggesting a dependence of the soil decomposer community on recently assimilated carbon in the maritime pine forest. This is consistent with results from other studies in pine forests (Hogberg *et al.*, 2001), but may not be general for all species of trees (Kowalski *et*

Table 2.5. Comparison of annual ecosystem fluxes between a clear-cut site in 2001 and a mature pine stand aged 28 years in 1997–98 and 31 years in 2001 (Kowalski et al., 2003).

Annual exchange	Clear-cut site	28-year-old stand	31-year-old stand
NEE (gC.m^{-2})	290	-575	-498
GPP (gC.m^{-2})	727	2255	2025
Reco (gC.m^{-2})	996	1680	1527

al., 2004). The greater relative reduction in GPP (versus Reco) indicated the clear-cut site to be a significant annual carbon source. It is clear from these data that intensively managed forests lose carbon to the atmosphere after disturbances. These studies showed the importance of taking into account the initial phase of stand settlement and resulting carbon loss in the accurate estimation of forest carbon sink over its full management cycle. The role of old-growth forests in the global carbon cycle is also a matter of interest because the old concept that their biogeochemical cycles are near equilibrium was recently challenged (Luyssaert *et al.*, 2008).

In the following sections, we summarize the mechanisms that can explain changes in stand productivity and carbon balance over the full management cycle in a temperate forest. First, we expand upon the results regarding decline in forest productivity with increasing stand age, pointing out the mechanisms responsible for this decline. Then, we present some preliminary research on the impact of a disturbance occurring at the early development stages on the carbon balance of a maritime pine ecosystem.

Evidence of a decline in growth with age in a maritime pine forest

In the Landes de Gascogne forest (south-western France), which consists of even-aged stands of maritime pine trees, empirical evidence for dates of the decline of forest production is well documented in inventories carried out by the National Forest Inventory. The decrease in wood production was about 70% between 24 and 54 year-old stands, whatever the fertility class, showing that age is one of the main factors inducing variability in forest productivity. Recent studies on growth, based on a chronosequence of even-aged stands of maritime pine, corroborated this trend and further investigated the physiological and structural causes of this decline. A rather clearer picture of maritime pine stand dynamics emerges from a chronosequence ranging between 10- and 91-year-old stands (Figure 2.16) (Delzon *et al.*, 2004; Delzon and Loustau, 2005).

The above-ground part of NPP, aNPP, decreased markedly with increasing stand age. This decline corresponded to 60% when comparing the 10- and 91-year-old stands. It was explained by a 40% decrease in stand leaf area, that is, LAI and plant area index (PAI), along the chronosequence (due to decline in tree density by thinning and crown abrasion). However, forest growth decline was only partly explained by the observed decline in LAI and light interception: productivity per unit of leaf area was also reduced (see Figure 2.16(b)). This result revealed that an altered photosynthesis may contribute to this decline therefore decreasing carbon fluxes and gains. In conclusion, these studies showed a significant decline in forest growth with age that can likely be explained by decreases both in leaf area at the stand level and in CO_2 assimilation rate at leaf level.

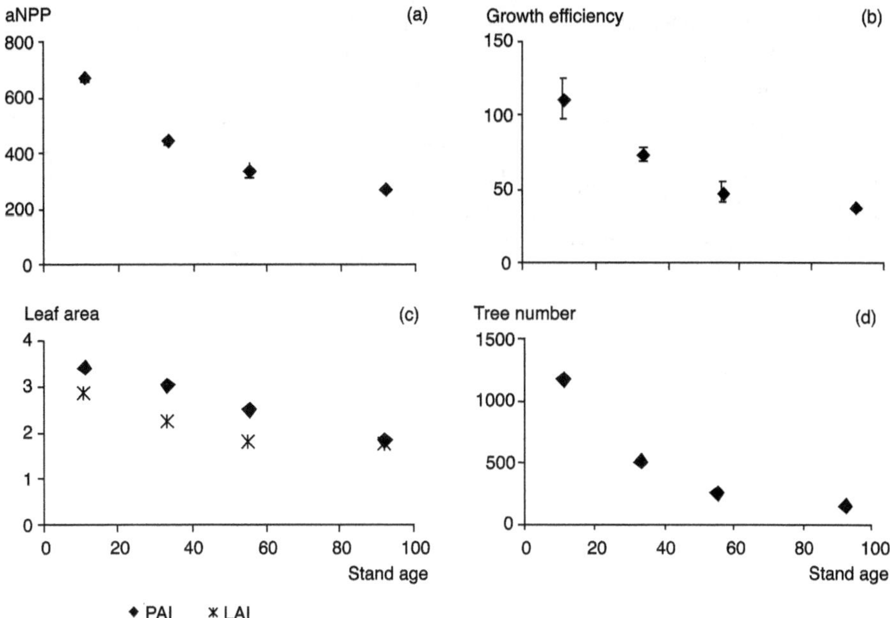

Figure 2.16. Changes in productivity as a function of stand age in a maritime pine chronosequence in the Landes de Gascogne forest of south-western France using various measures: (a) aNPP ($gC·m^{-2}·year^{-1}$) represents above-ground net primary production (NPP) estimated using biomass increments and litterfall measurements per stand; (b) growth efficiency ($gC·m^{-2}·year^{-1}$) corresponds to the above-ground annual tree biomass increment expressed per unit leaf area; (c) stand leaf area (LAI: leaf area index; PAI: plant area index; $m^2·m^{-2}$); (d) tree number (ha^{-1}).

The decline in photosynthesis with stand age

Most studies on photosynthesis in woody plants of different ages were not designed to isolate age-induced variation from all other sources of variation (e.g. that owing to size or environment). However, data from experimental studies generally indicate that both photosynthesis (Kull and Koppel, 1987; Yoder *et al.*, 1994; Kolb and Stone, 2000) and stomatal conductance are reduced as trees get older (Ryan *et al.*, 2000; Schafer *et al.*, 2000). Recent results suggest that the hypothesis of cellular or molecular ageing is not convincing for trees (Mencuccini *et al.*, 2005; Peñuelas, 2005): meristematic cellular senescence (age-related control) does not explain forest growth decline with age. Instead, extrinsic factors mediated by size, rather than irreversible senescence, have been proposed as driving decline in growth rates in old/tall trees.

Results from intensive measurements of biomass allocation, leaf water potential, photosynthetic capacity, sapflow and carbon discrimination across a maritime pine chronosequence in south-western France showed that the dramatic decline in growth with age was explained by increasingly negative hydraulic impacts as trees grew taller (Delzon *et al.*, 2004, 2005). This conclusion is consistent with the hydric homeostasis hypothesis of Magnani *et al.* (2000). We demonstrated further that growth decline results from structural compensatory adjustments to height-related hydraulic constraints and stomatal closure. Trees compensate for hydraulic limitation by reducing the amount of photosynthetic area relative to water-conductive area (the leaf to sapwood area ratio, where sapwood is the hydraulically active part of the stem) (Delzon *et al.*, 2004). This structural compensation is, however, not sufficient to counteract the height effect on hydraulic conductance, causing a decline in stomatal conductance with tree height and reducing CO_2 diffusion into the leaf (Figure 2.17).

This conclusion was also confirmed by a comparison of leaf photosynthetic capacity across the same chronosequence (Delzon *et al.*, 2005). These results are consistent with Mencuccini *et al.* (2005), showing that photosynthetic capacity does not decline with tree age and indicating that senescence of the photosynthetic apparatus does not explain reduced forest growth.

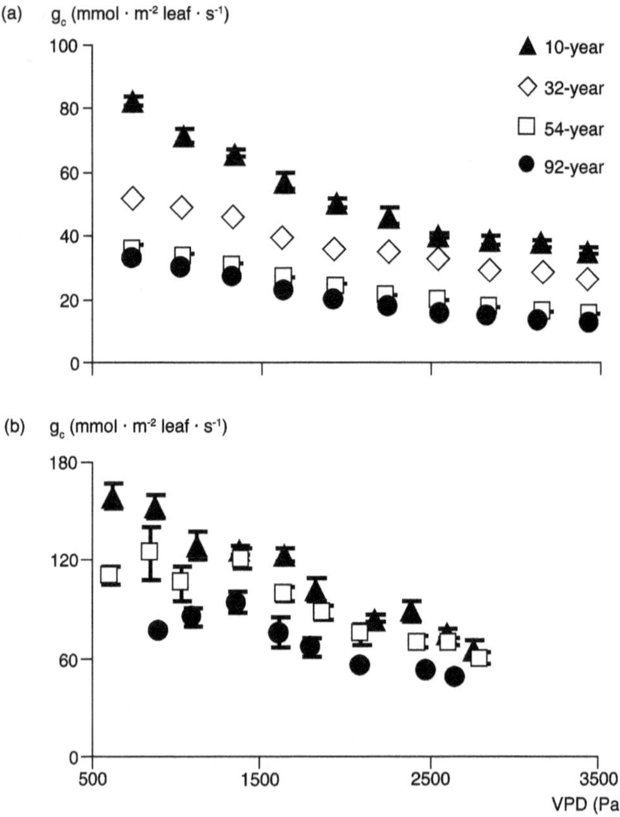

Figure 2.17. Mean values of canopy conductance (a) and stomatal conductance (b) versus air vapour pressure saturation deficit (VPD) for four maritime pine stands of different age. gc was determined from sapflow measurements on six to seven trees per stand and gs was measured on 1-year-old needles using a steady-state null balance porometer. (Redrawn from Delzon *et al.* (2005) with permission of *Plant, Cell and Environment*).

References

Aubinet M., Berbigier P., Bernhofer C., Cescatti A., Feigenwinter C., Granier A., Grünwald T., Havrankova K., Heinesch B., Longdoz B., Marcolla B., Montagnani L., Sedlak P., 2005. Comparing CO_2 storage and advection conditions at night at different CARBOEUROFLUX sites. *Boundary Layer Meteorology*, 116, 63–94.

Aubinet M., Grelle A., Ibrom A., Rannik Ü, Moncrieff J., Foken T., Kowalski A.S., Martin P.H., Berbigier P., Bernhofer C., Clement R., Elbers J. Granier A., Grünwald T., Morgenstern K., Pilegaard K., Rebmann C., Snijders W., Valentini R., Vesala T., 2000. Estimates of the annual net carbon and water exchange of forests: The EUROFLUX methodology. *Advances in Ecological Research*, 30, 113–176.

Berbigier P., Bonnefond J.-M., Mellmann P., 2001. CO_2 and water vapour fluxes for 2 years above Euroflux forest site. *Agricultural and Forest Meteorology*, 108, 183-197.

Bonal D., Bosc A., Goret J.-Y., Burban B., Gross P., Bonnefond J.-M., Elbers J., Ponton S., Epron D., Guehl J.-M., Granier A., 2008. The impact of severe dry season on net ecosystem exchange in the Neotropical rainforest of French Guiana. *Global Change Biology*, 14, 1917–1933.

Bond-Lamberty B., Wang C.K., Gower S.T., 2004. Net primary production and net ecosystem production of a boreal black spruce wildfire chronosequence. *Global Change Biology*, 10, 473–487.

Bosc A., De Grandcourt A., Loustau D., 2003. Variability of stem and branch maintenance respiration in a *Pinus pinaster* tree. *Tree Physiology*, 23 (4), 227–236.

Ceschia E., Damesin C., Lebaube S., Pontailler J.-Y., Dufrêne E., 2002. Spatial and seasonal variations in stem respiration of beech trees (*Fagus sylvatica*). *Annals of Forest Science*, 59 (8), 801–812.

Damesin C., 2003. Respiration and photosynthesis characteristics of current-year stems of *Fagus sylvatica*: from the seasonal pattern to an annual balance. *New Phytologist*, 158 (3), 465–475.

Delzon S., Bosc A., Cantet L., Loustau D., 2005. Variation of the photosynthetic capacity across a chronosequence of maritime Pine correlates with needle phosphorus concentration. *Annals of Forest Science*, 62, 537–543.

Delzon S., Loustau D., 2005. Age-related decline in stand water use: sap flow and transpiration in a pine forest chronosequence. *Agricultural and Forest Meteorology*, 129, 105–119.

Delzon S., Sartore M., Burlett R., Dewar R., Loustau D., 2004. Hydraulic responses to height growth in maritime pine trees. *Plant, Cell and Environment*, 27, 1077–1087.

Dufrêne É., Pontailler J.-Y., Saugier B., 1993. A branch bag technique for simultaneous CO_2 enrichment and assimilation measurements on beech (*Fagus sylvatica* L.). *Plant, Cell and Environment*, 16, 1131–1138.

Epron D., Le Dantec V., Dufrêne É., Granier A., 2001. Seasonal dynamics of soil carbon dioxide efflux and simulated rhizosphere respiration in a beech forest. *Tree Physiology*, 21, 2–3.

Falge E., Baldocchi D., Olson R., Anthoni P., Aubinet M., Bernhofer C., Burba G., Ceulemans R., Clement R., Dolman H., Granier A., Gross P., Grünwald T., Hollinger D.,

Jensen N.-O., Katul G., Keronen P., Kowalski A., Ta Lai C., Law B.E., Meyers T., Moncrieff J., Moors E., Munger J.W., Pilegaard K., Rannik Ü., Rebmann C., Suyker A., Tenhunen J., Tu K., Verma S., Vesala T., Wilson K., Wofsy S., 2001. Gap filling strategies for defensible annual sums of net ecosystem exchange. *Agricultural and Forest Meteorology*, 107, 43–69.

Falge E, Baldocchi D.D., Tenhunen J.D., Aubinet M., Bakwin P., Berbigier P., Bernhofer C., Burba G., Clement R., Davis K.J., Elbers J., Goldstein A.H., Grelle A., Granier A., Gudmundsson J., Hollinger D., Kowalski A.S., Katul G., Law B., Mahli Y., Meyers T., Monson R.K., Munger J.W., Oechel W., Paw U K.T., Pilegaard K., Rannik U., Rebmann C., Suyker A., Valentini R., Wilson K., Wofsy S., 2002(a). Seasonality of ecosystem respiration and gross primary production as derived from FLUXNET measurements. *Agricultural and Forest Meteorology*, 113, 53–74.

Falge E., Tenhunen J.D., Baldocchi D.D., Aubinet M., Bakwin P., Berbigier P., Bernhofer C., Bonnefond J.-M., Burba G., Clement R., Davis K.J., Elbers J., Falk M., Goldstein A.H., Grelle A., Granier A., Grunwald T., Gudmundsson J., Hollinger D., Janssens I.A., Keronen P., Kowalski A.S., Katul G., Law B., Mahli Y., Meyers T., Monson R.K., Moors E.J., Munger J.W., Oechel W., U K.T.P., Pilegaard K., Rannik U., Rebmann C., Suyker A., Thorgeirsson H., Tirone G., Turnipseed A., Wilson K., Wofsy S., 2002(b). Phase and amplitude of ecosystem carbon release and uptake potentials as derived from FLUXNET measurements. *Agricultural and Forest Meteorology*, 113, 75–95.

Finnigan J.J., Clements R., Malhi Y., Leuning R., Cleugh H., 2003. A re-evaluation of long-term flux measurement techniques. Part I: averaging and coordinate rotation. *Boundary Layer Meteorology*, 107, 1–48.

Gower S.T., McMurtrie R.E., Murty D., 1996. Aboveground net primary production decline with stand age: potential causes. *Trends in Ecology and Evolution*, 11, 378–382.

Granier A., Aubinet M., Epron D., Falge E., Gudmundsson J., Jensen N.O., Kostner B., Matteucci G., Pilegaard K., Schmidt M., Tenhunen J., 2003. Deciduous forests: Carbon and water fluxes, balances and ecophysiological determinants. *In: Fluxes of Carbon, Water and Energy of European Forests. Ecological Studies*, No. 163, *Analysis and Synthesis*. Springer-Verlag, New York, pp. 55–70.

Granier A., Bréda N., 1996. Modelling canopy conductance and stand transpiration of an oak forest from sap flow measurements. *Annals of Forest Science*, 53, 537–546.

Granier A., Bréda N., Longdoz B., Gross P., Ngao J., 2008. Ten years of fluxes and stand growth in a young beech forest at Hesse, North-eastern France. *Annals of Forest Science*, 65, 704, 13 p., doi: 10.1051/forest: 2008052.

Granier A., Biron P., Lemoine D., 2000(a). Water balance, transpiration and canopy conductance in two beech stands. *Agricultural and Forest Meteorology*, 100, 291–308.

Granier A., Ceschia E., Damesin C., Dufrêne É., Epron D., Gross P., Lebaube S., Ledantec V., Legoff N., Lemoine D., Lucot E., Ottorini J.M., Pontailler J.-Y., Saugier B., 2000(b). The carbon balance of a young Beech forest. *Functional Ecology*, 14, 312–325.

Granier A., Loustau D., 1994. Measuring and modelling the transpiration of a maritime pine canopy from sap-flow data. *Agricultural and Forest Meteorology*, 71, 61–81.

Granier A., Loustau D., Bréda N., 2000(c). A generic model of forest canopy conductance dependent on climate, soil water availability and leaf area index. *Annals of Forest Science*, 57, 755–765.

Granier A., Pilegaard K., Jensen N.O., 2002. Similar net ecosystem exchange of beech stands located in France and Denmark. *Agricultural and Forest Meteorology*, 114, 75–82.

Gu L.H., Falge E.M., Boden T., Baldocchi D.D., Black T.A., Saleska S.R., Suni T., Verma S.B., Vesala T., Wolfsy S.C., Xu L., 2005. Objective threshold determination for nighttime eddy flux filtering. *Agricultural and Forest Meteorology*, 128, 179–197.

Hogberg P., Nordgren A., Buchmann N., Taylor A.F.S., Ekblad A., Hogberg M.N., Nyberg G., Ottosson-Lofvenius M., Read D.J., 2001. Large-scale forest girdling shows that current photosynthesis drives soil respiration. *Nature*, 411, 789–792.

Hui D., Wan S., Su B., Katul G., Monson R., Luo Y., 2004. Gap-filling missing data in eddy covariance measurements using multiple imputation (MI) for annual estimation. *Agricultural and Forest Meteorology*, 121, 93–111.

Janssens I.A., Lankreijer H., Matteucci G., Kowalski A.S., Buchmann N., Epron D., Pilegaard K., Kutsch W., Longdoz B., Grunwald T., Montagnani L., Dore S., Rebmann C., Moors E.J., Grelle A., Rannik U., Morgenstern K., Clement R., Gudmundsson J., Minerbi S., Berbigier P., Ibrom A., Moncrieff J., Aubinet M., Bernhofer C., Jensen N.O., Vesala T., Granier A., Schulze E.D., Lindroth A., Dolman A.J., Jarvis P.G., Ceulemans R., Valentini R., 2001. Productivity and disturbance overshadow temperature in determining soil and ecosystem respiration across European forests. *Global Change Biology*, 7, 269–278.

Jarosz N., Brunet Y., Lamaud E., Irvine M., Bonnefond J.-M., Loustau D., 2008. Carbon dioxide and energy flux partitioning between the understorey and the overstorey of a maritime pine forest during a year with reduced soil water availability. *Agricultural and Forest Meteorology*, 148 (10), 1508–1523.

Kaimal J.C., Wyngaard J.C., Izumi Y., Cote O.R., 1972. Spectral characteristics of surface-layer turbulence. *Quarterly Journal of the Royal Meteorological Society*, 98, 563–589.

Kolb T.E., Stone J.E. 2000. Differences in leaf gas exchange and water relations among species and tree sizes in an Arizona pine-oak forest. *Tree Physiology*, 20, 1–12.

Kowalski A.S., Loustau D., Berbigier P., Manca G., Tedeschi V., Borghetti M., Valentini R., Kolari P., Berninger F., Rannik U., Hari P., Rayment M., Mencuccini M., Moncrieff J., Grace J., 2004. Paired comparisons of carbon exchange between undisturbed and regenerating stands in four managed forests in Europe. *Global Change Biology*, 10, 1707–1723.

Kowalski S., Sartore M., Burlett R., Berbigier P., Loustau D., 2003. The annual carbon budget of a French pine forest (*Pinus pinaster*) following harvest. *Global Change Biology*, 9, 1051–1065.

Kull O., Koppel A., 1987. Net photosynthetic response to light intensity of shoots from different crown positions and age in *Picea abies* (L.) Karst. *Scandinavian Journal of Forest Research*, 2, 157–166.

Longdoz B., Gross P., Granier A., 2008. Multiple quality tests for analysing CO_2 fluxes in a beech temperate forest, *Biogeosciences*, 5, 719–729.

Luyssaert S., Schulze E.D., Borner A., Knohl A., Hessenmoller D., Law B.E., Ciais P., Grace J., 2008. Old-growth forests as global carbon sinks. *Nature*, 455 (7210), 213–215.

Magnani F., Mencuccini M., Grace J., 2000. Age-related decline in stand productivity: the role of structural acclimation under hydraulic constraints. *Plant, Cell and Environment*, 23, 251–263.

Magnani F., Mencuccini M., Borghetti M., Berbigier P., Berninger F., Delzon S., Grelle A., Hari P., Jarvis P.G., Kolari P., Kowalski A.S., Lankreijer H., Law B.E., Lindroth A., Loustau D., Manca G., Moncrieff J.B., Rayment M., Tedeschi V., Valentini R., Grace J., 2007. The human footprint in the carbon cycle of temperate and boreal forests. *Nature*, 447 (7146), 848–850.

Mencuccini M., Martinez-Vilalta J., Vanderklein D., Hamid H.A., Korakaki E., Lee S., Michiels B., 2005. Size-mediated ageing reduces vigour in trees. *Ecology Letters*, 8, 1183–1190.

Misson L., Baldocchi D.D., Black T.A., Blanken P.D., Brunet Y., Yuste J.C., Dorsey J.R., Falk M., Granier A., Irvine M.R., Jarosz N., Lamaud E., Launiainen S., Law B.E., Longdoz B., Loustau D., McKay M., Paw U K.T., Vesala T., Vickers D., Wilson K.B., Goldstein A.H., 2007. Partitioning forest carbon fluxes with overstory and understory eddy-covariance measurements: A synthesis based on FLUXNET data. *Agricultural and Forest Meteorology*, 144 (1–2), 14–31.

Moffat A.M., Papale D., Reichstein M., Hollinger D.Y., Richardson A.D., Barr A.G., Beckstein C., Braswell B.H., Churkina G., Desai A.R., Falge E., Gove J.H., Heimann M., Hui D.F., Jarvis A.J., Kattge J., Noormets A., Stauch V.J., 2007. Comprehensive comparison of gap-filling techniques for eddy covariance net carbon fluxes. *Agricultural and Forest Meteorology*, 147 (3–4), 209–232.

Nellemann, C., Thomsen M.G., 2001. Long-term changes in forest growth: Potential effects of nitrogen deposition and acidification. *Water Air and Soil Pollution*, 128, 197–205.

Norman J.M., Garcia R., Verma S.B., 1992. Soil surface CO_2 fluxes and the carbon budget of a grassland. *Journal of Geophysical Research*, 97, 18845–18853.

Papale D., Reichstein M., Aubinet M., Canfora E., Bernhofer C., Kutsch W., Longdoz B., Rambal S., Valentini R., Vesala T., Yakir D., 2006. Towards a standardized processing of Net Ecosystem Exchange measured with eddy covariance technique: algorithms and uncertainty estimation. *Biogeosciences*, 3 (4), 571–583.

Papale D., Valentini R., 2003. A new assessment of European forest carbon exchanges by eddy fluxes and artificial neural network spatialization. *Global Change Biology*, 9, 525–535.

Peñuelas J., 2005. Plant physiology – A big issue for trees. *Nature,* 437, 965–966.

Raich J.W., Bowden R.D., Steudler P.A., 1990. Comparison of two static chamber techniques for determining carbon dioxide efflux from forest soils. *Soil Science Society of America Journal*, 54, 1754–1757.

Rayment M.B., Jarvis P.G., 1997. An improved open chamber system for measuring soil CO_2 effluxes in the field. *Journal of Geophysical Research* 102 (D24), 28779–28784.

Rayment, M.B., Loustau, D., Jarvis, P.G., 2002. Photosynthesis and respiration of black spruce at three organizational scales: shoot, branch and canopy. *Tree Physiology*, 22, 219–229.

Reichstein M., Falge E., Baldocchi D.D., Papale D., Aubinet M., Berbigier P., Bernhofer C., Buchmann N., Gilmanov T., Granier A., Grunwald T., Havrankova K., Ilvesniemi H., Janous D., Knohl A., Laurila T., Lohila A., Loustau D., Matteucci G., Meyers T., Miglietta F.,

Ourcival J.M., Pumpanen J., Rambal S., Rotenberg E., Sanz M., Tenhunen J., Seufert G., Vaccari F., Vesala T., Yakir D., Valentini R., 2005. On the separation of net ecosystem exchange into assimilation and ecosystem respiration: review and improved algorithm. *Global Change Biology*, 11 (9), 1424–1439.

Ryan M.G., Binkley D., Fownes J.H., 1997. Age-related decline in forest productivity: pattern and process. *Advances in Ecological Research*, 27, 213–262.

Ryan M.G., Binkley D., Fownes J.H., Giardina C.P., Senock R.S., 2004. An experimental test of the causes of forest growth decline with stand age. *Ecological Monographs*, 74, 393–414.

Ryan M.G., Bond B.J., Law B.E., Hubbard R.M., Woodruff D., Cienciala E., Kucera J., 2000. Transpiration and whole-tree conductance in ponderosa pine trees of different heights. *Oecologia*, 124, 553–560.

Schafer K.V.R., Oren R., Tenhunen J.D., 2000. The effect of tree height on crown level stomatal conductance. *Plant, Cell and Environment*, 23, 365–375.

Spiecker H., 1999. Overview of recent growth trends in European forests. *Water, Air, and Soil Pollution*, 116, 33–46.

Valentini R., Matteucci G., Dolman A.J., Schulze E.D., Rebmann C., Moors E.J., Granier A., Gross P., Jensen N.O., Pilegaard K., Lindroth A., Grelle A., Bernhofer C., Grunwald T., Aubinet M., Ceulemans R., Kowalski A.S., Vesala T., Rannik U., Berbigier P., Loustau D., Gumundsson J., Thorgeirsson H., Ibrom A., Morgenstern K., Clement R., Moncrieff J., Montagnani L., Minerbi S., Jarvis P.G., 2000. Respiration as the main determinant of carbon balance in European forests. *Nature*, 404, 861–865.

Vickers D., Mahrt L., 1997. Quality control and flux sampling problems for tower and aircraft data. *Journal of Atmospheric and Oceanic Technology*, 14, 512–526.

Vries W. d., Reinds G.J., Gundersen P., Sterba H., 2006. The impact of nitrogen deposition on carbon sequestration in European forests and forest soils. *Global Change Biology*, 12 (7), 1151–1173.

Watson D.J., 1947. Comparative physiological studies on the growth of field crops. *Annals of Botany*, 11, 41–76.

Webb E.K., Pearman G.I., Leuning R., 1980. Correction of the flux measurements for density effects due to the heat and water vapour transfer. *Quarterly Journal of the Royal Meteorological Society*, 106, 85–100.

Yang B., Hanson P.J., Riggs J.S., Pallardy S.G., Heuer M., Hosman K.P., Meyers T.P., Wullschleger S.D., Gu L.-H., 2007. Biases of CO_2 storage in eddy flux measurements in a forest pertinent to vertical configurations of a profile system and CO_2 density averaging. *Journal of Geophysical Research*, 112, D20123.

Yoder B.J., Ryan M.G., Waring R.H., Schoettle A.W., Kaufmann M.R., 1994. Evidence of reduced photosynthetic rates in old trees. *Forest Science*, 40, 513–527.

Chapter 3

Dynamics of soil carbon and moder horizons related to age in pine and beech stands

BERNARD GUILLET, JEAN-ROBERT DISNAR, DENIS LOUSTAU, JÉRÔME BALESDENT

Introduction

For a long time, the attention paid to soil organic matter (SOM) has been motivated by the well-recognized role that this constituent exerts on soil chemical, physical and biological properties (e.g. van Cleve and Powers, 1995; Robert, 1996; Karlen *et al.*, 1997). Increasing concern about climate change due to the release of greenhouse gases into the atmosphere now raises a question on the effective part played by soils, which are a major terrestrial reservoir of carbon (GIEC, 2000; Arrouays *et al.*, 2002). With the aim of understanding and forecasting variations of stock, the dynamics of forest soil organic carbon is currently described by compartmental models, for example, the Century model Parton *et al.* (1987), which is included in many ecosystem carbon models. These models describe the relationships between input fluxes and the distributions of residence times depending on soil and climate conditions. However, the following points can be formulated as to the use of these models.

– (i) If the input fluxes are easily accessible (excluding rhizodeposition), some values of the parameters can be called into question because they are largely inherited from model calibration on either cultivated or grassland areas, or on tropical forest ecosystems where natural C isotope tracing has been largely used (Jenkinson *et al.*, 1991, Trouvé *et al.*, 1994; Balesdent and Mariotti, 1996). Such isotopic approaches cannot be applied in temperate and boreal forest ecosystems. In these latter systems the ^{14}C isotope in natural abundance remains a valuable but insufficiently used tracer for measuring carbon residence times in forest soils (Balesdent and Guillet, 1982; Trumbore, 1996). Sites with a humus type moder, owing to their vertical stratification, are favourable for such a determination because deeper layers of humus may be hypothesized to be older. Therefore, analysis of ^{14}C according to layer depth is expected to provide an insight into carbon dynamics.

– (ii) The vertical stratification of carbon in holorganic O and organomineral A horizons (Elzein and Balesdent, 1995), together with its variations, is not often taken into account in models. On the one hand, it implies varied responses of the dynamics to microclimatic and edaphic factors and, on the other hand, it depends on variations of edaphic conditions and on management practices.

– (iii) Many models neglect the evolution of carbon stocks on the scale of a forest ecosystem life cycle, from disturbance to regeneration, growth, maturation and senescence, especially in the description of the initial states. In addition, models do not take into account the evolution of the vertical stratification in relation to the life cycle. However, such an evolution has been clearly proved by, for example, Jolivet (2000), Nys *et al.* (2002) or Huet *et al.* (2004; pers. comm.).

The aim of the present contribution to the French project CARBOFOR was to quantify the dynamics of carbon in typical temperate forest ecosystems, and more precisely, the build-up and turnover of the carbon stocks of horizons O and A_1. The work was carried out on two sites: the beech forest of Fougères (Brittany, France) and the pine forest of the Landes de Gascogne (Le Bray and Baudes stands, Aquitaine, France). In each case, we used the following approach to assess the temporal evolution of organic matter stocks in soil. Firstly, we established the relationship between the carbon content of humus and soil layers with stand age along chronosequences (Huet, 2004; Huet *et al.*, 2004; pers. comm.). Secondly, we related quantitative data on net primary production (NPP) and annual litterfall to the carbon stock in humus to deduce the apparent "turnover" of the organic matter. Thirdly, we developed and applied a simple model to date the soil organic matter by radiocarbon determinations. In the two study areas, stands of contrasting age were analysed.

Study sites and methods

The Fougères beech forest

The national forest of Fougères is located in the north-east of the Ille-et-Vilaine department in Brittany (France) (48° 20' N; 1° 10' E) at an elevation of 115–191 m above sea level. It covers a total surface of 1660 ha and is sub-divided into 130 stands. Beech (*Fagus sylvatica* L.), either alone or mixed with oak (pedunculate *Quercus robur* L. and sessile *Q. petraea* Liebl.) cover approximately 76% of the surface of the forest (Toutain, 1966). Coniferous trees, which have been planted for a few decades, occupy less than 60 ha. The climate is of the oceanic type, that is to say, temperate and wet. The average annual precipitation is 881 mm, with more marked rainfall in December and October. The mean annual temperature is around 10.3°C, with a monthly minimum and maximum of 3.9°C and 17.3°C, respectively.

The forest of Fougères is divided into even-aged stands and undergoes a high level of forest management. Harvesting consists of successive thinnings and clear-cutting: a first thinning is followed by one or more subsequent thinnings and final clear-cutting. The current management of the forest is based on the duration of a complete 150-year rotation. Four stands of different ages, FOU1 to FOU4, were sampled. In each stand, five or six sample plots, A–E or F, were selected depending on the nature and density of plant species in the understorey (Table 3.1). Nevertheless, the influence of these plants is negligible, the soil being frequently barren in most parts of the forest, except in the thicket.

In the largest part of the Fougères forest, the soils are of the slightly leached acid brown type, that is, Brunisol luvic according to the French nomenclature (Karroum, 2002) or alocrisol luvisol according to the FAO/UNESCO soil system (FAO, 1998). These soils developed on silty material, often mixed with sandy arena in the depth of the profiles. The soil cover is generally 1–2 m thick. According to whether they were formed on quasi-pure silts, or silts mixed with granitic arena, schists or quartzitic or alluvial material, the soils do not have the same texture or the same morphology, but their formation always followed the same evolutionary path. The soils are acidic and base saturation is low. Some indices of hydromorphy and/or moderate podzolization can sometimes also be observed at depth, when they present a thick humus of the dysmoder-type (the formation of which is favoured by acidifying species such as beech).

Table 3.1. Spatial variability and floristic heterogeneity of the studied plots in the Fougères forest.

Reference	FOU1: 10 years	FOU2: 27 years	FOU3: 87 years	FOU4: 145 years
Plot A	Gramineae (5) and fern (+)	Fern (+) and ivy (+)	Fern (1), ivy (+), carex (+) and wavy hair-grass (+)	Ivy (2)
Plot B	Gramineae (5), bramble (+) and moss (+)	Fern (1) and ivy (1)	Ivy (3) and polytric (1)	Barren with ivy (3)
Plot C	Gramineae (5), bramble (+) and moss (+)	Fern (+) and ivy (1)	Ivy (2)	Barren with ivy (1)
Plot D	Gramineae (5), fern (+) and bramble (+)	Fern (+) and ivy (+)	Ivy (2) and oak (2)	Wavy hair-grass (5)
Plot E	Fern (5)	Fern (1) and ivy (1)	Ivy (1) and polytric (+)	Ivy (5) and oxalis (1)
Plot F	Gramineae, bramble, honeysuckle, broom and moss			

(+) = presence; from 1 to 5 = increasing surface cover (Karroum, 2002 after Lebret, 2002).

The pine forest of the Landes de Gascogne: Le Bray and Baudes sites

The site of Le Bray (Jolivet, 2000) consists of one adult stand of maritime pine *(Pinus pinaster* Ait.) seeded in 1970. It is located in a wet zone of the Landes, with some localized hillocks of mesophilic moors. The understorey vegetation is primarily made up of purple heath grass *(Molinia coerulea* M.) in the wetlands and western brackenfern *(Pteridium aquilineum)* on the mesophilic hillocks. The soil, a humic podzol with A2 (E) and iron pan, displays hydromorphy at depth (podzosol duric). At the time of sampling (2003), the stand was 33 years old. Silviculture here includes some tillage of the soil before seeding. The pines at the Baudes site were 92 years old in 2003, and the soil is also a humic podzol with iron pan partly covered with tussocks of *Molinia*.

Soil sampling and analytical methods

Soil sampling and sample pre-treatment

At Fougères, three sample sets were used. In October 1997 all plots were sampled for carbon stocks. The 1997 sampling contributed to the ECOFOR programme (Guillet and Disnar, 2000, in Jabiol, 2000; Karroum, 2002; Karroum *et al.*, 2005). The objective of this project was to estimate the carbon stock of the soil and its variability in the holorganic layers, as well as in the upper organomineral horizons (A_1 sub-divided into A_{11} and A_{12} and occasionally E_1). Additional samples were taken at FOU3 and FOU4 in February 2000 (just after the severe storm in December 1999), and in March 2002 at FOU2 for additional carbon stocks and ^{14}C determination. The sampling procedure of the five replicates was the same for each plot and was also applied in the Landes de Gascogne sites.

At the Landes site, soil sampling was carried out in January 2003. At each site, five plots were selected at random inside a 20 m radius circle near the micrometeorological tower. The hologanic O horizon layers were taken inside a 0.25 m^2 square frame. The soil mineral horizons were extracted using a 15 cm long and 8 cm wide cylinder with a manual corer. The cores were separated into layers or horizons in the laboratory. The samples collected below the *Molinia* tussocks were treated separately taking care to measure the surface area that they occupied, their thickness and their wet weight. After separation of the roots from the soil, the intimately associated fine root fractions were weighed, processed and analysed separately. Annual litterfall values were estimated from measurements made on 24 litterfall traps collected monthly from 1996 onwards (data not shown).

Analysis

In the laboratory, the thickness of the horizons was measured and the various horizons or even layers were subsequently separated from one another "by scouring". After having been weighed wet, all the samples, as well as the roots and large woody remains, were dried at 55°C and reweighed. The five replicates were treated separately for total organic carbon (TOC) and bulk density and pooled for ^{14}C dating.

Soil bulk density was determined after oven drying at 55°C: the dry weight of the whole sampled horizon was divided by the surface area of either the frame (O horizons) or the corer (A and E), and then by the thickness, so that the stocks were not biased by uncertainty about thickness determination. The roots were separated from the soil samples and subsequently dried and weighed.

All the dry samples were submitted for organic carbon determination before any other treatment. In several cases, this analysis was a prerequisite for further analyses, especially the preparation of composite samples, such as those used for ^{13}C and ^{14}C isotopic measurements (see below). The organic carbon content of the Landes soils was determined by combustion using an elementary analyser (Flash EA, Thermofinnigan). The carbon and nitrogen content of the Fougères soil samples were determined either with a Leco CNS 2000 analyser, or by Rock-Eval® pyrolysis with a model 6 "Turbo" device (Vinci Technologies). The analyses were carried out on 100 mg of crushed sample under standard conditions (Disnar *et al.*, 2003). Since this latter technique underestimates the TOC of the soils, we applied the following corrections as recommended by Disnar *et al.* (2003): the contents of the TOC provided by the apparatus were multiplied by 1.166 if they came from samples of litter (OL and OF) rich in polysaccharides, and by 1.092 for all the others.

The stock of carbon in each horizon/layer was calculated as:

$$\text{C-stock (kgC·m}^{-2}) = \text{thickness (m)} \times \text{density (kg soil·m}^{-3}) \times \text{TOC (kgC·kg soil}^{-1}) \quad (1)$$

^{14}C and ^{13}C measurements were carried out on composite samples made of aliquots of the five individual samples of each layer. The ^{14}C activity measurements of the soil organic matter were made at the Centre of Radiocarbon Dating (Lyon, France) using liquid scintillation. The results are expressed as:

$$\Delta^{14}\text{C (‰)} = 1000 \times (A - Ao)/Ao \quad (2)$$

where A is the activity of the sample and Ao is $0.95 \times$ the activity of the NBS 1154 standard. The unit Δ^{14}C is corrected for isotopic fractionations.

The apparent age ^{14}C (unit BP, Before Present with "Present" = 1950) is conventionally defined as:

$$\text{apparent age} = -8035 \times \text{Ln (A/Ao)} \quad (3)$$

Principle of interpretation of ^{14}C activities

The measured radioactivity of a carbon sample is equal to the sum of the mean carbon activities of the various components of SOM, the age of which varies from zero (fresh leaves of the year) to several millennia. The ^{14}C activity of atmospheric CO_2 increased considerably after 1955 following the thermonuclear tests carried out in the atmosphere by the USA and USSR (Figure 3.1). After reaching a maximum in 1963, the production

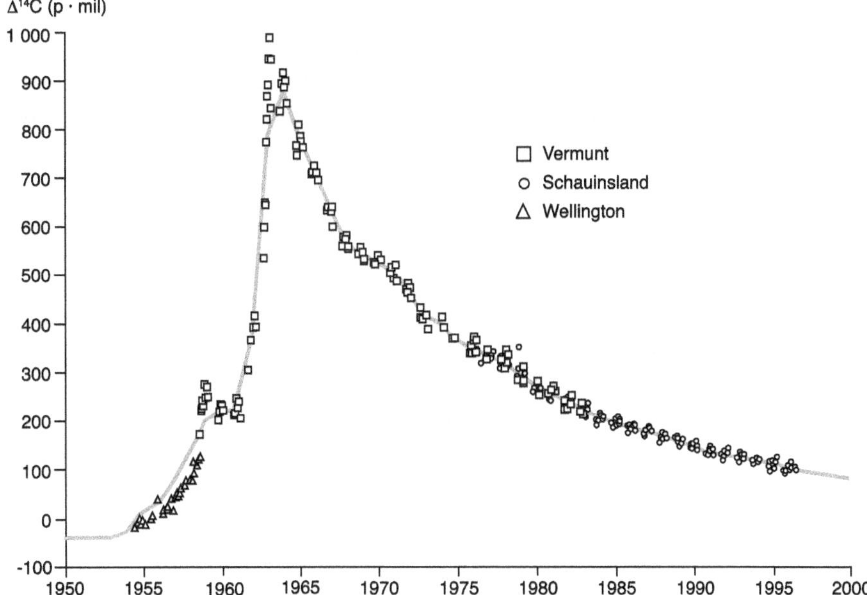

Figure 3.1. Database of tropospheric $^{14}CO_2$ made up for the study (curve) and values measured in spring and summer at the European observatories of Vermunt and Schauinsland (according to Levin and Kromer, 1997 and Levin *et al.*, 1994), and, for comparison, Wellington in 1953–59.

of thermonuclear ^{14}C decreased and is still decreasing. This [14]C "peak" has turned out to be a fortuitous isotopic labelling, firstly of plants that use CO_2 for their metabolism and, secondly, of SOM through the incorporation of ^{14}C-enriched litterfall material. This radiolabelling of SOM allows us to determine carbon residence times, provided realistic assumptions can be made about the age distribution of the various components of the considered soil sample (Balesdent and Guillet, 1982; and the review by Wang and Hsieh, 2002). In the present study, the activity data are interpreted in terms of age by means of the compartmental model described previously.

We built a database for the tropospheric ^{14}C activities in the northern hemisphere from the arithmetic mean of measurements made each year from April to September by Levin and Kromer (1997) from 1979 to 1996, and Levin *et al.* (1994) from 1954 to 1978. The data for the period 1996–2001 are extrapolated and fixed on the values of 1998–99.

Results and discussion: experimental data

Organic carbon stocks

Fougères beech forest

The distribution of TOC in the Fougères forest soils and its variation with tree age can be defined when referring to the two datasets sampled in 1997 and 2000–02. The TOC values of the first sample set represent the mean of TOC determinations made on samples taken from the different layers of five to six soil profiles from a given beech stand. In contrast, for the 2000–02 sampling dataset, TOC determinations were carried out on composite samples that were mainly prepared for ^{14}C analysis and were assembled from soil layer aliquots extracted in five different profiles from the same stand. The results obtained with the first sample set are presented in Tables 3.2 and 3.3 and are discussed below.

Table 3.2. Total organic carbon contents in the various layers or horizons of the soils of the Fougères forest sampled in 1997 (average values and variation coefficient for five to six samples of a given horizon in a given stand).

	FOU1: 10 years		FOU2: 27 years		FOU3: 87 years		FOU4: 143 years	
	Average TOC	Variation coefficient	Average TOC	Variation coefficient	Average TOC	Variation coefficient	Average TOC	Variation coefficient
	$(mg \cdot g^{-1})$	(%)	$(mg \cdot g^{-1})$	(%)	$(mg \cdot g^{-1})$	(%)	$(mg \cdot g^{-1})$	(%)
OL	338.0	25.1			488.5	2.3	479.8	1.2
OF					462.9	2.1	451.9	12.3
OL, OF			432.3	6.7				
OH			287.3	32.2	317.3	30.5	262.5	33.1
A11	115.1	33.6	138.9	44.7	87.8	30.4	75.3	44.9
A121	61.7	60.4	75.7	30.2	46.7	11.2	42.2	28.0
A13	29.7	18.7	43.6	24.7				

TOC: total organic carbon.

Table 3.3. Mean organic carbon stock values and coefficient of variation in the holorganic O and A_1 horizons of soils of the four stands that were studied at Fougères.

	FOU1 (thicket)		FOU2 (sapling)		FOU3 (young high forest)		FOU4 (old high forest)	
	Mean stock	Variation coefficient	Average stock	Variation coefficient	Average stock	Variation coefficient	Average stock	Variation coefficient
	kgC·m^{-2}	(%)	kgC·m^{-2}	(%)	kgC·m^{-2}	(%)	kgC·m^{-2}	(%)
OL					0.72	43.30	0.50	51.50
OF					0.59	30.40	0.83	47.30
OL + OF	0.43	28.20	1.20	47.00	1.31	31.60	1.33	45.10
OH			0.77	132.30	0.93	18.50	1.18	34.30
Total O	0.43	28.20	1.97	68.50	2.34	22.40	2.71	18.30
A1	4.03	32.40	5.48	30.70	1.96	24.00	2.05	26.90
Total soil	4.45	60.60	7.45	22.30	4.30	12.60	4.75	16.00

According to Guillet and Disnar (2000).

TOC values of 480–490 mg·g^{-1} in the OL soil layers of young and old mature forest stands (FOU3 and FOU4) are typical of recent dead plant material values, especially leaves (Guillet and Disnar, 2000). Slightly lower TOC values of 450–460 mg·g^{-1} in OF suggests a minor mineral contamination of this layer. The much lower TOC values in the OH layer (320 mg·g^{-1} in FOU3 and 260 mg·g^{-1} in FOU4) are consistent with the presence of organomineral aggregates as observed. The much greater variability in the OH (variation coefficient 30–33 mg·g^{-1}) than in the OL and OF, mostly reflects specific features of the plots, such as the presence of *Deschampsia flexuosa* in one FOU4 plot (Table 3.1) and another unidentified factor in FOU3. The greater variability observed in A_{11} than in A_{12} is mostly attributable to some mixing between A_{11} and the overlying OH layer.

In FOU2, the percentages of organic carbon in the OL are lower than those observed in FOU3 and FOU4. While this is due to a mixing with fragments of litter of the OF type, it is mainly the result of the introduction of mineral matter by earthworms. The OH layer, sometimes absent (FOU2B, FOU2C), presents large variability of content, probably because of an imperfect control of sampling caused by irregular boundaries between OH and A_{11}.

Variability is also large (45%, $N = 7$) in FOU2 horizons ($A_{11} + A_{12}$). On average, they are richer in organic carbon (140‰) than in FOU3 (90‰) and FOU4 (75‰). This is presumably due to a stronger biological stabilization of organic matter by fine mineral fractions caused by the action of earthworms, which are more abundant in FOU2 and FOU1.

In FOU1, the variability of TOC contents of the OL is the largest of all the analysed litters. As the TOC values are also much lower than in other sites (340‰ on average), it can be assumed that this variability arises from a non-homogeneous SOM dilution with mineral matter, undoubtedly brought up by the mesofauna. In the organomineral horizons, the percentages of organic carbon are comparable to those observed in FOU2.

The average stock of organic carbon and its variability within each station are presented in Table 3.3 and Figure 3.2. The following points about the stock of organic matter and its vertical distribution are worth mentioning: in the soil of the sapling stand (FOU2), an important organic carbon stock of moderate variability (22%) resulted from storage in the A_1 horizon. The cause of this is certainly to be found in the activity of earthworms, which acted as stabilizers of the humus. In FOU2, there was also high stock variability (68%) within the O horizons due to the presence of both mull- and moder-like humus forms in this stand, the stocks in the mull type being necessarily lower than those of the moder type. In contrast, lower variability (30%) was observed in the A_1 horizon at the same site.

In the high forest (FOU3 and FOU4), the organic carbon stocks were not significantly different in the OL + OF, OH and even the A_1 horizon. The relatively low values of the coefficients of variation of the carbon stocks in the dysmoder are consistent with the visual impression of homogeneity observed at these two sites, but especially so in FOU3.

Table 3.4. Annual litterfall, carbon stocks in humus layers and estimated carbon mean residence times at the Fougères sites as calculated from 1997 and 2002 measurements.

1997 measurements		**FOU1**	**FOU2**		**FOU3**	**FOU4**
Mean litterfall						
Mass	(kg dw·m^{-2}·year^{-1})	0.21	0.38		0.39	0.47
Carbon	(kgC·m^{-2}·year^{-1})	0.089	0.161		0.165	0.2
Horizon OL + OF:						
carbon stock	(kgC·m^{-2})	0.424	1.179		1.299	1.323
K = flux/stock	(year^{-1})	0.21	0.137		0.127	0.151
MTR = stock/flux	(year)	4.76	7.32		7.87	6.62

2002 measurements		**FOU2 mull**	**FOU2 moder**	**FOU3**	**FOU4**
Horizon OL + OF:					
carbon stock	(kg·m^{-2})	0.481	0.689	1.118	0.916
K = flux/stock	(year^{-1})	0.335	0.234	0.148	0.218
MTR = stock/flux	(year)	2.99	4.28	6.78	4.58

Annual litterfall from Lebret (2002). MTR: mean time of residence.

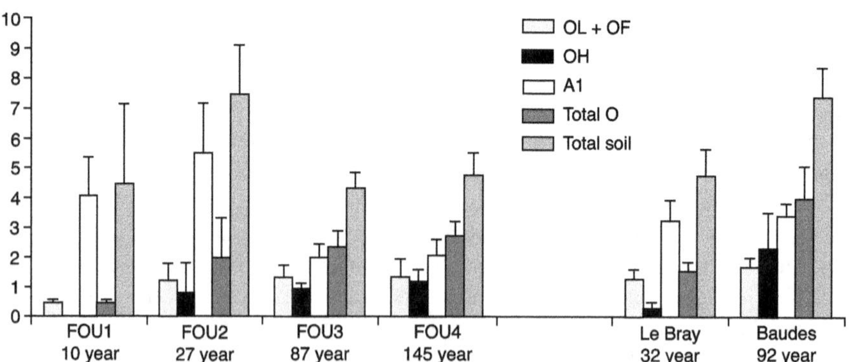

Figure 3.2. Carbon stocks in organic horizons of stands of different age at the Fougères (average and standard deviation for five or six profiles) and the Landes forests. All values in kgC·m^{-2}.

The Landes pine forest

The most striking differences between the profiles of Le Bray (32 years) and Baudes (92 years) are: (i) the absence of an OH horizon at Le Bray and the advanced development of the OH horizon at Baudes; (ii) a larger carbon stock in the 15-45 cm layer at Le Bray. Altogether, the carbon stock in the Baudes site soil + humus is 50% higher than that of Le Bray.

Vertical root distribution

At the FOU3 and FOU4 sites, the root biomass is distributed in a rather homogeneous way down to 15–20 cm (Figure 3.3). Although present in the O horizons, the roots mainly occur in horizons A (0–20 cm).

The soils of the sites of Bray and Baudes form a kind of mosaic made of two systems: humus with the traditional morphology of a moder and *Molinia* tussocks, the latter accounting for approximately 25% of the surface and below which *Molinia* roots are densely imbricated forming a mat of roots. The dry matter mass of this *Molinia* root mat accounts for 20% of litter at the two sites. The *Molinia* root system occurs in the OH horizon at the Baudes site and in the A_1 horizon at Le Bray, where the OH horizon is missing. Apart from this *Molinia* root mat, the majority of roots are found exclusively in the A horizons, where their volumetric density decreases exponentially with depth (Figure 3.3).

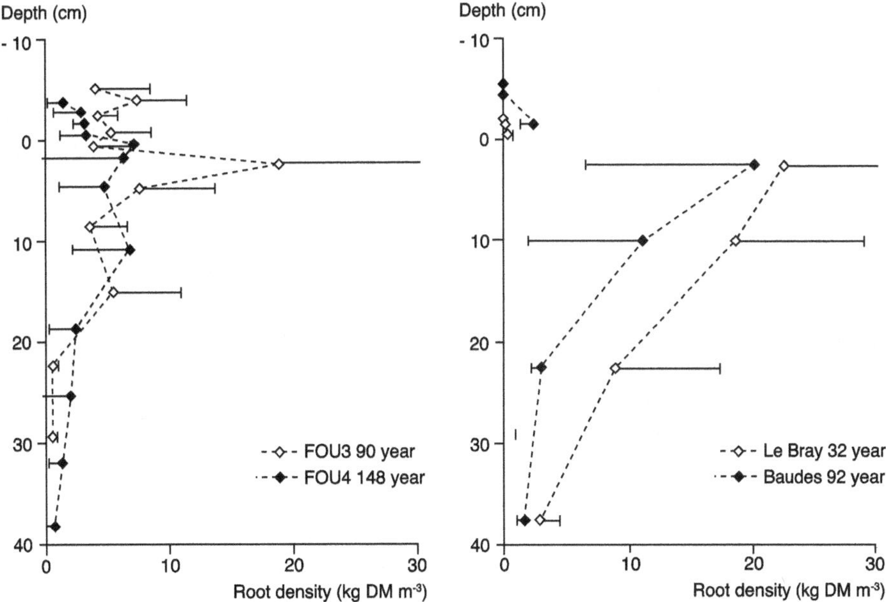

Figure 3.3. Depth profiles of root mass per unit volume of soil. The roots of the pine forest *Molinia* tussocks (Le Bray and Baudes) are not represented.

Vertical distributions of the ^{13}C and ^{14}C isotopes

The ^{13}C profiles (Figure 3.4) in the young and mature high forest stands FOU3 and FOU4 are typical of temperate forest soils developed on a light and thick material, comprising an argillaceous phase (Balesdent *et al.*, 1993). The enrichment of 2‰ to 3‰ with depth can be explained classically (Balesdent and Mariotti, 1996) by one or several of the following factors: (i) the evolution with increasing age of the photosynthesized CO_2, the present day CO_2 being depleted by 1.5‰ compared with pre-industrial CO_2; (ii) the predominance of lignin (depleted in ^{13}C) in the O horizons; (iii) the enrichment that occurs in the heterotrophic (microbial) trophic chain, the products of which become predominant in evolved matter; (iv) the downward transport of fulvic compounds rich in carboxyls and in ^{13}C; (v) the possible presence of a background of carbon rich in ^{13}C of periglacial times (Adams and Faure, 1996). The most noticeable fact is the low Δ^{13}C in the O horizons of FOU2 plots that cannot be solely explained by the young age of the carbon. This seems to be related to a particular and transitory state; either to the trophic status of the young trees or to a high proportion of litterfall and root exudates from the understorey depleted in ^{13}C.

In the Landes de Gascogne, the ^{13}C distribution is identical at the two sites. The podzols of this area differ from the leached soils of Fougères by having a lower enrichment at depth. This can be related to the absence of clay fractions likely to retain the isotopically enriched fractions (see factors (iii), (iv) and (v) above). The little data available on *Molinia* plant ^{13}C composition at the Bray site (unpublished data) suggests that the recycling of respired CO_2 via photosynthesis is not significant here.

The ^{14}C profiles (Figure 3.4) account for the enrichment in the thermonuclear ^{14}C isotope. The 1964 peak of radioactivity (36 years old at the time of sampling) is located at the top of the profile, in the OH_1 in the moder. This horizon is lacking in mull because of the incorporation of SOM in the A horizons by fauna. The profiles of FOU3 and FOU4 are very similar. Overall, the O horizons at the FOU2 site with moder are younger than those of the high forest sites. Activity may also be slightly lower at this site due to the more recent sampling date. Below 20 cm, the figures correspond to apparent ^{14}C ages greater than 1000 years.

In the two studied sites of the Landes, the ^{14}C profiles are very similar to those of equivalent age in the Fougères beech forest. In the Landes, the addition of post-bomb ^{14}C is also higher in the 92-year-old stand (Baudes), with a maximum in the OH horizon. For all the horizons, the apparent age of the carbon is greater in the 92-year-old stand (Baudes). It is more marked by the post-bomb C of the years 1960 in the O horizons, and is less active in the organomineral horizons. The *Molinia* root mat and the fine organic matter associated with it were dated separately. They have the same ^{14}C activities as those of the horizons from which they were separated, that is, OL and A_1 for the 32-year-old stand and OH for the 92-year-old stand. As a consequence, in the dynamic study, the material of these *Molinia* tussocks were regarded as pertaining to these horizons, respectively. The origins (*Molinia* or pine) of the O horizons will thus be merged. These distributions are interpreted in the last section with the modelling results.

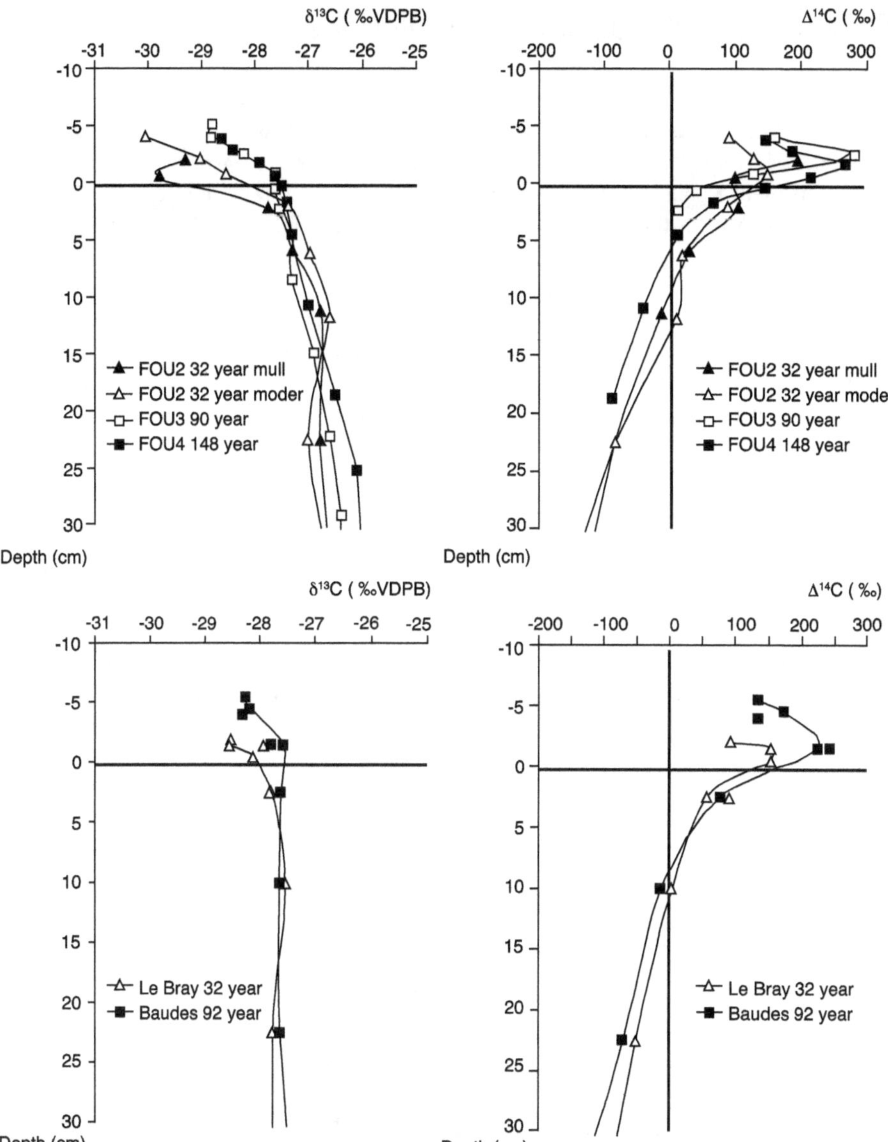

Figure 3.4. Vertical profiles of the ¹³C and ¹⁴C carbon isotopic composition of the humus and soil organic matter at the Fougères and Landes sites.

Results and discussion: modelled carbon dynamics

Model

The objective of the modelling was to determine the distributions of carbon residence times in the studied soils. We used here typically a regression procedure giving the best adjustment to observed stocks and ^{14}C activities, horizon by horizon. The horizons were pooled to form three layers: OL + OF called OLF, OH, and A_1 level 0–10 cm or 0–20 cm.

The equation system for horizon age distribution is given by a box model (Figure 3.5). Typically, the carbon is transferred from OLF towards OH, and then from OH towards A_1 and each compartment is assumed to be perfectly mixed. In the case of transient modes, the biodegradation and transfer equations follow a first order law ($dC/dt = -k \cdot C$). The ^{14}C activity at the date of sampling is calculated numerically accounting for the annual contributions of each year from infinite time until sampling.

In addition to the vertical transfer of carbon from litterfall, a sub-model for carbon injection into the soil from the roots accounts for the direct incorporation of young carbon at each depth. The following assumptions are incorporated:
– the carbon flux injected into the soil down to 20 cm should be equal to 40% of the litterfall input. This flux is directed towards the horizons in proportions equal to the respective measured root masses (Figure 3.3);
– this flux also follows the same laws as the OLF horizon: a proportion is allocated to a "resistant plant material" (RPM) compartment (Jenkinson and Rayner, 1977), corresponding to the root litter, with the same biodegradation rate value as that of the OLF horizon;

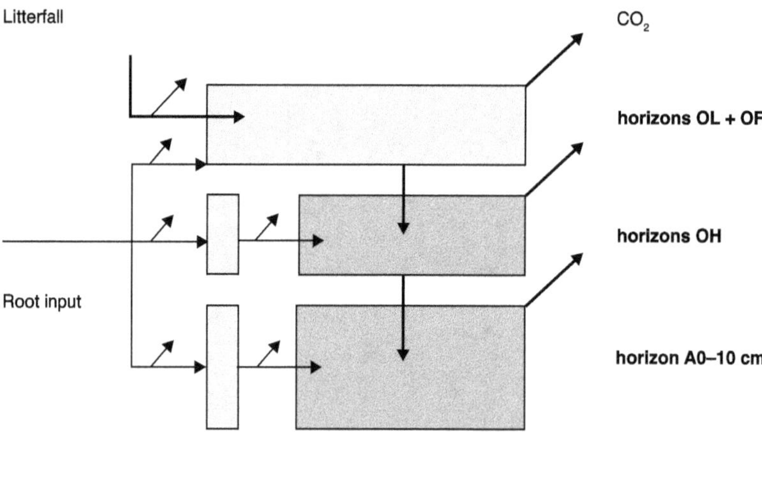

Figure 3.5. Model developed to describe the age distributions of organic carbon in the moder humus at the Fougères and Landes sites. RPM (resistant plant material) and HUM (humic and protected) refer to terms used by Jenkinson and Rayner (1977).

– a proportion of 10% is allocated to the compartments known as humic. This 10% is inferred from: (i) the flux of transfer from litterfall into the A horizons; and (ii) the measurement of long-term stabilization of plant material in agricultural systems (Jenkinson and Rayner, 1977; Balesdent, 1996). This assumption mostly concerns A horizons.

Steady-state age distribution of carbon

At Fougères, in the 90-year-old and 148-year-old stands (FOU3 and FOU4), stocks and ^{14}C activities are similar for all horizons. The system is thus considered to be in a stationary regime. This is valid for the OL, OF and OH horizons, whose mean times of residence (MTR) are largely lower than the time sequence age. The average flow of carbon at the two sites was thus used to adjust the model. An average annual litterfall-C flow value of 0.175 kgC·m^{-2}·year^{1} was retained, to which a flow of 0.07 kgC·m^{-2}·year^{1} was added from the root contribution. The adjustment with observed data is excellent because of the small variation between the two sites and the low number of degrees of freedom (Figure 3.6). The predicted values of the carbon MTR in the O horizons are close to the values estimated using ^{14}C activities. This is due to the strong thermonuclear ^{14}C signal and small root contribution. The model performance is weaker for the E horizons because of the weight of the assumption on root flow and the possible contribution of an old and stable carbon pool. The Baudes site may also be considered to be in a stationary state, the MTR of the O horizon being mostly lower than 92 years. There is no adjustment here, the number of degrees of freedom for the model is nil in this case. Data for the Bray and Baudes sites are given in Table A1 in the Appendix.

The model results (Table 3.5) account for the heterogeneity of the MTR of carbon in the various soil horizons and the partitioning between carbon mineralization and its transfer to subjacent horizons. The most outstanding features are as follows.

The age varies from 5 years (Fougères) to 8 years (Landes) in the OLF horizon and from 60 years (Fougères) to 64 years (Landes) in the A$_1$ horizon (0–10 cm).

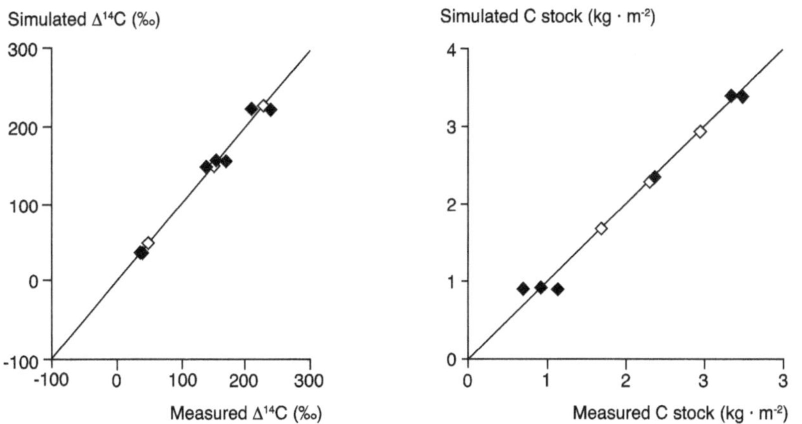

Figure 3.6. Adjustment of the simulated versus observed ^{14}C values and stocks for the FOU3 and FOU4 sites.

Table 3.5. Parameters of the dynamics of organic carbon in the moder of the old stands, in stationary regime. Forest of Fougères (A) and pine forest of the Landes (B). Each row provides input and output carbon flow together with stocks, decay rate, mean time of residence and age in a carbon compartment. In the OH and A horizons, compartment RPM corresponds to root litter only.

(A) Fougères beech forest

	Input flow	Stock	K	MTR	Mean age	^{14}C activity in 2000	Output flow		
							Mineralization (CO_2)	Transfer	Destination
	(kg·m^{-2}·year^{-1})	(kg·m^{-2})	(year^{-1})	(year)	(year)	(PMC)	(kg·m^{-2}·year^{-1})		
OL + OF:									
Instant mineralization				< 1			0.072		
OL + OF	0.107	0.91	0.116	9	9	115.6	0.012	0.096	OH
OH:									
Instant mineralization				< 1			0.002		
OH RPM	0.003	0.03	0.116	9	9	115.6	0.003	0.001	OH
OH HUM	0.096	2.32	0.042	24	33	122.4	0.077	0.019	A1
A1 0–10 cm:									
Instant mineralization				< 1			0.014		
A1 0–10 cm RPM	0.020	0.18	0.116	9	9	115.6	0.017	0.003	A1 HUM
A1 0–10 cm HUM	0.023	3.22	0.007	143	172	102.8	0.015	neglected	A1 10–20 cm
A1 10–20 cm:									
Instant mineralization				< 1			0.010		
A1 10–20 cm RPM	0.016	0.13	0.116	9	9	115.6	0.016		
A1 10–20 cm HUM	neglected	1.89	ND				neglected		
Total OL + OF	0.179	0.91	0.116	5	9	115.6	0.083	0.096	
Total OH	0.101	2.35	NA	23	32	122.3	0.082	0.019	
A1 0–10 cm total	0.053	3.39	NA	60	164	103.5	0.046		

... —

Table 3.5. (continued)

(B) Baudes Pine forest

	Input flow	Stock	K	MTR	Mean age	^{14}C activity in 2000	Output flow		
							Mineralization (CO_2)	Transfer	Destination
	(kg·m^{-2}·year^{-1})	(kg·m^{-2})	(year^{-1})	(year)	(year)	(PMC)	(kg·m^{-2}·year^{-1})		
OL + OF:									
Instant mineralization				< 1			0.0320		
OL + OF	0.168	1.682	0.100	10	10	115	0.0735	0.0945	OH HUM
OH:									
Instant mineralization				< 1			0.0023		
OH RPM	0.0011	0.019	0.100	10	10	115	0.0028	0.0001	OH HUM
OH HUM	0.0946	2.293	0.041	24	34	122.8	0.0765	0.0182	A1 HUM
A1 0–10 cm:									
Instant mineralization				< 10			0.0136		
A1 0–10 cm RPM	0.0208	0.210	0.100	10	10	115	0.0188	0.0021	A1 HUM
A1 0–10 cm HUM	0.0202	2.730	0.007	135	167	104.2	0.0150	neglected	A1 10–20 cm
E 10–20 cm:									
Instant mineralization				< 1			0.0034		
E 10–20 cm RPM	0.0135	0.130	0.100	10	10	115	0.0135	0.0013	A HUM
E 10–20 cm HUM	0.0013	1.890	ND				neglected		
Total OL + OF	0.200	1.68	0.100	8	10	115.0	0.106	0.095	
Total OH	0.096	2.30	NA	24	34	122.8	0.082	0.018	
A1 0–10 cm total	0.044	2.94	NA	64	156	105.0	0.047		

K = decay rate; MTR = mean time of residence; HUM = humic and protected; RPM = resistant plant material.

The ^{14}C activity in the OLF horizons is made compatible by partitioning the annual carbon input between two fractions:

– a fraction consisting of a short residence time (<1 year) carbon, 40% at Fougères and 20% in the Landes;

– a remaining fraction, 60% and 80% respectively, engaged in the construction of OLF horizons with an MTR of 8–10 years.

This partitioning agrees with litter mass loss rates generally observed in the first year.

The mean carbon age in OH horizons is 33 years (Fougères) or 34 years (Landes). It is thus expected that in a sequence or rotation in which this horizon was initially buried, its rebuilding could extend over many decades.

The average age of the A_1 horizon (0–10 cm) is nearly 160 years. This value is higher than that predicted by standard models like Century or RothC, but the mean age of O + A is more consistent with the latter. The E horizons are even older. It is assumed that the participation of these horizons in soil respiration is primarily due to the contribution of the roots.

The source of heterotrophic respiration can be specified as 35% from OL + OF, 35% from OH and 30% in A_{11} (0–20 cm) according to the model in both sites.

The heterotrophic respiration of the A horizons is mostly due to roots. However, the organic matter present in these horizons comes mainly from litterfall.

The results obtained from Landes and Fougères are very similar. This was expected since similar ^{14}C activities were found at both sites (Figure 3.7). The MTR in the OH and A horizons are almost identical. On the other hand, the Fougères site has a much higher proportion of litterfall with a short residence time (40%) than at Landes (20%). This is consistent with the larger OLF horizons and the biochemical composition of litterfall comprising mainly dead needles in Landes.

Detailed distribution of the ages of carbon
in the profile of the old beech stand (FOU4)

The sampling of thin layers from the moder-type humus profile at the FOU4 beech stand makes it possible to propose a detailed age distribution in the O horizons. The carbon content and stocks of the various layers distinguished in the profile, as well as the corresponding ^{14}C measurements, are presented in Table 3.6. The stock represented by all the layers of this 11 cm long section contains almost half of the "total carbon" stored in the top-soil. The positive ^{14}C activity from OL to A_{12}, which reveals an obvious contribution of organic matter marked "during their life" by the ^{14}C of atomic bombs, was converted to ages according to the following hypothesis: inputs enter solely into OL, each layer being a perfectly mixed compartment transferring part of its carbon to the next underlying layer. Such an interpretation generates an increasing dispersion of ages with depth. The strongest positive ^{14}C values were observed in OH_1 and OH_2. Negative activity values in A_{13} to E_1 correspond to older C (Figure 3.4), the age calculated for the carbon stock of E_1 at a depth of 17 cm is 740 ± 30 years BP. The excessive age in OL may be attributed to an underestimation of the residence time of carbon within living trees between photo-assimilation and senescence. The results of this age model in cascade remain in general agreement with those of the generic model described above.

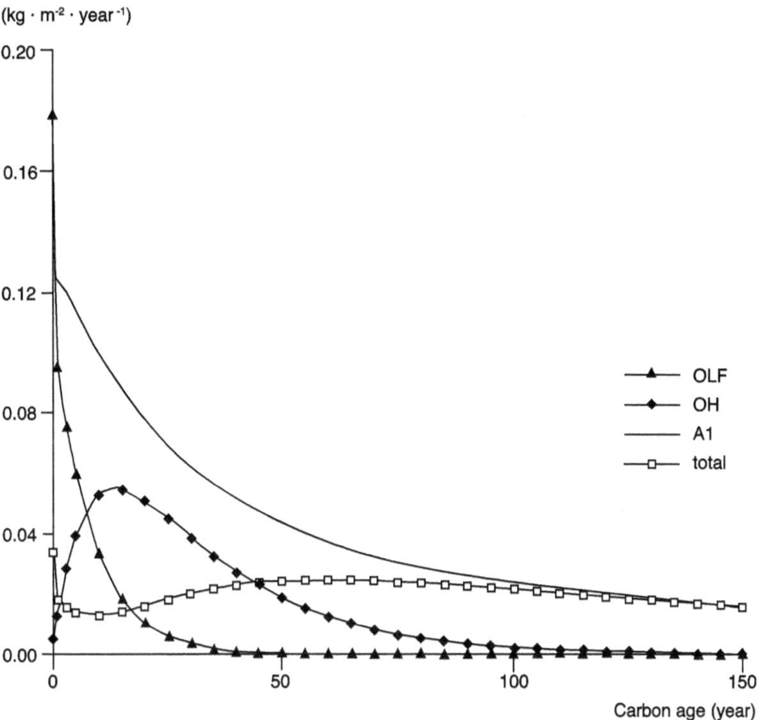

Figure 3.7. Distribution of the ages of carbon in moder of the oldest stands of the forest of Fougères in stationary regime. The distribution is almost identical for the Landes site (not represented).

Evolution of stocks with increasing stand age

As discussed in the previous sections, the differentiation of soil profiles with stand age is dominated by the progressive formation of the OH horizon of the moder. In order to relate this phenomenon to the MTR of carbon, we calculated the kinetics of formation of the O horizons starting from an initial absence of O horizon and following the predictions of the dynamics modelling (Figure 3.8). The observed values of stocks were used to calibrate the model, except those of the 32-year-old stand. Although lower than expected, these latter values remain compatible with the model. This weakness can be explained by the fact that the model supposes that the litterfall flux is already at a maximum from the very start of tree growth, thus neglecting the progressive increase in primary production with increasing tree age. In fact, this latter factor must not be overemphasized, the amount of annual litterfall increases notably between the thicket and sapling stages (and thus before the beginning of moder formation), but tends to level off subsequently as trees age (Lebret *et al.*, 2001 and references therein; see Chapter 2 in this volume). The appearance of OH horizons is progressive and spreads out over nearly 100 years. Our model can be used for predicting the O horizons temporal dynamics even if several issues still need to be addressed:

Table 3.6. Carbon stock and ^{14}C dating of soil horizons in the oldest site FOU4 in Fougères.

FOU4 Horizon	Depth (cm)	TOC (%)	Stock (kgC·m^{-2})	Δ^{14}C (‰)	MTR* (year)	Mean age (year)	Stock/MTR (kgC·m^{-2}·year^{-1})	Cumulated stock (kgC·m^{-2})
OL	(−4.3 – 3.3)	44.80	0.358	145 ± 5	6	6	0.060	0.358
OF	(−3.3 – 2.3)	42.88	0.557	185 ± 5	4	10	0.139	0.915
OH1	(−2.3 – 1.1)	34.20	1.067	267 ± 5	7	17	0.152	1.982
OH2	(−1.1 – 0.0)	29.77	1.277	214 ± 5	26	44	0.049	3.259
A11-OH3	(0.0 0.7)	12.42	0.522	144 ± 5	12	56	0.044	3.781

MTR: mean time of residence; TOC: total organic carbon.
* MTR and average age values were determined by means of a model said to be "in cascade" (see text).

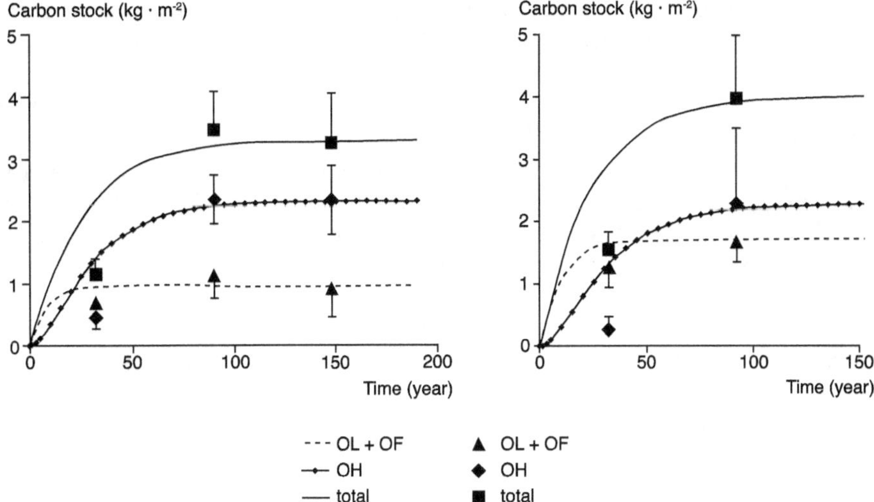

Figure 3.8. Kinetics of formation of the O horizons at the Fougères (left) and Landes (right) sites, as predicted by model ^{14}C carbon dynamics, starting with an absence of initial O horizon. The carbon input to the soil is supposed to be constant from time zero. Ordinates: carbon stock (kg·m⁻²). Lines: model prediction. Symbols: stock data measured.

– the sensitivity of these dynamics to the initial state (unknown here);
– the role of the progressive conversion of the mull-type humus (observed in FOU1, 10 years, and FOU2, 30 years) to moder-type humus (FOU2 moder, FOU3 and FOU4).

The latter can be related to the progressive disappearance of earthworms (Deleporte, 2001; Ponge, 2003), together with the export of cations and the resulting acidification of beech litter and O horizons. This process thus appears central to the simulation of carbon dynamics in the O horizons in relation to the age of the stands.

Concluding remarks

The model we used to interpret ^{14}C values and describe carbon dynamics was of the simplest, using only two compartments in each of the three layers investigated. The same model reproduced both the age distribution under steady-state regime (mature stands) and the kinetic build-up of stocks and layers. This means that, on the one hand, a model with first-order hypotheses appears sufficient for a general description of bulk carbon dynamics, and that, on the other hand, most of the dynamics can be described by the same parameters throughout the sequence. The other surprising finding is that the carbon dynamics appeared very similar in sequences with contrasting vegetation and soil type (i.e. beech Fougères forest versus pine Landes forest). This might result from a coincidence of factors acting in various directions, for example, acidity and aluminium in the Landes site reducing decay rates and compensating higher temperature than in the Fougères site. However, it strengthens confidence in the observed magnitude of carbon residence time. The magnitude of these residence times is compatible with those of models such as Century (Parton *et al.*, 1987) or RothC (Jenkinson *et al.*, 1991). The

OL and OF horizons are of an age close to that of structural or resistant plant material compartments of these models. In contrast, the OH horizon clearly belongs to the slow compartments. Root-derived carbon input has to be taken into account to explain carbon ages in the A horizons, and probably represents the major part of the heterotrophic respiration in the A horizon of moder-type humus.

During the first century of a rotation, the development of the O horizons will be determined by the previous history of the stand and the initial state of humus layers and A_1 horizon. If these horizons have been removed or buried, their rebuilding may extend over more than 50 years. The rate of this process will be also determined by the pedobiological differentiation of humus into mull or moder. In order to determine the impact on the total stock (O + A horizons), it would be necessary to analyse the mineralization rate of the O horizon material that has been incorporated in A_1. It can be concluded from our results that the non-steady-state of O horizons must be taken into account when modelling the dynamics of forest soil carbon. The transition of humus from a mull-type to a moder-type appears determining for the simulation of carbon dynamics in relation to the age of the broadleaved forest studied, whereas accumulation of carbon in O horizons and thickening were the main age-related trends observed in the coniferous stand.

Acknowledgements

The authors would like to thank Mark Bakker for his contribution to the selection and sampling of the sites in Landes, Delphine Derrien for her help in the field and Christine Marol for the ^{13}C isotopic analyses. The study was supported by the program CARBOFOR, INRA, CNRS and the University of Orléans.

References

Adams J., Faure H., 1996. Changes in moisture balance between glacial and interglacial conditions: influence on carbon cycle processes. *In: Global Continental Changes: The Context of Palaeohydrology* (J. Brandson, A.G. Brown, K.J. Gregory, eds). *Geological Society Special Publication*. Geological Society, London, pp. 27–42.

Arrouays D., Balesdent J., Germon J.C., Jayet P.A., Soussana J.F., Stengel P., 2002. *Stocker du carbone dans les sols agricoles en France ? Expertise scientifique collective*. INRA Éditions, Paris.

Balesdent J., 1996. The significance of organic separates to carbon dynamics and its modelling in some cultivated soils. *European Journal of Soil Science*, 47, 485–493.

Balesdent J., Girardin C., Mariotti A., 1993. Site-related $\Delta^{13}C$ of tree leaves and soil organic matter in a temperate forest. *Ecology*, 74 (6), 1713–1721.

Balesdent J., Guillet B., 1982. Les datations par le ^{14}C des matières organiques des sols. Contribution à l'étude de l'humification et du renouvellement des substances humiques. *Science du Sol*, 2, 93–112.

Balesdent J., Mariotti A., 1996. Measurement of soil organic matter turnover using ^{13}C natural abundances. *In: Mass Spectrometry of Soils* (T.W. Boutton, S.I. Yamasaki, eds), Marcel Dekker Inc., New York, 83–111.

Deleporte S., 2001. Changes in the earthworm community of an acidophilous lowland beech forest during a stand rotation. *European Journal of Soil Biology*, 37, 1–7.

Disnar J.R., Guillet B., Keravis D., Di-Giovanni C., Sebag D., 2003. Soil organic matter (SOM) characterization by Rock-Eval pyrolysis: scope and limitations. *Organic Geochemistry*, 34, 327–343.

Elzein A., Balesdent J., 1995. A mechanistic simulation of the vertical distribution of carbon concentrations and residence times in soils. *Soil Science Society of America Journal*, 59, 1328–1335.

FAO, 1998. *World Reference Base for Soil Resources. World Soil Resources Reports*, N° 84. Food and Agriculture Organization of the United Nations, Rome.

GIEC, 2000. *Land Use, Land-use Change and Forestry (LULUCF)*. Cambridge University Press/Organisation Météorologique Mondiale, Cambridge.

Guillet B., Disnar J.R., 2000. Évolution des constituants organiques majeurs, polysaccharides et lignine, dans les humus et horizons A1 des sols des hêtraies de la forêt de Fougères (I et V). Rapport final de la Convention de recherche ECOFOR, N° 97. ECOFOR, Paris, 33 p.

Huet S., 2004. Contribution à l'étude du fonctionnement carboné d'une hêtraie de plaine atlantique, incluant sa végétation de sous-bois, dans une chronoséquence. Thèse de doctorat, université de Rennes 1.

Huet S., Forgeard F., Nys C., 2004. Above- and belowground distribution of dry matter and carbon biomass of Atlantic beech (*Fagus sylvatica* L.) in a time sequence. *Annals of Forest Science*, 61 (7), 683–694.

Jabiol B., 2000. Évolution de la diversité et du fonctionnement des humus au cours d'une révolution forestière en futaie régulière de hêtre. Site-atelier de la forêt de Fougères (I et V.). Rapport de synthèse de la Convention de recherche ECOFOR, N° 97. ECOFOR, Paris.

Jenkinson D.S., Adams D.E., Wild A., 1991. Model estimates of CO_2 emissions from soil in response to global warming. *Nature*, 351, 304–306.

Jenkinson D.S., Rayner J.H., 1977. The turnover of soil organic matter in some of Rothamsted classical experiments. *Soil Science*, 123 (5), 298–305.

Jolivet C., 2000. Le carbone organique des landes de Gascogne. Variabilité spatiale et effet des pratiques agricoles et sylvicoles. Thèse de doctorat, université de Bourgogne, 313 p.

Karlen D.L., Mausbach M.J., Doran J.W., Cline R.G., Harris R.F., Schuman G.E., 1997. Soil quality: a concept, definition and framework for evaluation (a guest editorial). *Soil Science Society of America Journal*, 61, 4–10.

Karroum M., 2002. Devenir des biopolymères (lignine et polysaccharides) et constitution d'humus (*mull/moder*) dans une chronoséquence de hêtre (*Fagus sylvatica*), en forêt de Fougères (France). Thèse de doctorat, université d'Orléans, 121 p.

Karroum M., Guillet B., Laggoun-Defarge F., Disnar J.R., Lottier N., Villemin G., Toutain F., 2005. Morphological evolution of beech litter (*Fagus sylvatica* L.) and biopolymer transformation (lignin, polysaccharides) in a mull and a moder, under temperate climate (Fougères forest, Brittany, France*). Canadian Journal of Soil Science*, 85 (3), 405–416.

Lebret M., Nys C., Forgeard F., 2001. Litter production in an Atlantic beech (*Fagus sylvatica* L.) time sequence. *Annals of Forest Science*, 58, 755–768.

Lebret M., 2002. Les humus forestiers en hêtraies de plaine: analyse des facteurs de l'évolution au cours d'une chronoséquence. Thèse de doctorat, spécialité biologie, université de Rennes 1, 292 p.

Levin I., Kromer B., 1997. $^{14}CO_2$ records from Schauinsland. *In: Trends: A Compendium of Data on Global Change.* Carbon Dioxide Information Analysis Center, US Department of Energy, Oak Ridge National Laboratory, Oak Ridge.

Levin I., Kromer B., Schoch-Fischer H., Bruns M., Münnich M., Berdau D., Vogel J.C., Münnich K.O., 1994. $^{14}CO_2$ record from Vermont. *In: Trends: A Compendium of Data on Global Change.* Carbon Dioxide Information Analysis Center, US Department of Energy, Oak Ridge National Laboratory, Oak Ridge.

Nys C., Arrouays D., Dupouey J.L., Richter C., Forgeard F., Gelhaye L., Hossann C., Huet S., Lebret M., Le Tacon F., 2002. Effets de la sylviculture sur le stockage de carbone dans les sols forestiers. Données pour une validation des paramètres du modèle d'évolution des stocks de carbone. Rapport scientifique GICC-99_02. Ministry of the Environment, Paris.

Parton W.J., Schimel D.S., Cole C.V., Ojima D.S., 1987. Analysis of factors controlling soil organic matter levels in Great Plains Grasslands. *Soil Science Society of America Journal*, 51, 1173–1179.

Ponge J.F., 2003. Humus forms in terrestrial ecosystems: a framework to biodiversity. *Soil Biology and Biochemistry*, 35, 935–945.

Robert M., 1996. Le sol: Interface dans l'environnement, ressource pour le développement. Masson, Paris, 244 p.

Toutain F., 1966. Étude du sol et des eaux de la forêt de Fougères. Thèse de doctorat, université de Paris XI, 192 p.

Trouvé C., Mariotti A., Schwartz D., Guillet B., 1994. Soil organic carbon dynamics under *Eucalyptus* and *Pinus* planted on savannas in the Congo. *Soil Biology and Biochemistry*, 26, 287–295.

Trumbore S.E., 1996. AMS measurement of ^{14}C and ^{41}Ca in soils. *In: Mass Spectrometry of Soils* (T.W. Boutton, S.I. Yamasaki, eds). Marcel Dekker Inc., New York, pp. 52–81.

Van Cleve K., Powers R.F., 1995. Soil carbon, soil formation, and ecosystem development. *In: Carbon Forms and Functions in Forest Soils* (W.W. McFree, M.J. Kelly, eds). Soil Science Society of America Inc., Madison, pp. 155–200.

Wang Y., Hsieh Y.P., 2002. Uncertainties and novel prospects in the study of the soil carbon dynamics. *Chemosphere*, 49, 791–804.

Appendix

Table A1. Distributions of carbon and ^{14}C activities in soil profiles of two sites at Landes de Gascogne (average and standard deviation of five replicates). ^{14}C measured on composite samples and analytical error.

Horizon/ depth (cm)	Concentration (g·kg^{-1})		δ^{13}C (‰)		Stock by horizon (kg·m^{-2})		Cumulated stock (kg·m^{-2})		Δ^{14}C (‰)	
Bray site										
Mat of *Molinia*	455.4	11.3	− 27.9	0.3	0.40	0.18	0.40	0.18		
OL	483.8	5.4	− 28.5	0.3	0.47	0.23	0.86	0.25	92.2	4.5
OF	428.8	67.8	− 28.6	0.2	0.66	0.26	1.52	0.43	152.8	4.6
OH	264.6	65.7	− 28.1	0.4	0.28	0.19	1.80	0.35	152.3	5.1
0–5 (*Molinia*)	37.4	7.5	− 27.8	0.3	0.33	0.09	2.13	0.29	91.3	4.9
0–5 (bare soil)	32.6	14.4	− 27.8	0.5	1.24	0.68	3.37	0.62	58.4	4.5
5–15	26.6	9.0	− 27.5	0.3	3.30	1.01	6.67	1.35	1.9	4.9
15–30	17.2	5.9	− 27.8	0.4	3.72	1.09	10.39	1.56	− 51.5	4.2
30–45	11.3	6.3	− 27.7	0.5	2.63	1.50	13.02	2.58	− 103.4	4.0
Baudes site										
Mat of *Molinia*	453.7	3.7	− 28.3	0.5	0.62	0.24	0.62	0.24	133.5	4.7
OL	468.8	6.5	− 28.3	0.4	0.94	0.28	1.56	0.14	133.1	5.2
OF	441.2	33.2	− 28.2	0.4	0.74	0.41	2.30	0.47	171.4	4.8
OH (bare soil)	299.7	71.6	− 27.6	0.6	1.83	1.10	4.13	1.14	224.0	5.2
OH (*Molinia*)	400.2	56.8	− 27.8	0.3	0.46	0.21	4.60	1.22	241.9	5.6
0–5	43.4	10.8	− 27.6	0.1	2.04	0.29	6.63	1.16	76.7	4.7
5–15	20.0	0.7	− 27.7	0.1	2.72	0.24	9.35	1.18	− 12.2	4.4
15–30	13.1	2.4	− 27.6	0.1	2.91	0.55	12.26	1.29	− 70.9	4.1
30–45	6.7	2.8	− 27.4	0.2	1.47	0.56	13.73	1.70	− 153.1	3.7

Chapter 4

Estimating carbon stocks in forest stands:
1. Methodological developments

Laurent Saint-André, Patrick Vallet, Gérôme Pignard, Jean-Luc Dupouey,
Antoine Colin, Denis Loustau, Christine Le Bas, Céline Meredieu,
Yves Caraglio, Annabel Porté, Nabila Hamza, Antoine Cazin,
Yann Nouvellon, Jean-François Dhôte

Introduction

Biomass assessment has long been an important research topic in ecology (Boysen-Jensen and Müller, 1927; Burger, 1929; or see, for example, the XVth IUFRO (The Global Network for Forest Science Cooperation) Congress in 1971, the working group on forest biomass studies and the emergence of complete tree utilization during the 1970s to 1980s), which is being renewed by the present-day carbon and bio-energy issues linked to the role of carbon in the composition of most anthropogenic greenhouse gases.

Estimating carbon stocks in forest biomass can be based on forest inventories or on remote sensing. The inventory-based method includes two main approaches: a direct one using allometric equations, and an indirect one using volume equations and biomass expansion factors. The latter is mainly used in national forest inventories, whereas the direct one is, for example, used at a stand level and is recommended by Nabuurs *et al.* (2004) 2003 for afforestation/reforestation carbon projects. Both approaches are presented in this chapter. They require: (i) an inventory of the plants present; (ii) models for assessing carbon stocks based on the size of the individuals measured (models for biomass or models for volume multiplied by biomass expansion factors); and (iii) an assessment of the carbon contained in the dead mass (standing dead wood) and understorey vegetation. The models for estimating carbon stocks in forest biomass set out to link a factor that is difficult to measure (such as tree volume, mass or mineral content) with factors that are easier to assess, such as diameter at 1.30 m or tree height.

This chapter sets out the state of development of models used for estimating the standing carbon stock above- and below-ground. The results presented, for example, the volume tables developed in France by LERFOB (Laboratoire d'Études Forêt-Bois) and the expansion factors obtained from the literature form the basis for a revised estimate of national forest carbon stocks per species and per region, based on data from the National Forestry Inventory (NFI) (see Chapter 5).

Standing carbon stock evaluation models

The form and degree of complexity of the models used to estimate the carbon biomass of standing trees may vary depending on the data used to construct them and/or the data with which they are used. This part successively presents:

(i) the methodological aspects inherent in estimating standing carbon stocks: despite their apparent simplicity, the equations used have to be adjusted carefully, using adequate regression techniques;

(ii) a bibliographical analysis of expansion factors (branches and roots) and leaf biomass for the main French species: the contribution made by architectural models such as those developed by the AMAP (botAnique et bioInforMatique de l'Architecture des Plantes) joint research unit on volume expansion factors is also covered;

(iii) five examples of biomass or volume estimates with error calculations in different forest systems (biomass for maritime pine, volume for the main French species).

Constructing biomass and volume equations

General considerations

According to Cailliez (1980), volume tables are tables of figures, formulae or graphs that give estimates of the volume of a tree or group of trees as a function of different variables that serve as entry values. Such tables are referred to as "individual" when they serve to estimate the volume of a single tree. In general, the correlations are good and the most commonly used functions are polynomial, logarithmic or power. The structure of the equations may be:

linear
$$Y_i = a_s + b_s \cdot X_i + \varepsilon_i \tag{1}$$

or non-linear
$$Y_i = a_s + b_s \cdot X_i^{C_s} + \varepsilon'_i \tag{2}$$

Y_i is the biomass or volume of the tree for the compartment in question; X_i the principal independent variable ($d_{1.3}$ or $r_{1.3}^2 \cdot h$; where $d_{1.3}$ and $r_{1.3}$ are the diameter and radius of the tree at 1.30 m and h its total height); a_s, b_s and c_s are the parameters to be estimated by the model, s refers to the stand and ε_i and ε'_i are the residual variances not explained by the model.

Despite their apparent simplicity, volume and biomass equations have to be fitted with care. Without trying to propose a standard method for building this type of model, we discuss the main difficulties that can occur with biomass/volume equations and we list a collection of key papers which give some solutions.

The first difficulty is linked to data structure. First, the variance of biomass increases with tree age or diameter at breast height (DBH) leading to heteroscedasticity and, sometimes, non-normally distributed data. Application of ordinary least square regression under this variance heterogeneity generates: (i) estimated parameters that do not have the minimum variance; (ii) biased estimators of the variances of the estimated parameters; and (iii) false estimation of the residual variance (Cunia, 1964; Grégoire and Dyer, 1989; Parresol, 1993). Fortunately, regression coefficients are unbiased and converged to the true parameters as sample size increases (Kelly and Beltz, 1987), meaning that average biomass is little affected by heteroscedasticity. Conversely, variances may be badly estimated and this is a major drawback when the user wishes to give a valid confidence interval of the tree or stand biomass estimates. There are several solutions to deal with heteroscedasticity. Log-log transformations can be a way to stabilize the variance as tree size increases. However, this kind of transformation is not always suitable and non-linear forms of log-log regressions have also been reported (Návar *et al.*, 2002). Furthermore, a conversion factor should be applied when calculating back-transformed values (see Dhôte and Hervé, 2000; Parresol, 2001; Zianis and Mencuccini, 2004; Joosten *et al.*, 2004; or the discussion given by Cienciala *et al.*, 2006). The other way is to consider that residual variances are proportional to variable X in Equations (1) and (2):

$$E\left[\varepsilon_i^2\right] = \sigma_i^2 \cdot X_i^{2k} \tag{3}$$

where σ^2 is the residual variance of the weighted model and k the weighting value. The main objective is to find the optimum value for k. This can be achieved using three different approaches: (i) calculating the biomass variance for each DBH class, and k is the slope of the relationship between the log of the diameter and the log of the variance (this method is the simplest but requires a large number of trees, which is rarely achievable in biomass studies); (ii) the use of the Furnival's index of fit (FI) (Furnival, 1961) for testing different possible weightings (i.e. different k values), the best model will be that with the lowest FI_k (Williams and Gregoire, 1993; Uzoh and Ritchie, 1996; Bi, 1999; Bi and Hamilton, 1999); (iii) the use of maximum likelihood techniques to fit simultaneously a model for the mean (e.g. $Y_i = a_s + b_s \cdot X_i$ for the linear model) and a model for variance ($\sigma_i^2 = \sigma^2 \cdot X_i^{2k}$). The latter solution is the most flexible because any model formulation can be used for the variance.

The second difficulty is related to data sampling. Trees are generally felled from the same plots meaning that observations are correlated between each other. To deal with these correlated observations, linear and non-linear mixed-effects modelling approaches can be applied (e.g. see Zhang and Borders, 2004; Wirth *et al.*, 2004; Levy *et al.*, 2004; Sharma and Parton, 2007). Furthermore, biomass or volume equations are basically non-generic and should be applied with care to other trees or silvicultural regimes. The way the sampling is carried out may have a great influence on the coefficient estimators, their confidence interval and the residual variance (Cunia, 1964; Madgwick, 1971). If genericity is a desired property for the biomass equation, it is necessary, therefore, to take all the "between stand" and "within stand" variability into account (e.g. species, age, silviculture regimes, sites, etc.) for the equation building (Rennie, 1966; Wirth *et al.*, 2004; Joosten *et al.*, 2004; Saint-André *et al.*, 2005; Sicard *et al.*, 2006; Cienciala *et al.*, 2006; Peichl and Arain, 2007; Antonio *et al.*, 2007). Some general equations were developed from a biome-wide database and can be useful for first approximate estimations (e.g. Cairns *et al.*, 1997 for below-ground biomass; Brown, 1997 for tropical species in dry, wet and moist climatic zones; Niklas and Enquist, 2002 or Enquist and Niklas, 2002 for angiosperm and conifer species; Zianis and Mencuccini, 2003 for beech trees, Zianis and Mencuccini, 2004 for different species spanning the world, Chave *et al.*, 2005 for tropical forests in America, Asia and Oceania; Muukkonen, 2007 for five tree species in Europe; Nabuurs *et al.* (2004) provide a list of biomass equations that can be used for both tropical and temperate species (in their tables 4.A1, 4.A2, 4.A3 and 4.A4, pages 4.114–4.116); some monographs such as the one produced by Zianis *et al.*, 2005 for temperate species are also useful). Nabuurs *et al.* (2004) recommend verifying the equation by destructively harvesting a few trees of different sizes. They fixed a 10% admitted error (their page 4.101, Step 3 of the "direct approach"). We also recommend using modelling efficiency introduced by Mayer and Butler (1993) and checking for bias (a simultaneous test on the intercept, which should be null, and the slope of the regression, which should be equal to unity, between the measured biomass and the simulated one). The ideal case would be a modelling efficiency close to 1 and no bias between the measured and simulated biomass.

The third difficulty is linked to tree splitting into several compartments. When building models for each of them, the sum of the estimated biomass with the "compartments equations" might not lead to the estimated biomass from the "total equation" (see Kozak, 1970 for an illustration; Cunia and Briggs, 1984; Reed and Green, 1985; Návar *et al.*, 2002). Parresol (1999) indicates three different methods to ensure additivity in biomass equations: (i) the total biomass equation computed by the sum of individually best fitted equations for each compartment; (ii) the total biomass and all compartment equations are of the same form and have the same weighted function, the regression coefficient of the total biomass equations being simply the sum of compartment coefficients; (iii) the total biomass and all compartment equations are all different, the additivity is ensured by setting constraints on coefficients (seemingly unrelated regressions, e.g. Parressol 1999, 2001; Návar *et al.*, 2002; Carvalho and Parresol, 2003; Saint-André *et al.*, 2005; Antonio *et al.*, 2007).

Developing models

Once trees are sampled in the different stands, there are two main ways of processing data. The simplest approach consists of using site variables such as site index, dominant height, altitude (used as a proxy for climate), age, etc., as additional variables to tree DBH or tree height in the equation (Joosten *et al.*, 2004; Wirth *et al.*, 2004). A second approach consists of fitting one or several candidate models to each sample unit (stand-level), then studying and modelling the parameter variations with site variables, and finally fitting a stand-dependent allometric relationship if significant variations among sites were found (Saint-André *et al.*, 2005; Cienciala *et al.*, 2006; Antonio *et al.*, 2007).

Model selection can be performed as follows: if the two equations to be compared are nested, then a classical F-test can be applied (Brown and Rothery, 1994), otherwise, maximum likelihood-based criterion such as Akaike information criterion or Bayesian information criterion should be applied. These statistical tests can be used in combination with graphical analysis of residuals, as well as with several measures of residual dispersion and normality.

Bibliographical overview of the situation regarding conversion factors

It has now been recognized that well-designed and statistically sound NFIs are useful tools for estimating carbon stocks and their dynamics at large scale. A conversion factor applies to forest inventories when the commercial volume is obtained either directly (derived from stem diameter measured at different heights on standing trees with instruments such as a Bitterlich relascope or laser dendrometer) or indirectly by volume or stem taper equations. It should be noted that such equations require the same sound procedure as biomass equations (genericity and heteroscedasticity). Once the commercial volume is known, it is converted to biomass using a density factor (tonnes of dry matter per cubic metre of green volume) and a biomass expansion factor (BEF) (to shift from commercial to whole tree biomass). Values of tree density for several tropical tree species can be found in Brown (1997, Appendix 1) and/or in Nabuurs *et al.* (2004) (Table 3A.1.9-1 for temperate and boreal species, Table 3A.1.9-2 for tropical species). Default values of BEFs can be found in Nabuurs *et al.* (2004) (Table 3A.1.10, page 3.178). However, both quantities may vary considerably with species, stand age or/and site (Fang and Wang, 2001; Fang *et al.*, 2001; Brown, 2002; Lehtonen *et al.*, 2004; Van de Walle *et al.*, 2005; Tobin and Nieuwenhuis, 2007). In this part of the work, we also explored the solution of using mean expansion coefficients for the branches ("branch biomass + stem biomass divided by trunk biomass" ratio) and roots ("above-ground biomass + root biomass divided by above-ground biomass" ratio), instead of one single biomass expansion factor. The results presented here are based on two bibliographical synopses by Cannell (1982) and Vogt *et al.* (1996). We analysed leaf biomasses separately, and proposed mean values per area unit. The calculated mean values are given in Tables 4.1–4.3. From this analysis,

Table 4.1. Mean and median (in brackets) branch expansion factors for 159 stands in temperate, boreal and Mediterranean zones. The number of samples used is indicated between dashes.

	Broadleaved species	**Conifers**	**All combined**
Young < = 30 years	1.27 (1.24) – 6 –	1.43 (1.43) – 36 –	1.41 (1.39) – 42 –
Old > 30 years	1.27 (1.25) – 49 –	1.17 (1.16) – 68 –	1.21 (1.18) – 117 –
All combined	1.27 (1.25) – 55 –	1.26 (1.18) – 104 –	1.26 (1.19) – 159 –

Data: Cannell (1982).

Table 4.2. Mean and median (in brackets) root expansion factors for 239 stands in temperate, boreal and Mediterranean zones. The number of samples used is indicated between dashes.

All combined	**Broadleaved species**	**Conifers**	**Total**
Root expansion factor: (above-ground + large + fine roots)/above-ground part	1.28 (1.24) – 80 –	1.30 (1.24) – 159 –	1.29 (1.24) – 239 –
< = 30 years	1.48 (1.29) – 14 –	1.37 (1.25) – 36 –	1.40 (1.28) – 50 –
> 30 years	1.24 (1.21) – 66 –	1.28 (1.24) – 123 –	1.26 (1.24) – 189 –

Data: Cannell (1982) and Vogt *et al.* (1996).

a single value of the branch expansion factor can be applied to broadleaved species (1.25 irrespective of age based on median values), whereas a strong variation with stand development was found for coniferous species (1.43 for young stands and 1.16 for old ones). Average values of the root expansion factor also varied with species and stand age, whereas median values were relatively more stable (between 1.21 and 1.29).

Table 4.3. Mean and median values (in brackets) of leaf biomass for 92 stands in temperate, boreal and Mediterranean zones. The number of samples used is indicated between dashes.

	Conifers	Broadleaved species
Leaf biomass (t DM·ha⁻¹)	9.1 (7.8) – 121 –	3.4 (3.1) – 92 –
< = 30 years	7.3 (6.4) – 36 –	2.4 (2.7) – 19 –
> 30 years	9.8 (9.0) – 84 –	3.6 (3.2) – 73 –

Data: Cannell (1982).

As we do not have BEFs for the main species growing in France, we improved the evaluation of carbon stocks from the French NFI by: (i) building a set of total volume equations (including trunk and branches) from French forestry research archives (data gathered between 1920 and 1950), thus avoiding uncertainties from the branch expansion factor (especially for coniferous species); (ii) compiling literature to calculate wood specific gravity values per species in France; and (iii) using the root expansion factor calculated from the Cannell (1982) and Vogt *et al.* (1996) database.

Examples of the development of specific models

Case study 1. Total aerial biomass for maritime pine on the heathlands of Gascony

In this example, we look at two biomass models developed for maritime pine on the heathlands of Gascony. The two equations considered can be used to model total above-ground dry biomass. The two forms chosen are as follows:

$$W_i = a_s \cdot D_i^{bs}, + \varepsilon_i \tag{4}$$

$$W_i = a_s \cdot D_i^{bs} \cdot A_i^{cs} + \varepsilon'_i \tag{5}$$

W_i is total aerial biomass (kg); D_i, the principal independent variable, is tree diameter at 1.30 m (cm); A_i, the second independent variable, is age, counted from germination; a_s, b_s and c_s are the parameters to be estimated by the model, with s referring to the complete sample considered; and ε_i and ε'_i are the residual variances not explained by the models. The two models are allometric ($y = a\,x^b$). Introducing the age variable makes it possible to take account of variations in time.

The set of data came from four plots of maritime pine in Gironde (France), and was split into five age categories (one site was sampled twice). For each stand, a set of allometric relationships was established per compartment (Porté, 1999; Porté *et al.*, 2000, 2002). The total biomass of each tree was obtained by adding the estimates for each compartment. The adjustment made using model 2 is shown in Figure 4.1. The results of the two non-linear adjustments are shown in Table 4.4. The model including age as an independent variable was better than that using diameter alone.

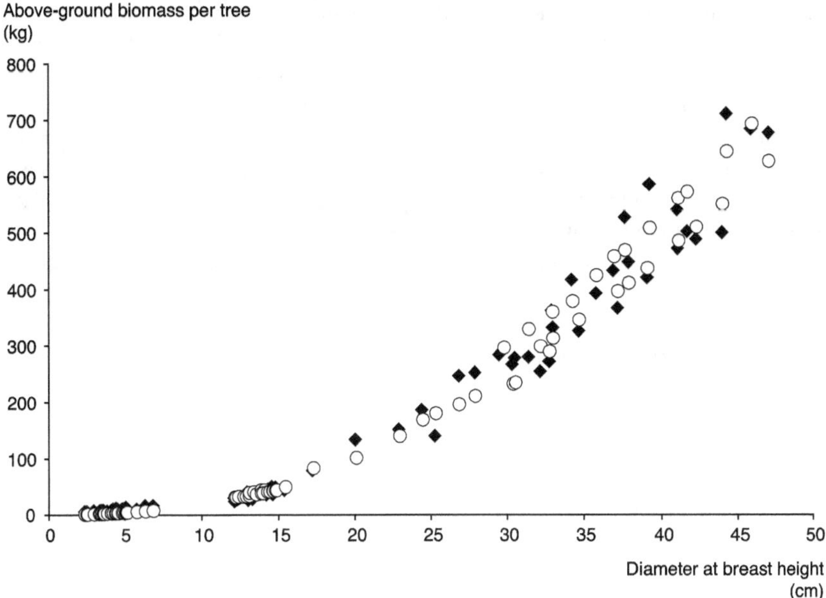

Figure 4.1. Adjustment of total aerial biomass of maritime pine trees using model 2. Empty circles are the predicted values.

Table 4.4. Parameters of the allometric models used to estimate total aerial biomass per tree.

Model	Parameter a	Parameter b	Parameter c	Sum of squares of residuals
1	0.137 (0.036)	2.217 (0.072)	–	77801.2
2	0.094 (0.020)	1.992 (0.062)	0.330 (0.034)	41108.1

NB: Estimate of parameters and (standard error values).

Continuing with this work should lead to a revision of the weightings used for these total aerial biomass models and a calculation of the confidence intervals of the estimates. These models can be used routinely in tools for estimating carbon stocks throughout the life of stands. A study has been launched to check the validity of these models under a range of fertility conditions.

Case study 2. Biomass tables for eucalyptus plantations in Congo

A complete set of equations for below-ground and above-ground biomass of a natural eucalyptus hybrid in Congo was constructed and validated by Saint-André *et al.* (2005). Heteroscedasticity of data was taken into account by fitting weighted non-linear regressions using maximum likelihood estimates and seemingly unlikely regressions. Stand age was introduced as a complementary variable in the equations allowing the estimation of biomasses for 5- to 135-month-old stands. This system of equations includes: (i) model parameters for the mean; (ii) model parameters for the variance; (iii) the correlation matrix between parameters (within a given compartment and between compartments); and (iv) the correlation matrix between residuals. Using this information, we used Monte Carlo simulations to calculate both the mean and confidence interval for the stand biomass estimates (see Sicard *et al.*, 2006 for a detailed description of the applied methodology). In this study, we present the results obtained for the six plots described in Saint-André *et al.* (2005), where tree growth was monitored in permanent plots with manual band dendrometers over 2 years. The objective was to apply biomass models to the successive inventories, so as to update the biomass histograms for the plots, calculate the increments (total and per compartment for the years 2001 and 2002) and determine the confidence intervals for these estimates (Figure 4.2). The standard errors were small in terms of standing biomass: for instance, for stand G3B, total standing biomass at 100 months was 128 ± 1.9 t·ha^{-1}, split between 104 ± 1.8 t·ha^{-1} of aerial and 24 ± 1.0 t·ha^{-1} of underground biomass. These results tallied with what we expected insofar as the material was clonal and the model parameters were relatively stable.

Standing carbon stock (aerial and underground) increment was calculated for the years 2001 and 2002 (Table 4.5). The standard errors around the predictions represented 7% to 25% of simulated biomass increment. The figures increased with age, from 1 t (G3B) to 3 t for the old stands (G3D). In terms of carbon, for the plot monitored through eddy correlations (plot G3B), standing tree carbon increment was around 6 t·ha^{-1}·year^{1} $\pm$ 500 kg.

Table 4.5. Standing biomass increment for the years 2001 and 2002 (mean $\pm$ 1 standard error).

Plot	Age (months)	Year	Start date	End date	Total biomass increment (t·ha^{-1})
G3A	9–22	2001	23/03/2001	23/03/2002	19.5 ± 1.6
G3A	22–34	2002	23/03/2002	02/04/2003	25.6 ± 3.3
G3B	26–38	2001	08/01/2001	27/12/2001	12.0 ± 0.9
G3B	38–49	2002	27/12/2001	28/11/2002	12.9 ± 1.0
G3C	49–61	2001	12/01/2001	11/01/2002	9.6 ± 1.9
G3C	61–74	2002	11/01/2002	31/01/2003	15.3 ± 2.4
G3D	74–86	2001	13/01/2001	11/01/2002	14.7 ± 2.9
G3D	86–99	2002	11/01/2002	31/01/2003	10.9 ± 2.8

As a conclusion, this study in the Congo revealed an age effect on "biomass/tree size" relations, and enabled the development of comprehensive biomass models. They were validated, in terms of the aerial part, using an independent sample. Using data from inventories, it was then possible to calculate biomass increments (aerial and underground) for the 2 years of the study, with their confidence interval at 95%. These intervals took account of the correlations between compartments and were still relatively large.

Case study 3. Volume tables for the total above-ground volume of the main French species (LERFOB tables)

Drawing up volume tables for the total above-ground volume of trees avoids multiplying the errors induced by branch expansion factors. As the aim of the study was to estimate the national resource, it was crucial that the sample be truly representative. The trees chosen constituted a database of 4543 individuals. The seven species for which we established volume tables were sessile oak, beech, Douglas fir, spruce, European

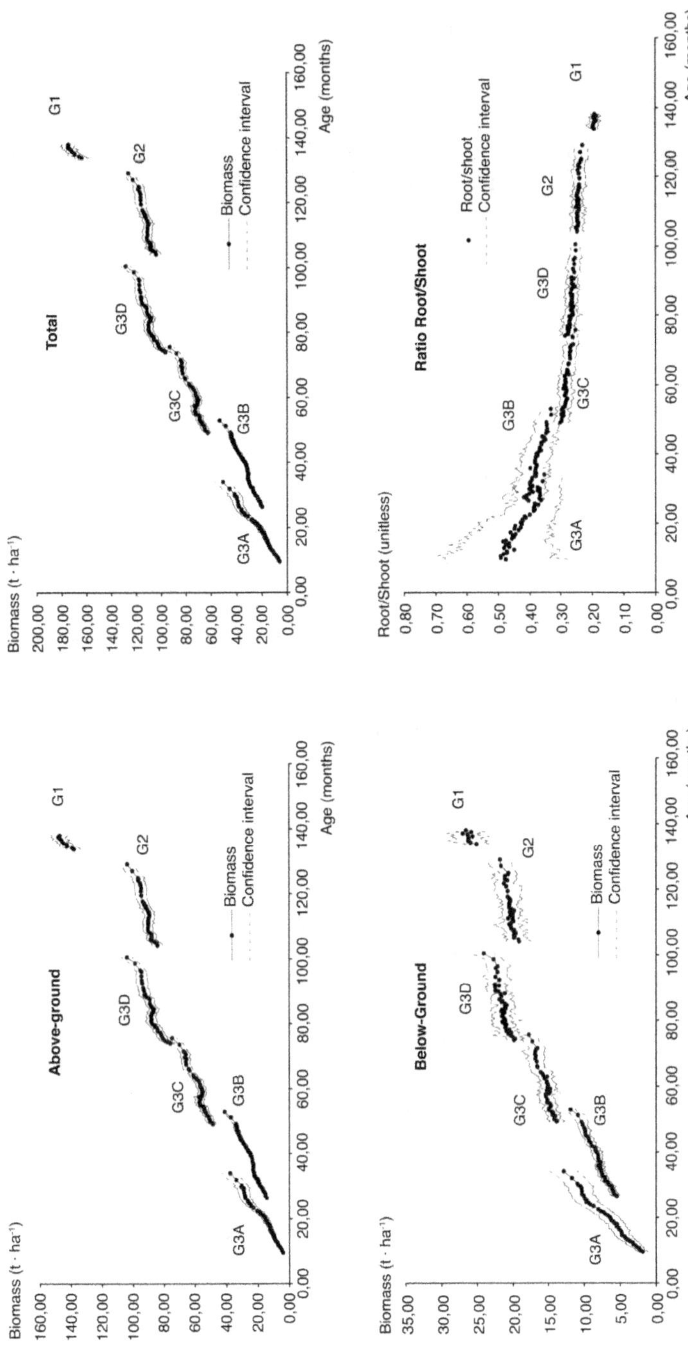

Figure 4.2. Total, aerial and underground biomass as a function of age for a chronosequence of eucalyptus stands in Pointe-Noire (Congo) simulated with biomass equations (Saint-André *et al.*, 2005). The letters refer to the stand number given in Table 4.5. The confidence intervals calculated by the Monte Carlo procedure correspond to the limits at 95% of individual values (around 2 standard errors).

silver fir, pines (maritime, Corsican, Scots and black) and larch. Detailed information can be found in Vallet *et al.* (2006). We recall the main results here.

The heteroscedasticity of volume led us to work on the form factor. Volume was thus expressed as:

$$v_{tot} = \frac{1}{4\pi} \cdot c_{130}^2 \cdot h_{tot} \cdot f \tag{6}$$

where f was the form factor (no unit of measurement), c_{130} and h_{tot} girth and height in metres, and v_{tot} volume in m³. We modelled variable f. The overall form was:

$$f(c_{130}, h_{tot}) = \left(\alpha + \beta \times c_{130} + \gamma \times \frac{\sqrt{c_{130}}}{h_{tot}} \right) \cdot \left(1 + \frac{\delta}{c_{130}^2} \right) + \varepsilon \tag{7}$$

where α, β, γ and δ were the parameters to be adjusted.

The results for each species are summarized in Table 4.6 hereafter. We observed similar changes in form factors as a function of c_{130} between some species. The "dark" conifers (i.e. spruce, Douglas fir and silver fir) showed a sharp reduction in the form factor in line with girth, that is, the larger they were, the more conical the shape of the tree. Conversely, the form factors for oak and beech both increased in line with the robustness of the tree, meaning that for a given girth, the stockier the tree, the more its volume was cylindrical in shape. The pines performed in a similar way to beech.

The aim of the study was to provide tools for estimating total aerial biomass on a national scale, but the available data corresponded to a mere sample taken from that resource some time ago. The work undertaken provided the following information:
– there is geographical variability, but we did not identify any clear structure;
– the available data on the total volume of individuals are old, but we did not observe any temporal trends;
– we tested our tables on 35 coppice-with-standard reserves; the results were not conclusive, but we used the same tables for coppice-with-standard;
– according to NFI data, the species for which we gave a specific volume table account for around 60% of the volume of the national resource. In particular, there were major gaps for various broadleaved species (except beech and sessile oak).

We drew up a table for "miscellaneous broadleaved species" using the data for beech and sessile oak, which gave good results with a set of 63 ash trees.

Case study 4. What modelling can contribute to plant architecture modelling to evaluate expansion factors

Analysing and modelling tree architecture make it possible to obtain highly botanically accurate three-dimensional plant structures (Fourcaud *et al.*, 1997) and then offer the opportunity for developing branch expansion factors for the studied species. These models rely upon a detailed representation of canopy architecture. However, because such data are long and costly to collect, the number of sampled stands is generally limited (especially for the oldest stands). Hence making studies of architecture variability with site index or stand density is very complex. Conversely, growth and yield models have been calibrated using a larger range of situations and silvicultural trials. Simulations from these later models should then help in better tuning the architectural model parameters and then render BEFs appropriate for a large range of situations.

Table 4.6. Estimate of parameters of Equations (6) and (7) for different tree species.

Tree species used to calibrate the equation	Species for which the equation was used	Parameters values			
		α	β	χ	δ
Quercus petraea	*Quercus petraea, Q. robur, Q. pubescens, Q. rubra*	0.471	− 0.345	0.377	0
Fagus sylvatica	*Fagus sylvatica*	0.395	0.266	0.421	45.4
Quercus petraea and Fagus sylvatica	Other broadleaved species	0.428	− 0.191	0.456	0
Pseudotsuga menziesii	*Pseudotsuga menziesii*	0.534	− 0.530	0	56.6
Pinus spp.	*Pinus* spp.	0.311	0.405	0.340	191
Larix spp.	*Larix* spp.	0.550	− 1.350	0.322	0
Abies alba	*Abies* spp.	0.550	− 0.749	0.277	0
Picea abies	*Picea abies* and other coniferous species	0.631	− 0.946	0	0

The functions that modify tree architecture are sequential: first the size (elongation) of internodes is considered, then the number of internodes, and finally the number of leaves. Implicitly, the variation generated for the internodes has an impact on lateral branch and shoot distribution and, therefore, on canopy and tree size.

To this end, we used and developed the combination of a plant architecture simulation software (AMAPsim) (Barczi *et al.*, 2008) and a platform that manages different types of growth and yield models (Capsis) (Coligny *et al.*, 2003), testing datasets on various conifers and eucalyptus. Applying constraints to the simulation through the modifying functions resulted in a good restitution of tree size (Figure 4.3), that tallied with the pattern of height and diameter values for a virtual 12-year-old maritime pine stand.

(a)

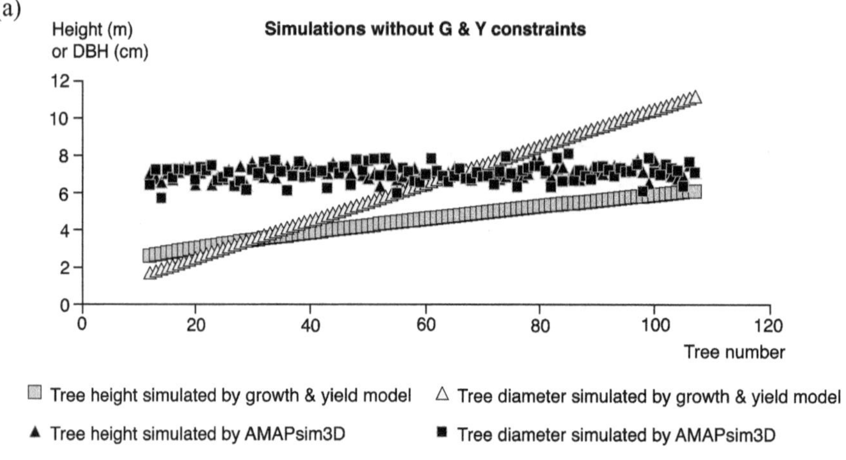

(b)

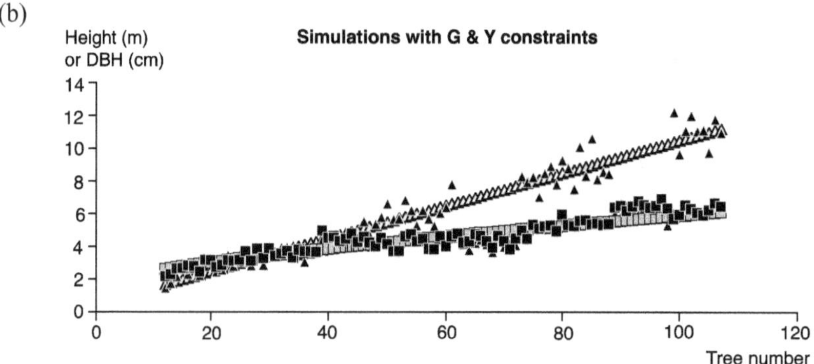

Figure 4.3. Tree characteristics (height and diameter) obtained using the growth and yield model (Lemoine, 1991; Salas-Gonzales *et al.*, 2001) and AMAPsim 3D simulations (Barczi *et al.*, 2008), shown for 96 individual maritime pine trees ranked in increasing order of diameter. (a) AMAPsim 3D simulations done without constraints; (b) AMAPsim 3D simulations done under constraint using the modifying functions.

Using the topological and geometrical information related to each simulated tree with AMAPsim, it is possible to determine volume or leaf area at a given age, and also to split total volume into trunk volume and crown volume, while within the crown, volume can be sub-divided in line with branching order. Using the range of modifying functions and data obtained from the growth and yield model, average information on expansion factors can be obtained. For instance, the expansion factor of interest to the NFI is that which can be used to assess the total volume of a tree (trunk + branches), TrVol + BrVol, given the volume of a trunk with a top diameter of 7 cm, TrVol +7. The factor calculated using our simulation changed considerably for small diameters and tended to stabilize for larger values. The mean values for expansion coefficient were 2.39 for the 12-year-old stand and 1.44 for the 40-year-old stand, and tallied closely with those in the bibliography.

We have presented some results, but in order to fully achieve the objectives set, we need to confirm certain points, and particularly: (i) to fine-tune the assumptions made about the modifying functions (work is under way on Scots pine and Austrian pine); and (ii) to validate not just the volume or biomass aspects but also the leaf areas or crown recession, so as to reflect the effects of stand density, for instance.

Discussion and prospects

General comments

Establishing a system of equations for a forest gives good results in terms of accuracy, provided the calibration sample is representative of the variations in tree age and size. The compartments with the greatest evaluation errors are those with a high turnover (such as leaves), probably because the size of those compartments is highly dependent on carbon distribution within the tree, which itself depends on the climatic conditions. That said, those compartments account for only a small share of standing biomass (5% to 6% on average). The greatest uncertainty surrounds the roots, since, because of sampling difficulties, very few data can be recorded. In terms of carbon increment (solely that contained in biomass), the errors are much the same as those observed for stocks.

The examples presented in this chapter tend to show that it is possible to establish generic equations, or at least an initial outline. Variable r^2h is particularly interesting, since it is equivalent to tree volume, apart from the form coefficient. Introducing age into the equations showed that proportionality relationships change over time. Like the volume table established by LERFOB, while it is possible to hope to obtain a form of equation that works for all species, the adjusted parameters differ fundamentally from one species to another. The main difficulty lies in data collection. To obtain sufficiently robust equations, it is necessary to cover a sufficiently contrasting range of situations in terms of tree growth. Databases, or data cooperatives, are valuable tools that make it possible to increase the number of trees sampled little by little, and thus to cover the existing variability more effectively. At the same time, silvicultural trials, or instrumented sites, serve to test hypotheses (e.g. effects of age, fertility, stand density or stand structure) and thus to fine-tune the form of the equations.

Validation of the aerial volume tables compiled by LERFOB

Although it comprised over 4500 trees, the sample used by LERFOB was not wholly representative:

– many species are absent; of the broadleaved species, only *Quercus petraea* and *Fagus sylvatica* were considered. The aerial volume of the other broadleaved species was evaluated using an equation compiled by combining the figures for these two species (Table 4.6);

– the forests from which the samples were taken were primarily in northern France;

– the figures are relatively old (1920–55), and several factors may have modified the shape of the trees (e.g. silvicultural changes, genetic improvement, environmental changes, etc.);

– the sample plots were exclusively set up in regular high forest stands.

The last point is particularly important, as a result, on the one hand, of the differences in shape between high forest trees and standards from coppice-with-standards, and, on the other hand, the proportion of mixed high forest and coppice in the broadleaved forest resource (the 8.8 million ha of productive broadleaved stands in France are split between 36% high forest, 24% simple coppice and 40% mixed high forest-coppice).

As trees in mixed high forest-coppice are characterized by their large crown development, one could think that the equations compiled based on high forest trees underestimate the total volume. However, the chosen model gives an important role to the $\sqrt{}$ (girth at breast height)/total height variable, representing the robustness of the tree, which could ensure a coherent performance outside the range in which it was developed.

The elements available to verify this are very sketchy. The results of a test on a limited sample, which LERFOB judged unrepresentative, were not conclusive. We feel that the most comprehensive synthesis concerning the national broadleaved resource is the one produced by Bouchon *et al.* (1981) as shown in Table 4.7. They give the results of recent work on simple coppice on the one hand and high forest on the other. For coppice, the following values were chosen: branch expansion factor for recordable coppice: 1.82; wood specific gravity: 0.55.

Branch expansion factors were calculated per diameter class for high forest oak and beech indicating mean values from 1.23 to 1.25 for oak and around 1.30 for beech. It is worth noting that the samples used by Bouchon *et al.* (1981) to study broadleaf high forest and by LERFOB to compile allometric equations were primarily subsets of the same overall sample, that of trees measured in forestry research sample plots from 1920 to 1955.

As regards coppice-with-standards, for want of recent usable data, the authors give a synopsis of the old technical literature. In particular, they recall the ratio used by foresters until the mid-20th century to assess reserve oak branch wood in coppice-with-standards, which they say roughly corresponds to 0.67 m³ of firewood per m³ of stemwood, to which bundles of sticks should be added to obtain the total wood figure. They primarily based their work on the *Expériences forestières entreprises de 1871 à 1874 par M. Le Chauff, Inspecteur des forêts à Moulins, Allier*, observing that the inspector in question was known to be highly conscientious and that Hüffel on occasion referred to his work and confirmed its merits. After confirming the close correlation between Le Chauff's estimates and those calculated by the Station de Sylviculture et de Production for (old) high forest oaks, the authors proposed adopting Le Chauff's data and opting for a branch expansion factor of 1.66 (rounded down to 1.6 for the sake of caution) for coppice-with-standards oaks with a diameter of more than 40 cm. Having observed a lack of similar data for beech and the existence of a certain consensus of opinion that beech crowns in coppice-with-standards are denser than those of oaks, they chose a value of 1.7 for large beech trees in coppice-with-standards (diameter of more than 37.5 cm at 1.30 m).

To take account of the existence of stands in mid-conversion in NFI "mixed high forest-coppice", they recommended using the following values for the branch expansion factor: 1.5 for oak and 1.6 for beech. Table 4.7 compares the branch expansion factors resulting from using the LERFOB allometric equations with the values proposed by Bouchon *et al.* (1981). The results obtained with the allometric equations were calculated on a national scale, but also for Lorraine, the region that undoubtedly has forest characteristics most similar to those of the samples used to compile the LERFOB equations and by Bouchon *et al.* (1981). It transpires that:

– generally speaking, the factors calculated for France as a whole are higher than those for Lorraine: this could reflect the fact that stands were converted and broadleaved species began to be managed as regular high forest longer ago than in other regions;

– the expansion factors resulting from the LERFOB volume allometric equations for regular high forest are significantly higher than those proposed by Bouchon *et al.* (1981): the difference, which is large for France as a whole, is less marked for Lorraine;

– as regards coppice-with-standards, the values are similar for oak, while the expansion factors obtained with the allometric equations are a little higher for beech (although Bouchon *et al.* (1981) were less certain about the figure for this species than about that for oak);

– the figures tallied closely for coppice.

Table 4.7. Branch expansion factors for broadleaved species: comparison of the result of applying the LERFOB allometric equations with the values proposed by Bouchon *et al.* (1981).

	Bouchon *et al.* (1981)			LERFOB equations France			LERFOB equations Lorraine		
	HF	CWS	SC	HF	CWS	SC	HF	CWS	SC
Quercus robur and *Q. petraea*	1.23–1.25	1.5–1.66		1.5	1.57		1.39	1.5	
Fagus sylvatica	1.3	1.6–1.7		1.61	1.86		1.54	1.8	
Other broadleafs				1.73	1.99		1.57	1.68	
All broadleafs			1.82			1.72			1.71

HF: regular high forest; CWS: coppice-with-standards (or NFI mixed high forest-coppice); SC: simple coppice.

It is difficult to draw any definite conclusions from these comparisons, but it is worth noting that the stands classified as regular high forest by the NFI include stands that have been managed as regular high forest from the outset and others resulting from the more or less recent conversion of coppice-with-standards. The trees in this second class of high forest have an intermediate form, somewhere between high forest stem and standards. It is thus logical that the expansion factors calculated for regular high forest in the NFI sense should fall between the values for true regular high forest calculated by Bouchon *et al.* (1981) and those calculated for standards.

The comparison was reassuring as regards the validity of the LERFOB allometric equations insofar as the results obtained for mixed high forest-coppice and for simple coppice, that is, a field not covered by the sample, tallied with the available evaluations. There are two points that are even more delicate, since the available data are largely inadequate for a satisfactory study: the geographical validity of the volume allometric equations (a question prompted by the differences observed between the estimates for France and Lorraine); and the validity of the estimates for broadleafs other than oak and beech.

As regards conifers, a few rough comparisons were made for Lorraine, between the LERFOB allometric equations and a series of allometric equations in a European database. In practice, only the spruce was really studied, based on references from Germany, Denmark and Scandinavia. The mean branch expansion factor calculated from the LERFOB equations for Lorraine (1.26) is lower than the values calculated from the other European allometric relations (1.29, 1.31, 1.32, 1.33, 1.38, 1.40, 1.65 and 1.67). However, it is difficult to interpret these comparisons with any accuracy for want of detailed information on the origin of the equations and given the obvious errors in the database.

From the LERFOB allometric equations, it is possible to calculate average branch expansion factors, particularly the values they take on each successive inventory. Table 4.8 presents the average branch expansion factors obtained for the eight species or group of species for which an allometric equation is available. A systematic increase of this factor is observed between the two inventories for all groups but the Douglas fir.

Table 4.8. Average branch expansion factors based on the LERFOB allometric equations, by species or group of species.

Species	1984	1996
Quercus spp.	1.51	1.55
Fagus sylvatica	1.58	1.66
Other broadleaved species	1.70	1.73
Pseudotsuga menziesii	1.18	1.18
Picea abies	1.18	1.22
Abies alba	1.24	1.29
Larix spp.	1.29	1.31
Pinus spp.	1.28	1.33

To what extent could this increase be explained by changes in the diameter distribution of the trees between the two inventories? Figures 4.5 and 4.6 show how the average expansion factor varies with diameter for broadleaved and coniferous species (last inventory). For all broadleaved species and pines, the branch expansion factor increases with diameter, above diameters of 25–30 cm. This somewhat unexpected pattern is certainly due to the way volumes are estimated by the NFI and to the underlying nature of the French forest resource: these species are characterized by an abundance of large branches, which implies removal of more wood during sawing and a subsequent reduction of the merchantable volume in comparison to the volume of a tree with more regular shape. The increase in average branch expansion factor between 1984 and 1996 is thus in agreement with observed changes in forest characteristics: for the eight tree species or group of species, the average diameter increased by 1–2 cm between the two inventories. Only the group composed of spruce and other coniferous species displayed stable diameters.

Other coniferous species (i.e. Douglas fir, spruce, larch and silver fir) display a more classical decrease of the branch expansion factor with increasing diameters. In these cases, the observed increase in average diameter with time cannot explain the increase of the average branch expansion factor. A more in-depth analysis is required to understand this phenomenon.

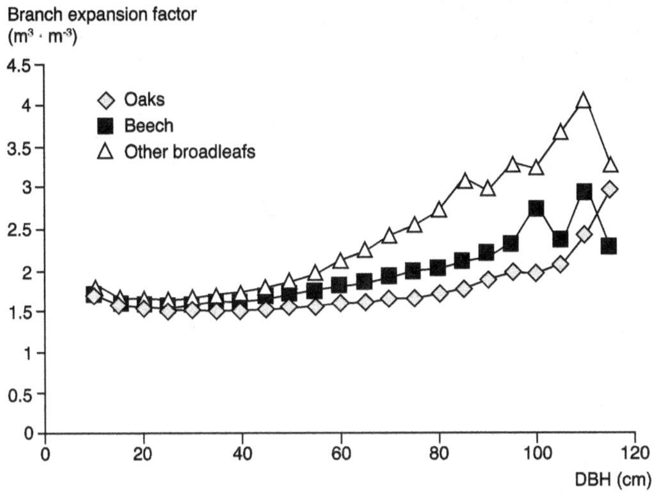

Figure 4.4. Variation of branch expansion factor according to diameter at breast height (DBH) for broadleaved species.

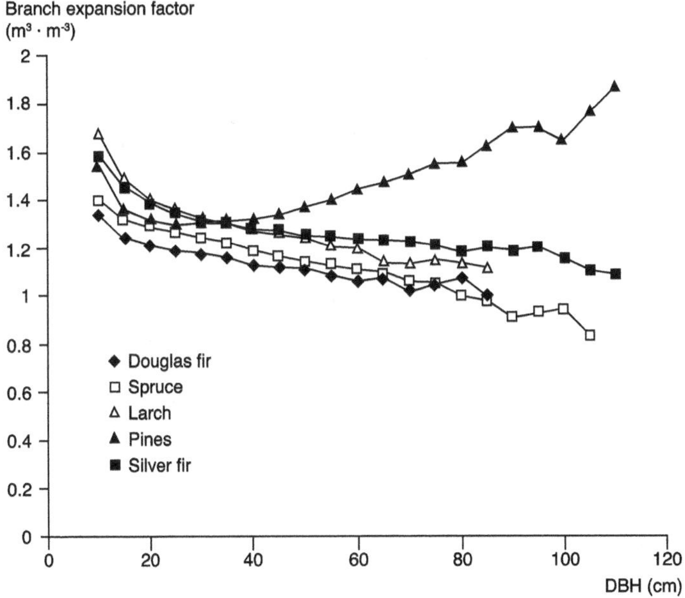

Figure 4.5. Variation of branch expansion factor according to diameter at breast height (DBH) for coniferous species.

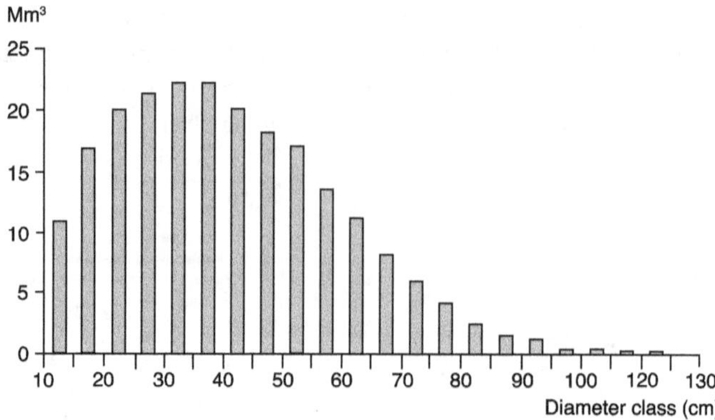

Figure 4.6. Distribution of merchantable volume of beech by diameter classes.

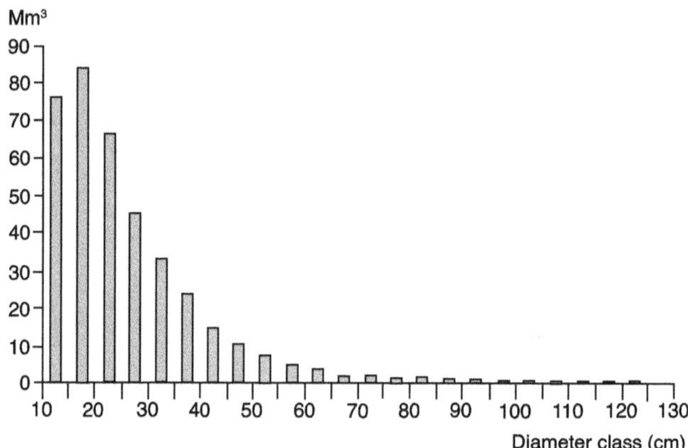

Figure 4.7. Distribution of merchantable volume of other broadleaved species by diameter classes.

Figure 4.4 suggests that the expansion factors used for the largest beech or other broadleaved trees over 50 cm in diameter are perhaps too high. However, Figures 4.6 and 4.7 show that the corresponding total volumes are very limited. Thus, the possible positive bias introduced into the calculations of carbon stocks due to this overestimation of branch volumes for the biggest trees by directly using the LERFOB equations probably remains low.

References

Antonio N., Tomé M., Tomé J., Soares P., Fontes L., 2007. Effect of tree, stand, and site variables on the allometry of *Eucalyptus globulus* tree biomass. *Canadian Journal of Forest Research*, 37, 895–906.

Barczi J.-F., Rey H., Caraglio Y., de Reffye P., Barthélémy D., Dong Q., Fourcaud T., 2008. AMAPsim: an integrative whole-plant architecture simulator based on botanical knowledge. *Annals of Botany*, 101 (8), 1125–1138.

Bi H., 1999. Predicting stem volume to any height limit for native tree species in southern New South Wales and Victoria. *New Zealand Journal of Forest Science*, 29 (2), 318–331.

Bi H., Hamilton F., 1999. Stem volume equations for native tree species in southern New South Wales and Victoria. *Australian Forestry*, 61 (4), 275–286.

Bouchon, J., Ottorini, J.M., Pardé, J., 1981. *Contribution à une meilleure connaissance des potentialités ligneuses totales en France à partir des données de l'inventaire forestier national.* Rapport contrat CEE N° 470.78.7.ESF. French National Institute for Agricultural Research, Nancy, 40 p.

Boysen-Jensen P., Müller D., 1927. Undersøgelser over stofproduktionen i yngre bevoksninger af ask og bøg. *Der Forstlige Forsøgsvæsen*, 9, 221–268.

Brown D., Rothery P., 1994. *Models in Biology: Mathematics, Statistics and Computing.* John Wiley, Chichester, 688 p.

Brown S., 1997. *Estimating Biomass and Biomass Change of Tropical Forests: a Primer.* FAO Forestry Paper No, 134. Food and Agriculture Organization of the United Nations, Rome. [http://www.fao.org/docrep/W4095E/w4095e00.HTM] (Consulted 31 December 2008.)

Brown S., 2002. Measuring carbon in forests: current status and future challenges. *Environmental Pollution*, 116, 363–372.

Burger H., 1929. Holz, Blattmenge und Zuwachs, I. Mitteilung, Die Weymounthsföhre. *Mitteilungen der Schweizerischen Centralanstalt für das forstliche Versuchswesen*, 15 (2), 243–292.

Cailliez F., 1980. *Estimation des volumes et accroissements des peuplements forestiers avec référence particulière aux forêts tropicales.* Vol. 1. *Estimation des volumes. Étude FAO Forêt*, N°. 22/1. Food and Agriculture Organization of the United Nations, Rome, 99 p.

Cairns M.A., Brown S., Helmer E.H., Baugardner G.A., 1997. Root biomass allocation in the world's upland forests. *Oecologia*, 111, 1–11.

Cannell M.G.R., 1982. World forest biomass and primary production data. *In: World Forest Biomass and Primary Production Data.* Academic Press, London, 391 p.

Carvalho J.P., Parresol B.R., 2003. Additivity in tree biomass components of Pyrenean oak (*Quercus pyrenaica* Willd.). *Forest Ecology and Management*, 179, 269–276.

Chave J., Andalo C., Brown S., Cairns M.A., Chambers J.Q., Eamus D., Foïster H., Fromard F., Higuchi N., Kira T., Lescure J.-P., Nelson B.W., Ogawa H., Puig H., Riera B., Yamakura T., 2005. Tree allometry and improved estimation of carbon stocks and balance in tropical forests. *Oecologia*, 145, 87–99.

Cienciala E., Černý M., Tatarinov F., Apltauer J., Exnerová Z., 2006. Biomass functions appicable to Scots pine. *Trees Structure and Function*, 20, 483–495.

Coligny F.d., Ancelin P., Cornu G., Courbaud B., Dreyfus P., Goreaud F., Gourlet-Fleury S., Meredieu C., Saint-André L., 2003. CAPSIS: computer-aided projection for strategies in silviculture: advantages of a shared forest-modelling platform. *In: Modelling Forest Systems. Workshop on the Interface Between Reality, Modelling and the Parameter Estimation Processes, Sesimbra, Portugal, 2–5 June 2002*, pp. 319–323.

Cunia T., 1964. Weighted least squares methods and construction of volume tables. *Forest Science*, 10 (2), 180–191.

Cunia T., Briggs D., 1984. Forcing additivity of biomass tables: some empirical results. *Canadian Journal of Forest Research*, 14, 376–384.

Dhôte J.F., Hervé J.C., 2000. Changements de productivité dans quatre forêts de chênes sessiles depuis 1930 : une approche au niveau du peuplement. *Annales des Sciences Forestières*, 57, 651–680.

Enquist B.J., Niklas K.J., 2002. Global allocation rules for patterns of biomass partitioning in seed plants. *Science*, 295 (5559), 1517–1520.

Fang J.-Y., Chen A., Peng C., Zhao S., Ci L., 2001. Changes in forest biomass carbon storage in China between 1949 and 1998. *Science*, 292, 2320–2322.

Fang J.-Y., Wang Z.M., 2001. Forest biomass estimation at regional and global levels, with special reference to China's forest biomass. *Ecological Research*, 16, 587–592.

Furnival G.M., 1961. An index for comparing equations used in constructing volume tables. *Forest Science*, 7 (4), 337–341.

Gregoire T.G., Dyer M.E., 1989. Model fitting under patterned heterogeneity of variance. *Forest Science*, 35 (1), 105–125.

Joosten R., Schumacher J., Wirth C., Schulte A., 2004. Evaluation tree carbon predictions for beech (*Fagus sylvatica* L.) in Western Germany. *Forest Ecology and Management*, 189, 87–96.

Kelly J.F., Beltz R.C., 1987. *A comparison of tree volume estimation models for forest inventories. USDA Forest Service Research Paper* SO-233. USDA Forest Service, Washington, 9 p.

Kozak A., 1970. Methods for ensuring additivity of biomass components by regression analysis. *Forestry Chronicle*, 46 (5), 402–404.

Lehtonen A., Cienciala E., Tatarinov F., Mäkipää R., 2004. Uncertainty estimation of biomass expansion factors for Norway spruce in the Czech Republic. *Annals of Forest Science*, 64 (2), 133–140.

Lemoine B., 1991. Growth and yield of maritime pine (*Pinus pinaster* Ait): the average dominant tree of the stand. *Annals of Forest Science*, 48, 593–611.

Levy P.E., Hale S.E., Nicoll B.C., 2004. Biomass expansion factors and root: shoot ratios for coniferous tree species in Great Britain. *Forestry*, 77 (5), 421–430.

Madgwick H.A., 1971. The accuracy and the precision of estimates of the dry matter in stems, branches and foliage in an old-field *Pinus virginiana* stand. *In: Forest Biomass Studies, XVth IUFRO Congress, Gainesville, Florida, 15–20 March*, IUFRO-Section 25, pp. 103–112.

Mayer D.G., Butler D.G., 1993. Statistical validation. *Ecological Modelling*, 68, 21–32.

Muukkonnen P., 2007. Generalized allometric volume and biomass equations for some tree species in Europe. *European Journal of Forest Research*, 126, 157–166.

Návar J., Méndez E., Dale V., 2002. Estimating stand biomass in the Tamulipan thornscrub of northeastern Mexico. *Annals of Forest Science*, 59 (8), 813–821.

Niklas K.J., Enquist, B.J., 2002. Canonical rules for plant organ biomass partitioning and annual allocation. *American Journal of Botany*, 89 (5), 812–819.

Parresol B.R., 1993. Modeling multiplicative error variance: an example predicting tree diameter from stump dimensions in Bald cypress. *Forest Science*, 39 (4), 670–679.

Parresol B.R., 1999. Assessing tree and stand biomass: a review with examples and critical comparisons. *Forest Science*, 45 (4), 573–593.

Parresol B.R., 2001. Additivity of nonlinear biomass equations. *Canadian Journal of Forest Research*, 31, 865–878.

Peichl M., Arain M.A., 2007. Allometry and partitioning of above- and belowground tree biomass in an age-sequence of white pine forests. *Forest Ecology and Management*, 253, 68–80.

Porté A., 1999. Modélisation des effets du bilan hydrique sur la production primaire et la croissance d'un couvert de pin maritime (*Pinus pinaster* Ait.) en lande humide. Thèse de doctorat, université de Paris XI, Orsay, 160 p.

Porté A., Bosc A., Champion I., Loustau D., 2000. Estimating the foliage biomass and area of maritime Pine (*Pinus pinaster* Ait.) branches and crowns with application to modelling the foliage area distribution in the crown. *Annales des Sciences Forestières*, 57 (1), 73–86.

Porte A., Trichet P., Bert D., Loustau D., 2002. Allometric relationships for branch and tree woody biomass of Maritime pine (*Pinus pinaster* Ait.). *Forest Ecology and Management*, 158 (1–3), 71–83.

Reed D.D., Green E.J., 1985. A method forcing additivity of biomass tables when using non-linear models. *Canadian Journal of Forest Research*, 15 (6), 1184–1187.

Rennie P.J., 1966. A forest-sampling procedure for nutrient uptake studies. *Commonwealth Forestry Review*, 45 (2), 119–128.

Saint-André L., Thongo M'Bou A., Mabiala A., Mouvondy W., Jourdan C., Roupsard O., Deleporte P., Hamel O., Nouvellon Y., 2005. Age-related equations for above- and below-ground biomass of a *Eucalyptus* hybrid in Congo. *Forest Ecology and Management*, 205, 199–214.

Salas-González R., Houllier F., Lemoine B., Pignard G., 2001. Forecasting wood resources on the basis of national forest inventory data. Application to *Pinus pinaster* Ait in southwestern France. *Annals of Forest Science*, 58, 785–802.

Sharma M., Parton J., 2007. Height-diameter equations for boreal tree species in Ontario using mixed-effects modelling approach. *Forest Ecology and Management*, 249, 187–198.

Sicard C., Saint-André L., Gelhaye D., Ranger J., 2006. Effect of initial fertilisation on biomass, nutrient content and nutrient-use efficiency of even-aged Norway spruce and Douglas-fir stands planted in the same ecological conditions. *Trees Structure and Function*, 20, 229–246.

Tobin B., Nieuwenhuis M., 2007. Biomass expansion factors for Sitka spruce (*Picea sitchensis* (Bong.) Carr.) in Ireland. *European Journal of Forest Research*, 126 (2), 189–196.

Uzoh F.C.C., Ritchie M.W., 1996. *Crown Area Equations for 13 Species of Trees and Shrubs in Northern California and Southwestern Oregon*. USDA Forest Service, Washington, PSW-RP-227-Web, 13 p.

Vallet P., Dhote J.-F., Moguedec G.L., Ravart M., Pignard G., 2006. Development of total aboveground volume equations for seven important forest tree species in France. *Forest Ecology and Management*, 229 (1–3), 98–110.

Van de Walle I., Van Camp N., Perrin D., Lemeur R., Verheyen K., Van Wesemael B., Laitat E., 2005. Growing stock-based assessment of the carbon stock in the Belgian forest biomass. *Annals of Forest Science*, 62 (8), 853–864.

Vogt K.A., Vogt D.J., Palmiotto P.A., Boon P., O'Hara J., Asbjornsen H., 1996. Review of root dynamics in forest ecosystems grouped by climate, climatic forest type and species. *Plant and Soil*, 187 (2), 159–219.

Williams M.S., Gregoire T.G., 1993. Estimating weights when fitting linear regression models for tree volume. *Canadian Journal of Forest Research*, 23, 1725–1731.

Wirth C., Schumacher J., Schulze E.D., 2004. Generic biomass functions for Norway spruce in Central Europe – a meta-analysis approach towards prediction and uncertainty estimation. *Tree Physiology*, 24, 121–139.

Zhang Y., Borders B.E., 2004. Using a system mixed-effects modeling method to estimate tree compartment biomass for intensively managed loblolly pines – an allometric approach. *Forest Ecology and Management*, 194, 145–157.

Zianis D., Mencuccini M., 2003. Aboveground biomass relationships for beech (*Fagus moesiaca* Cz.) trees in Vermio Mountain, Northern Greece, and generalised equations for *Fagus* sp. *Annals of Forest Science*, 60, 439–448.

Zianis D., Mencuccini M., 2004. On simplifying allometric analyses of forest biomass. *Forest Ecology and Management*, 187, 311–332.

Zianis D., Muukkonen P., Mäkipää R., Mencuccini M., 2005. Biomass and stem volume equations for tree species in Europe. *Silva Fennica Monographs*, 4, 63 p.

Chapter 5

Estimating carbon stocks and fluxes in forest biomass: 2. Application to the French case based upon National Forest Inventory data

JEAN-LUC DUPOUEY, GÉRÔME PIGNARD, NABILA HAMZA, JEAN-FRANÇOIS DHÔTE

Introduction

The carbon balance of terrestrial ecosystems is still a matter of debate (Ito *et al.*, 2008). Even in Europe, estimates for the forest sector vary by a factor of 10 (Nabuurs *et al.*, 2003). Although they were originally more oriented towards wood production issues, today the National Forest Inventories (NFIs) are valuable tools for estimating the carbon balance of forests owing to the quantity and quality of data they have accumulated on forest resources (Goodale *et al.*, 2002).

Among the various methods of evaluation of carbon fluxes, they have two main advantages:
– the possibility of defining precisely the spatial extent of measurements of fluxes on varied scales from the stand to biome level. This aspect is particularly important when the assessments of fluxes have to be used for national carbon accounting;
– the impacts of silviculture or other disturbances of the forest ecosystems are directly taken into account and integrated over large scales.

The French NFI is based on the measurement of about 110,000 temporary plots every 12 years since the early 1960s. Plots are randomly selected on the nodes of a regular grid according to a stratified sampling scheme in order to represent all major forest types in France. The sampling intensity is high: one point for every 130 ha of productive woodland on average.

Using these data and a new set of values for branch volume and wood density, we present here estimates of carbon stocks and fluxes in living biomass (above- and below-ground) of French forests, which significantly modify our own former calculations (Dupouey *et al.*, 1999; Pignard *et al.*, 2000).

Comparison of various estimates

We define the net carbon flux to the forest living biomass as the balance between photosynthesis, gross primary production (GPP), on the one hand, minus autotrophic respiration, R_a, fellings and natural losses, on the other hand. It corresponds to the net increment in carbon of the forest biomass. According to definitions listed in Chapter 1, the net carbon flux defined here is the net primary production (NPP) minus carbon losses through harvest, mortality and natural disturbances. Several estimates of carbon stocks and fluxes in forest biomass in France have been published in the past 5 years. The results obtained fall within a relatively broad range of estimates, although most were based on the same NFI data.

There are many factors behind the variations observed among the different estimates.

• The estimates concerned different dates or periods. Strong trends in terms of accumulation of standing material and increased productivity over the past several decades have meant similar variations in carbon stocks and fluxes in the biomass.

• The compartments considered were not strictly identical: some, for which few data are available and whose quantitative importance is generally limited, were in some cases overlooked or evaluated on an inclusive basis. This category includes the foliage, herbaceous and shrubby strata, dead wood and also trees smaller than the measurement thresholds set by the NFI (diameter of less than 7.5 cm at 1.30 m). The share of these compartments in the total biomass of forest stands is relatively small on a national level, but varies considerably depending on the type of stand (shade/light species, young/ mature stands) or region (e.g. the relative contribution of the herbaceous and shrubby strata is considerably greater in the Mediterranean zone).

• Different methods and parameters were used to assess tree biomass based on NFI data:

– for each sample tree, the NFI measures girth at 1.30 m (C_{1300}) and total height (HTOT), data which are used to calculate the individual volume of the tree (VIFN). In NFI jargon, the estimated volume, VIFN, is the standing timber volume over bark, that is, the stem volume to a top diameter of 7 cm, also called hereafter the "merchantable volume". This general definition may be adapted in the event of very substantial stem deformation that would otherwise prevent the tree from being used for timber; such deformations are primarily seen in broadleaved species and some conifers (Figure 5.1). In this occurrence, the volume taken into account is practically restricted to the volume below the deformation, up to a section cut according to its form (secondary logs on the stem, above the saw point, may sometimes be taken into account in the protocol of volume calculation provided they satisfy such stringent size and form criteria that they generally only represent a minor proportion of the stem volume above the saw point). Adopting this definition is warranted by socio-economic considerations, since the NFI's primary objective when it was set up was to assess the production potential of French forests with a view to supplying the timber and paper pulp industries. The volume estimated by the NFI, therefore, represents only part of the total aerial volume of a tree; in particular, it excludes the branches.

– two different approaches were used:

    i. calculating the total aerial volume of the tree by applying mean branch expansion factors to the stem volume estimated by the NFI;

    ii. more precise calculation based on C_{1300} and HTOT of the total aerial volume of the tree, or its biomass, using adapted allometric relationships.

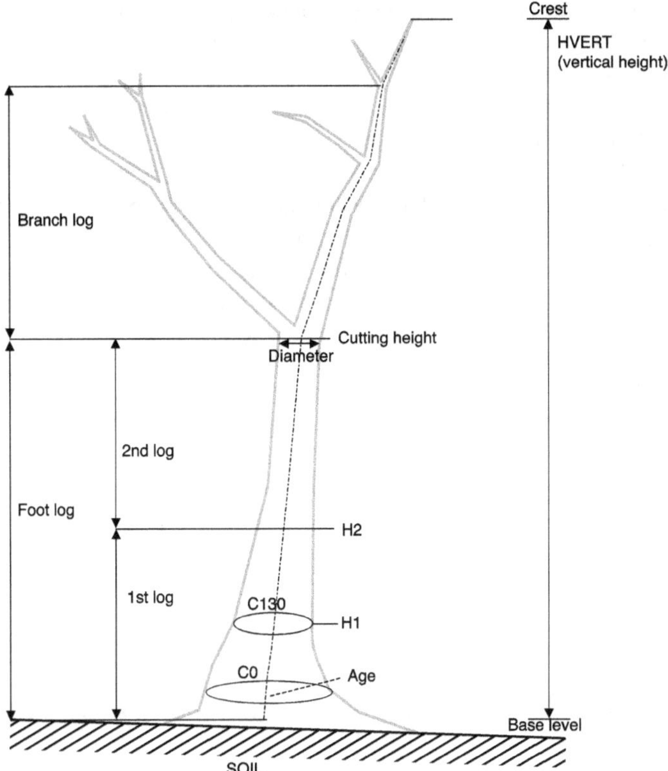

Figure 5.1. Tree measurement by the National Forestry Inventory (NFI).

• The mean branch expansion factor is a convenient way of comparing different estimates. It often differs between broadleaved species and conifers, and is defined as the ratio of total aerial volume to merchantable volume. The factor varies substantially depending on the species, site conditions, age and silvicultural methods used and is particularly difficult to evaluate without taking complex, destructive measurements. The branch expansion factor may either be one of the key parameters used in the calculation of carbon stocks, or be calculated *a posteriori*, when allometric relations are used (Somogyi *et al.*, 2007), as a simple informative parameter.

• Root expansion factors can then be used to integrate the below-ground part of the wood volume or tree biomass; these values are difficult to ascertain, and are another source of variability.

• Calculating biomass (excluding foliage) means multiplying the total volume of the trees by the wood specific gravity, which also varies depending on the species, tree growth rate, position within the tree (for instance, the bark is less dense), etc.

• Lastly, multiplying the total biomass of the tree by the average carbon concentration gives carbon mass. This carbon concentration is apparently less variable than expansion factors or wood densities. It is generally between 0.45 and 0.5.

Calculating an overall ratio, defined as the ratio of total carbon mass to merchantable volume, which is equal to the product of the branch and root expansion factors, the wood specific gravity and the carbon concentration, makes it possible to condense each of the estimates considered into a few key values (mean overall ratio for the main groups of species such as broadleaved species and conifers).

In addition to the previous sources of variation, two main methods were used for evaluating net carbon fluxes:
– through stock variations, comparing two successive inventories; or
– through the difference between influx (growth) and outflux (wood removal and mortality).

The latter approach is certainly more difficult, since it relies on calculating the difference between data from different sources, which are often heterogeneous. The NFI evaluates volume growth and mortality, although it has some difficulties with mortality due to accidental causes (e.g. windfall, fires, etc.), as a result of the low inventory frequency (approximately every 12 years). Wood removal figures are often both diverse and partial: the volume felled but left in the forest (sometimes burned) is not taken into account, the cubage method varies depending on the end use of the timber (e.g. sawmill, paper pulp or other such as poles, etc.), and the estimates for the self-consumed volume (mainly for fuelwood) are very imprecise, which amount to 25% to 30% of the total volume removed in France. However, this approach is recommended by the Intergovernmental Panel on Climate Change and is used by the Interprofessional Technical Centre for Studies on Atmospheric Pollution (CITEPA), the French national organization responsible for the annual inventory of greenhouse gas emissions (Fontelle *et al.*, 2003).

Table 5.1 enables a comparison of different estimates including that proposed on completion of the present CARBOFOR project. This last estimate, the results of which are given in "Methodology used for revised estimates of carbon stocks and fluxes within French forest biomass" hereafter, is based on:
• using the total aerial volume allometric equations compiled by Vallet *et al.* (2006) from French forestry research archives (data gathered 1920–50) and presented below;
• the root expansion factors and wood specific gravity values based on an analysis of existing literature (see "Comparaison of various estimates" for wood density; Chapter 4 for root expansion factors).

The overall ratio values corresponding to the preceding estimates are relatively uniform in terms of broadleaved species; for conifers, the CITEPA estimate is around 20% higher than the other two. The CARBOFOR estimate is based on a markedly higher overall ratio value for broadleaved species (around +25%) and a value towards the top of the range for conifers. The branch volume allometric equations used are the main reason for this re-evaluation, which will be discussed below.

Table 5.2, from Löwe *et al.* (2000), summarizes the values used for the annual inventory of greenhouse gas emissions in various European countries. Given that it is left to the signatories of the convention to decide whether to take account of below-ground biomass, there is some doubt surrounding the expansion factors quoted. The values obtained and used in the present CARBOFOR project (Table 5.1) are in the highest part of the European range (Table 5.2). However, the European values used in 2000 for biomass expansion were not always based on truly representative samples of adequately sampled trees. Instead, they often resulted from literature analysis. It is interesting to note that Levy *et al.* (2004), after a reassessment of expansion factors using a large dataset of tree measurements, obtained a mean branch expansion factor of 1.43 and a mean root expansion factor of 1.36 for coniferous trees in the UK, higher than our values (1.34 and 1.30, respectively, Table 5.1).

Table 5.1. Comparison of parameter values used for different estimates of carbon stocks and fluxes in French forests.

	AGRIGES 1999[1]		CITEPA 1999[2]		IGD 2000[3]		CARBOFOR 2004[4]	
	BRO.	CON.	BRO.	CON.	BRO.	CON.	BRO.	CON.
Branch expansion factor (aerial wood volume/merchantable volume)	1.30	1.13	1.25	1.25	1.40	1.30	1.61	1.34
Root expansion factor (total wood volume/aerial wood volume)	1.19	1.24	1.28	1.28	1.14	1.15	1.28	1.30
Overall expansion factor (total wood volume/merchantable volume)	1.55	1.39	1.60	1.60	1.60	1.50	2.06	1.74
Wood specific gravity	0.56	0.41	0.54	0.43	0.53	0.40	0.55	0.44
Carbon concentration	0.5	0.5	0.5	0.5	0.5	0.5	0.475	0.475
Overall ratio ($tC \cdot m^{-3}$ NFI) (carbon mass/merchantable volume)	0.431	0.285	0.432	0.344	0.422	0.296	0.535	0.361
"Secondary" compartments ($tC \cdot ha^{-1}$) (foliage, herbaceous and shrubby strata, non-recordable trees and dead wood)	2.3	6.5	0	0	0	0	0	0

1 AGRIGES 1999: estimate calculated by the NFI and French National Institute for Agricultural Research (INRA) at the end of the AGRIGES project (Dupouey *et al.*, 1999). The expansion factors and wood specific gravities (per species) were drawn from a bibliographical analysis. Allometric relations were used for coppice stands. All-inclusive estimates per hectare were used for the foliage (1.7 tC·ha-1 for beech, 2 tC·ha-1 for other broadleaved species, 6 tC·ha-1 for conifers) and herbaceous stratum (1 tC·ha-1, except for maritime pine, 4 tC·ha-1); the dead wood measured by the NFI (partial: mortality in the previous 5 years) was taken into account.

2 CITEPA 1999: parameters used to calculate emissions from the forest sector in annual greenhouse gas inventories (Rivière, 1999). Origin: MIES/INRA, pers. comm. The method is based on the difference between influx and outflux.

3 IGD 2000: calculation of national carbon storage in the tree stratum and forest soils (IFN/MAP, 2000). Parameters drawn from Morrin and Janz (1986).

4 CARBOFOR 2004: present project, see "Methodology used for revised estimates of carbon stocks and fluxes within French forest biomass". Figures are given for the last inventory.

BRO.: broadleafs; CON.: conifers.

Table 5.2. Factors used to estimate the forest carbon sink in the inventory of greenhouse gas emissions (from Löwe *et al.*, 2000).

Country	Species	Branch expansion factor: above-ground biomass volume/ stem volume	Overall expansion factor: total biomass volume/ stem volume	Wood specific gravity	Carbon concentration	Overall ratio (tC·m⁻³)
Austria	Conifers		1.45	0.39	0.49	0.28
	Broadleafs		1.46	0.53	0.48	0.37
Denmark	Conifers		2.00[1]	0.38	0.50[4]	0.38
	Broadleafs		2.00[1]	0.58	0.50[4]	0.58
Finland	Pine		1.527[2]	0.39	0.519	0.31
	Spruce		1.859[2]	0.39	0.519	0.38
	Broadleafs		1.678[2]	0.49	0.501	0.41
France (CITEPA 1999)	Conifers		1.60	0.43	0.50[4]	0.34
	Poplars		1.60	0.35	0.50[4]	0.28
	Broadleafs		1.60	0.54	0.50[4]	0.43
Germany	Pine, larch	1.14		0.43	0.50[4]	0.25
	Spruce, fir, Douglas fir	1.14		0.37	0.50[4]	0.21
	Oaks, beech	1.24		0.56	0.50[4]	0.35
	Other broadleafs	1.24		0.55	0.50[4]	0.34
Ireland	Conifers		1.30	0.37	0.43	0.21
	Broadleafs		1.30	0.55	0.45	0.32
Netherlands	All species	1.20		0.50[4]	0.50[4]	0.30
Portugal	Conifers		1.247	0.38	0.45	0.21
	Broadleafs		1.237	0.70	0.45	0.39
Spain	All species		1.60[3]	0.50[4]	0.45	0.36
Sweden	Conifers	1.30		0.42	0.45	0.25
	Broadleafs	1.50		0.58	0.45	0.39
UK	Conifers		1.39	0.35	0.50[4]	0.24
	Broadleafs		1.52	0.55	0.50[4]	0.42

1 Includes "some carbon in the undergrowth and soil".
2 "Also contains foliage".
3 Includes "part of the surrounding shrub vegetation".
4 IPCC default values.

They pointed to the need for an upward revision of estimated carbon fluxes in UK forests. Tobin and Nieuwenhuis (2007) reached the same conclusion after a re-evaluation of biomass expansion factors of *Picea sitchensis* in Ireland.

Methodology used for revised estimates of carbon stocks and fluxes within French forest biomass

Annual carbon stocks and fluxes in French forests have now been reassessed based on the last two sets of departmental inventories. The calculation method was as follows:

– application of total aerial volume allometric equations of Vallet *et al.* (2006) to merchantable volume of NFI sample trees. In each of around 110,000 plots, all trees above a threshold diameter (7.5 cm within a 6 m radius circle, 22.5 cm within a 9 m radius circle and 37.5 cm within a 15 m radius circle), that is, 11 trees on average were measured and subsequently used for the calculation of total aerial volume. Table 5.3 presents the set of eight parameter values used for applying the equation used (Equation 1), the tree species used to calibrate the equation and the species for which the equation was used;

– application of constant root expansion factors of 1.28 for broadleaved species and 1.30 for conifers;

– application of a carbon concentration in the wood of 0.475;

– the foliage, herbaceous and shrubby strata, trees below the inventory threshold diameter and dead wood were not taken into account;

– trees outside the forest (e.g. hedges and tree rows, parks and gardens) are not accounted for. However, they can represent a significant proportion of the wood stock in some regions of France, for example, in Brittany or in the Paris area (Dupouey and Pignard, 2001). Poplar plantations have also been excluded from our calculations for several reasons. They have a shorter rotation length (10–20 years) and a very specific management system, which induce different dynamics to the remaining French forest. Moreover, there was no branch allometric equation available for this species and lastly, it represents only 220,000 ha (less than 1.5% of the total French forest area).

Table 5.3. Parameters and species used for the branch allometric equations.

Tree species used to calibrate the equation	Species for which the equation was used	Parameters of the equations			
		a	b	c	d
Quercus petraea	*Quercus petraea, Q. robur, Q. pubescens, Q. rubra*	0.471	– 0.345	0.377	0
Fagus sylvatica	*Fagus sylvatica*	0.395	0.266	0.421	45.4
Quercus petraea and *Fagus sylvatica*	Other broadleafs	0.428	– 0.191	0.456	0
Pseudotsuga menziesii	*Pseudotsuga menziesii*	0.534	– 0.530	0	56.6
Pinus spp.	*Pinus* spp.	0.311	0.405	0.340	191
Larix spp.	*Larix* spp.	0.550	– 1.350	0.322	0
Abies alba	*Abies* spp.	0.550	– 0.749	0.277	0
Picea abies	*Picea abies* and other coniferous species	0.631	– 0.946	0	0

Equation:

$$\text{total above-ground volume} = \text{shape factor} \times \left(\sqrt{C_{130}}\right) \times \text{HTOT}/(40{,}000 \times \Pi)$$

with:

C_{130}:   girth at breast height (cm)

HTOT: total height (m)

rob: robustness $= \left(\sqrt{C_{130}}\right)/\text{HTOT}$

and shape factor $= \left(a + b \times C_{130}/1000 + c \times \text{rob}\right) \times \left(1 + d/\sqrt{C_{130}}\right)$

Results

The French forest carbon stock and its temporal change

The carbon stock in forest biomass (above- and below-ground) was assessed at 71.2 tC·ha^{-1} on average at the time of the last inventory in 1996, the merchantable volume was 154 m^{-3}·ha^{-1}. This value per hectare was estimated for the area of productive woodland, that is, 13.8 million ha in 1996, which gives a total stock of 984 MtC. "Non-productive" forests, where wood production is not the main target (e.g. protected areas, peri-urban woodlands, unexploited forests on steep slopes of the mountains), are not inventoried by French NFIs. If the previous stock per hectare is extrapolated to the whole woodland area, that is, 14.9 million ha, this gives an overall estimate of 1059 MtC in 1996.

The carbon stock in forest biomass (above- and below-ground) was assessed at 59.4 tC·ha^{-1} on average at the time of the previous inventory in 1984, the merchantable volume was 133 m^{-3}·ha^{-1}. The stock was estimated for the area of productive woodland, that is, 13.4 million ha in 1984, which gives a total stock of 795 MtC. If this stock is extrapolated to the whole woodland area, that is, 14.0 million ha, this gives an overall estimate of 830 MtC in 1984.

By calculating the difference in stocks, these two inventories were used to calculate the net annual carbon flux in forest biomass: 15.6 MtC·year^{-1} based on productive woodland and 18.7 MtC·year^{-1} if the whole woodland area is considered. It is important to note that the results quoted here barely take account of the impact of the December 1999 storms as the departmental inventories, conducted since 2000, have concerned regions that were not severely affected.

Without even considering the methods used to evaluate the carbon mass in the trees, these results were affected by several types of bias resulting from the forest inventory system:

– it is debatable whether it is wise to extrapolate the estimate to the forest area as a whole, since those forests that the NFI classes as non-productive are primarily stands on steep slopes, and to a lesser extent, stands with considerable recreational uses. It is reasonable to assume that stocks and primary biomass production in such forests are lower, but the same can doubtless also be said for wood removal. The relative value therefore seems to be valid;

– some changes over time in the inventory methods resulted in bias when calculating the variations in carbon stocks; around 250,000 ha, primarily Mediterranean heathlands, were included in the previous survey but not in the last one because of their low economic impact. Such stands had a low merchantable volume: around 52 m^{-3}·ha^{-1} (32 tC·ha^{-1}) at the time of the previous inventory. The mean stock per hectare was thus overestimated in the last inventory, since the biomass of these stands was not counted along with the productive areas, and the variation in the stock was also overestimated. However, the impact was relatively limited: around 0.5 tC·ha^{-1} for the stock estimate and 0.5 MtC·year^{-1} for the estimate of mean annual storage;

– some areas, abandoned by farmers, that became woodland between 1984 and 1996 as a result of natural colonization already had some trees in 1984, and this biomass, which is difficult to assess, needs to be deducted from the variation in stocks calculated above. Although such natural woodland now accounts for over 80% of forest recolonization in

France, it still has only a limited impact on carbon storage estimates. If we assume that 80% of the areas corresponding to forest increase over the period 1984–96 contained an initial volume of 10 m^{-3}·ha^{-1}, the annual storage estimate will be reduced by just 0.3 MtC·year^{-1};
– the values for net annual carbon storage in forest biomass quoted above may, therefore, be overestimated by around 0.8 MtC·year^{-1}, which gives an estimate of the forest sink of around 14.8 MtC·year^{-1} for productive woodland and 17.9 MtC·year^{-1} for French forests as a whole.

These estimates are a significant increase on previous evaluations, notably those proposed by Dupouey *et al.* in 1999 (Figure 5.2). The total stock was estimated at 54.8 tC·ha^{-1} in 1991 and 48.5 tC·ha^{-1} in 1979, considering only the biomass compartment of recordable trees. The main reasons for this re-evaluation are the use of new volume allometric equations and the adoption of higher root expansion factors.

The upward revision of net annual storage (calculated as stock variation) is much more marked, since the overall carbon sink for the period 1979–91 was estimated at 10.5 MtC·year^{-1}. This is a result of the combined effects of adjusting the method (an increase of around 25% in the overall carbon mass to NFI volume ratio) and a real increase in net annual storage. The latter is in turn the result of several factors:
– the volume growth per area unit has been increasing, for reasons that have yet to be determined, by an average of 1% per year for at least two to three decades (see IFN/ MAP, 2000, for a more in-depth analysis of this phenomenon), while wood removal has progressed less rapidly, and has been stagnating since the early 1990s (until the 1999 storms);
– since the mid-20th century, the area of forest has been growing steadily and this increase has remained strong despite a reduction in artificial reforestation: an average of +73,000 ha·year^{-1} between 1984 and 1996, that is, 0.5% per year, primarily through natural recolonization of land abandoned by farmers. This spread has also contributed to

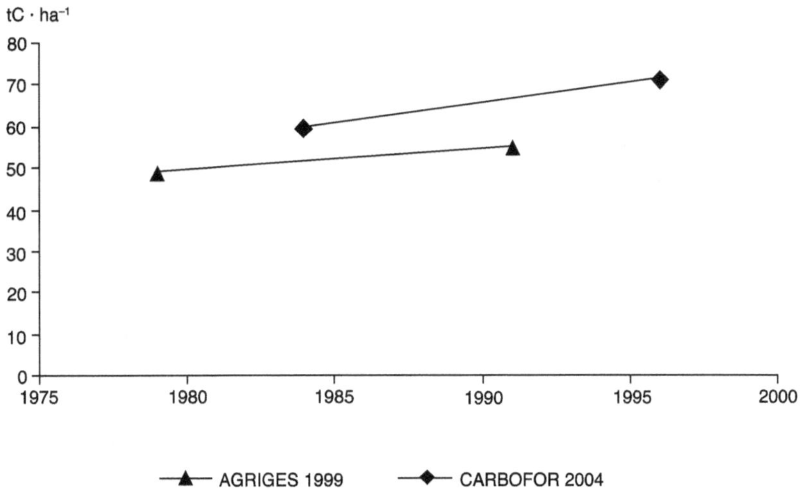

Figure 5.2. Comparison of carbon stock estimates per area unit in forest biomass in two successive studies.

the increase in total growth in volume. The impact of an increase in forest area on carbon storage and biological production, assessed using NFI data, is not generally noticeable for some 10 to 20 years (excluding poplars), which corresponds to the time taken for young trees to reach the threshold diameter for inventory. Any increase in area between two successive inventories thus has only a limited impact on the stock variations calculated between these two inventories. Moreover, if the increase in area is considerable, it will reduce the mean stock calculated for the forests as a whole (mean of stock in existing woodlands and the lower stock in new woodlands). On the other hand, the increases in forest area that occurred before this period have a significant impact on the carbon stock and flux, because the stands of intermediate age (20 to 40 years) have a high production and weak wood removal. Large areas have been afforested in France in the second half of the 20th century, mainly with coniferous species (Figure 5.3). They contribute positively to the current carbon sink, but this impact should gradually decrease when these stands are harvested.

Table 5.4 summarizes some of the key changes in French forests previously discussed.

Table 5.4. Main data related to French forests (except poplar plantations). The overall branch expansion factor is the ratio of total aerial volume calculated for the whole of France to the total merchantable volume measured by the NFI for the whole of France.

	Unit	1984	1996
Total area	1000 ha	13,988	14,875
Growing stock	$m^3 \cdot ha^{-1}$	133	154
Gross increment	$m^3 \cdot ha^{-1} \cdot year^{-1}$	5.7	6.5
Part of regular high forest (area)	%	43.5	48.5
Average age of high forests	year	64.4	64.6
Carbon stock (biomass)	$tC \cdot ha^{-1}$	59.4	71.2
Branch expansion factor (overall)		1.46	1.50

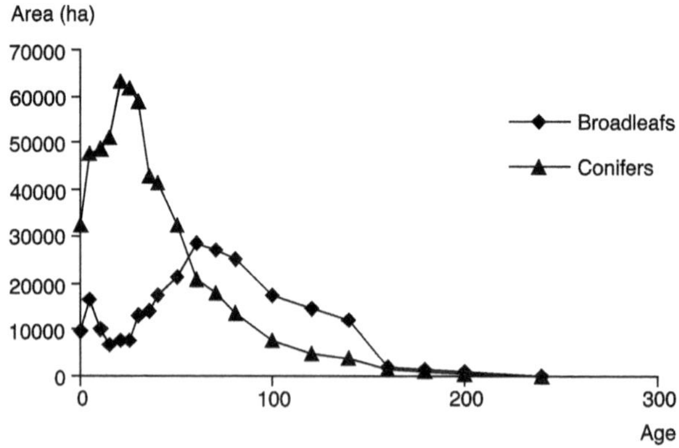

Figure 5.3. Area of forest stands managed as regular high forest by age and species group (average date: 1996).

The change over time of carbon stocks and fluxes in the forest biomass can be analysed more precisely. A third inventory was carried out between 1971 and 1980 and is available for nearly half of France (55 departments). We applied the method described above to all inventories and carried out a linear interpolation to calculate in each department the carbon stock for each year between two successive inventories; we also used this linear equation to evaluate the carbon stock before the oldest inventory and beyond the most recent inventory. These data made it possible to calculate the national carbon stock for each year, and we only retained the years for which fewer than 15 departments (out of 90) had extrapolations of more than 5 years. The annual carbon flux was assessed by the difference between two successive estimates of the carbon stock. The results are presented in Figures 5.4 and 5.5. The latter clearly highlights the increasing trend in annual storage, which increases from about 14.6 MtC·year^{-1} in 1976 to 18.7 MtC·year^{-1} in 1996, which means an increase in the sink of 4.1 MtC·year^{-1} in 20 years. However, because of the method used, there is the same biases here as for the average flux in the period 1984–96 and the variation of the sink in 20 years is certainly lower and nearer to +3.3 MtC·year^{-1} than to +4.1 MtC·year^{-1}. There is a strong increase in annual biomass storage during the decade 1980–90, and two periods of stagnation before 1980 and after 1990. The stagnation at the end of the 1970s could be partly a consequence of the drought of 1976, which had a strong and lasting effect on the productivity of stands.

The curve in Figures 5.4 and 5.5 is, however, to be analysed carefully because of the method of interpolation. For example, the similarity between the evolution of the sink and that of the wood harvest (Figure 5.6) is surprising: it implies that the increase in harvest during the 1980s would have been more than compensated for by the increase of the gross increment.

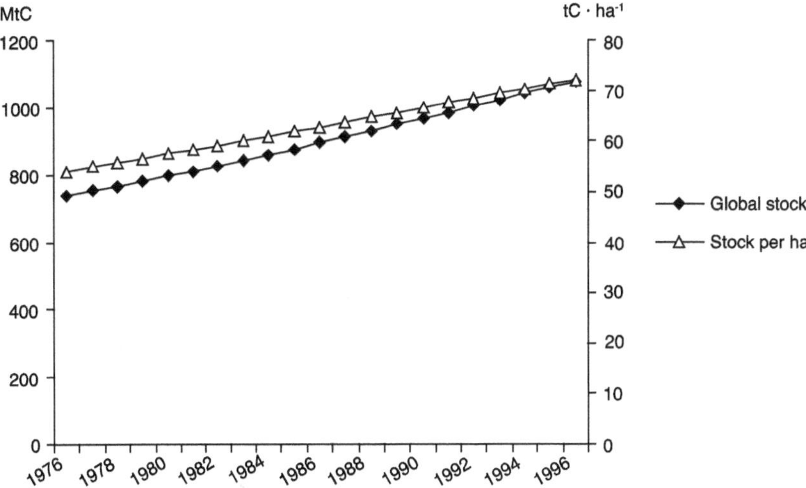

Figure 5.4. Carbon stock of the biomass in French forests from 1976 to 1996.

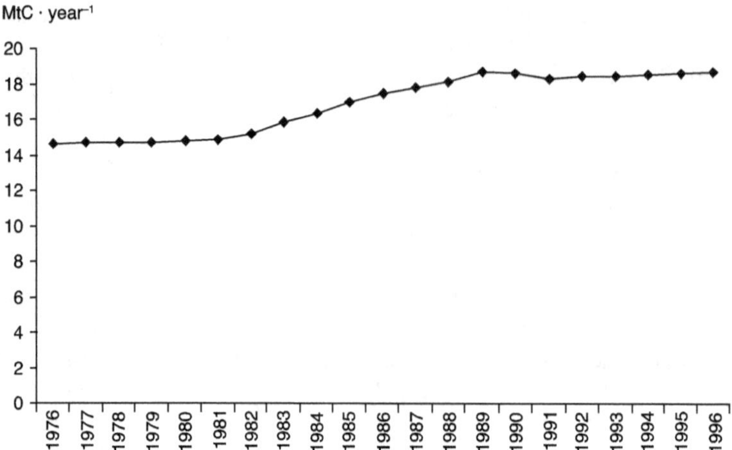

Figure 5.5. Annual change in the carbon stock of the biomass in French forests from 1976 to 1996.

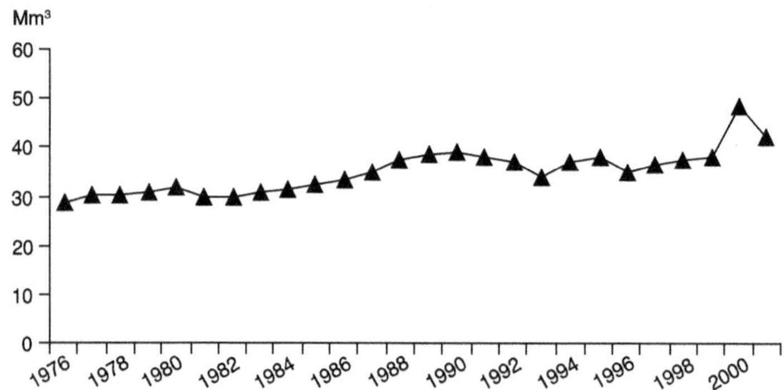

Figure 5.6. Volume of wood harvested and marketed in France between 1976 and 2001 (volume in million·m^{-3} over bark).

A detailed analysis of the results reveals a number of points:

– despite a lower NFI volume, which can be attributed to the existence of coppices (143 m^3·ha^{-1} compared with 173 m^{-3}·ha^{-1} for conifer stands), broadleaved species generally have a higher carbon stock per hectare: 76 tC·ha^{-1} compared with 62 tC·ha^{-1} for conifers. This situation results from their higher wood density and proportion of branches. The stock variation for the two groups of species is much the same, that is, around +1 tC.ha^{-1}·year^{-1};

– as regards distribution per ownership category, the carbon stock ranking is the same as for merchantable volume: 85 tC·ha^{-1} in state-owned forests (191 m^{-3}·ha^{-1}), 81 tC·ha^{-1} in other state-managed woodlands (181 m^{-3}·ha^{-1}) and 67 tC·ha^{-1} in private woodlands (143 m^{-3}·ha^{-1}). However, the gaps are shrinking, and the variations for the period

1984–96 were in the reverse order: +6 tC·ha⁻¹ in state-owned forest, +9 tC·ha⁻¹ in other state-managed woodlands, and +13 tC·ha⁻¹ in private woodlands. This ranking is closely correlated to the relative extent of reforestation and to wood mobilization rate in the three ownership categories, 83%, 67% and 62% over the 1984–96 period, respectively. Overall, storage in productive stands was + 1MtC·year⁻¹ in state-owned forest (1.4 million ha), +2.5 MtC·year⁻¹ in other state-managed woodlands (2.2 million ha) and +12.1 MtC·year⁻¹ in private woodlands (10.1 million ha);

– the carbon stock per hectare is 80 t on average in regular high forest and mixed high broadleaved forest-coppice and only 39 t in simple coppice. Mixed high conifer forest-coppice and irregular high forest have an intermediate stock per hectare, calculated to be 70 tC·ha⁻¹ and 52 tC·ha⁻¹, respectively. These average values are highly dependent on stand distribution in terms of age category;

– figure 5.7 shows the mean carbon stock in forest biomass per area unit, department by department. It shows that the highest stocks on a departmental level are in north-eastern France: Alsace, Lorraine and Franche-Comté, in the northern Alps, and in the western Pyrenees. Mediterranean regions have the lowest values for this estimate, which only takes account of the biomass of recordable trees. A table (see Table A1 in the Appendix) summarizes the main figures, region by region;

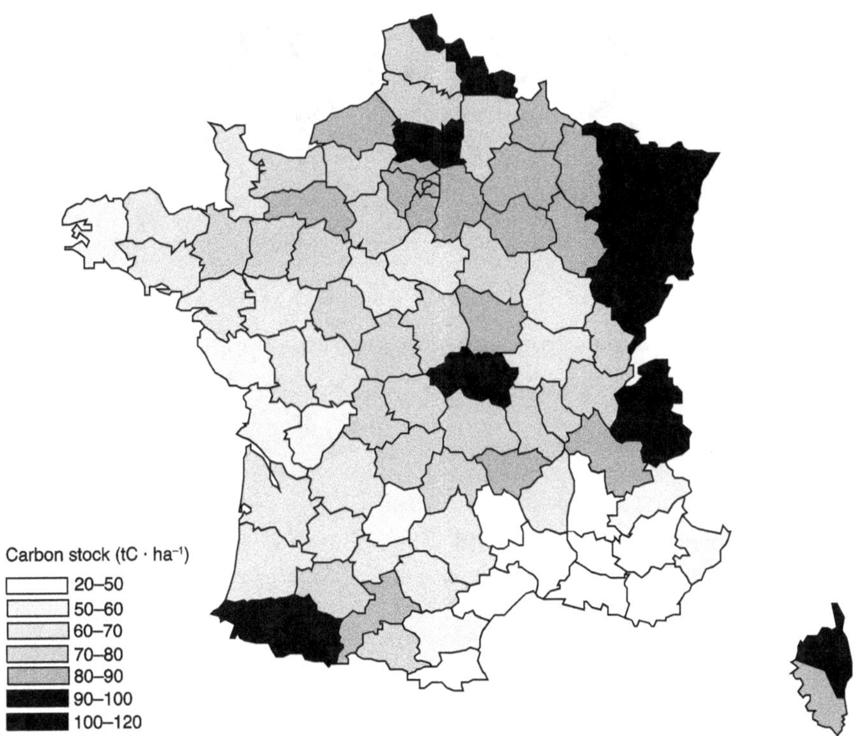

Figure 5.7. Mean carbon stock per area unit in forest biomass per department.

– figure 5.8 shows net annual carbon storage in forest biomass. On a departmental scale, the most significant sinks follow a broad diagonal line from the south-west to the north-east, with Aquitaine, Burgundy and Auvergne prominent. This situation was doubtless considerably disrupted by the December 1999 storms, which caused severe damage across a large part of the zone. It must be pointed out that these figures are global sequestration accumulated per department and that, at this scale of integration, sequestration depends first on the forest area of each department.

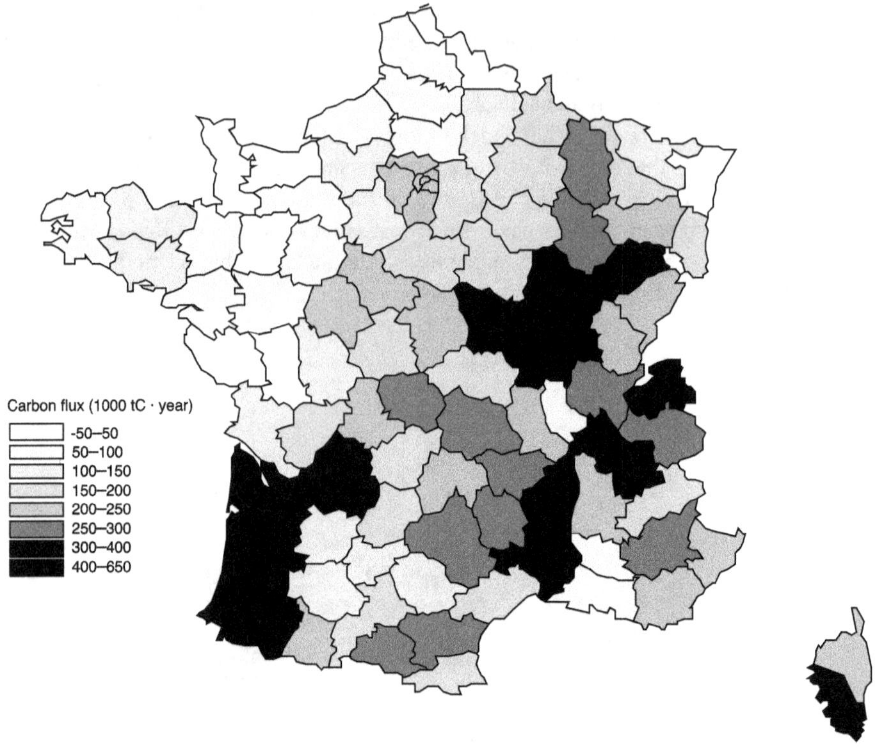

Figure 5.8. Annual variation of carbon stock in forest biomass per department averaged over the period 1984–96.

Impact of the storms of December 1999

The total volume of wood damaged by the storms of December 1999 is not known precisely. The conclusion of all works carried out by the French NFI (IFN/AFOCEL, 2004) is that this volume could reach 170 million m^3, which is about 8% of the growing stock, or two times the annual gross increment, or three times the annual harvest (the annual harvest is about two-thirds of the gross increment). The use of a reduction factor of 0.92 allows the assessment of carbon stock of the forest biomass after the storms. However, the implicit model, which only takes into account the pool of living biomass, is too simple to describe the phenomenon correctly.

As of 1 January 2000, this pool was actually reduced by 8%, but the total carbon stock in French forests had not changed. The carbon contained in the damaged trees was simply transferred to dead wood or litter.

A fraction of the carbon stock contained in the damaged trees has been exported subsequently from the damaged stands during the 2 years following the storms. These removals have partly replaced the usual harvest, which has been temporarily avoided and delayed. According to the annual survey on harvested and marketed wood the harvests in 2000 and 2001 were respectively, 1.28 and 1.12 times the harvest in 1999. It can be supposed that in 2000 and 2001, a volume equivalent to two times the harvest of 1999 was taken in damaged wood. About 2 to 3 years after the storms, the main impact on carbon stock in the forest biomass, with the above assumption, is a loss of approximately one annual harvest, which is approximately two-thirds of the annual current production or average annual storage for 2 years). This loss is evaluated to be approximately 1.8 $tC \cdot ha^{-1}$. It will result in an emission to the atmosphere through the decay or burning of the residual biomass, which is more or less delayed, according to whether the residual damaged wood is burnt or left in the forest.

Another impact of the storms is the re-setting of highly damaged stands dynamics to a younger age class of the same or a new growth trajectory. Since about 450,000 ha (3% of the forest area) suffered a loss of more than 50% of forest cover, this impact must also be taken into account. A more complete calculation of the impact of the 1999 storm on the carbon budget is presented in Pignard *et al.* (2009).

Discussion: results validation and fine-tuning

Validation of the aerial volume calculated with allometric equations

Calculating the ratio of total above-ground volume/stem volume showed that the present estimations, based on new allometric equations (Vallet *et al.*, 2006), produced a significantly higher figure for branch volumes in comparison with previous estimates, particularly with regard to broadleafs (Table 5.1). The mean value of this ratio for the last inventory in the present study is 1.61 for broadleaved species, whereas it was only 1.30 in our previous work (Dupouey *et al.*, 1999). This re-evaluation is commented in depth in Chapter 4. In short, although it comprised over 4500 trees, the sample used to build the new allometric equations was perhaps not fully representative of the entire French forests because:
– many species are absent; of the broadleaved species, only *Quercus petraea* Lieb. and *Fagus sylvatica* L. were considered. The aerial volume of the other broadleaved species was evaluated using an equation compiled by combining the figures for these two species (Table 5.3);
– the forests from which the samples were taken were primarily in northern France;
– the figures are relatively old (1920–55), and several factors may have modified the allometric relationships within trees (e.g. silvicultural changes, genetic improvement, environmental changes, etc.);
– the sample plots were exclusively set up in regular high forest stands.

However, as far as we could check these different points, no real biases have been found (see Chapter 4) and this set of allometric equations is currently the best available way of calculating total tree aerial volume in France.

Using an allometric approach for the calculation of branch volumes, instead of constant branch expansion factors, lead to an increase of the overall mean branch expansion factor for the whole of France between the two last inventories, from 1.46 to 1.50 (Table 5.4). The causes for such an increase are discussed in Chapter 4. We calculated that this apparent change in the shape of trees was responsible by itself for 13% of the 17.9 MtC·year^{-1} carbon flux into forest biomass between 1984 and 1996.

Impact of wood density estimates

Two sets of wood specific gravity values per species were used for France. They were compiled from bibliographical references. One older reference (Mathieu, 1877), proposes much higher values for most species. These differences may be due to some type of systematic bias in measuring the samples used, in which case it would no doubt be best not to use those measurements. This reasoning resulted in an initial set of values, including corrected values for Mathieu samples, which was used for the present CARBOFOR estimates presented above. There is another possibility: Mathieu's sample may be more representative of the French forest resource. Woods from mountain regions or the Mediterranean zone, which are characterized by higher density, are more common in Mathieu's work than in the other sources, which were primarily published by authors from northern Europe. One way to shed light on this discrepancy would be to look again at Mathieu's samples, which are still available at l'École Nationale du Génie Rural des Eaux et des Forêts (ENGREF) (Nancy, France). Extensive campaigns of wood density measurements should also be launched in order to obtain new and more representative estimates at the national scale for a large range of species. In the meantime, a second set of wood specific gravity values has been compiled, primarily based on raw Mathieu's results.

This second set was used to calculate a second estimate of the carbon stock in French forest biomass. This produced a mean carbon stock figure of 77 tC·ha^{-1}, that is, 8.1% more than the first estimate. Mean wood specific gravity values were 0.593 for broadleafs (+8.6%) and 0.467 for conifers (+6.7%).

Accuracy of carbon stock and flux estimates

Evaluating the error made in estimating carbon stocks and fluxes in forest biomass is a complex task. It involves not only sampling and modelling errors (applying the allometric equations to NFI sample trees), but also some forms of bias difficult to pinpoint:
– applying equations compiled from data on oaks and beech to other broadleaved species;
– applying volume equations outside the field for which they were developed (e.g. types of stand, regions, species, etc.);
– using data for which the degree of error is not known (e.g. wood specific gravity, carbon concentration, etc.).

We attempted to estimate the uncertainty in the global carbon flux by three different ways: (i) recalculating this flux based on different, selected sets of parameters for wood density, branch volume and carbon concentration; (ii) building a random distribution of possible input values by Monte-Carlo simulation; and (iii) using an input/output budget approach, instead of measuring the difference between carbon stocks on two different dates.

Table 5.5 presents eight estimates of total carbon stock in 1996 and annual flux during the 1984–96 period, by combining two different sets of wood specific gravity values (see above), two different estimates of branch volume (allometric equations on one hand and branch expansion factors obtained from a literature analysis on the other hand, that is, 1.27 for broadleafs and 1.26 for conifers (see Chapter 4), and two different values for carbon concentration. The highest estimates differ from the lowest by 36% for stocks and 51% for fluxes. The estimates based on allometric equations, the lowest specific gravity values and the 0.475% carbon concentration seem to be the most adequate according to our present knowledge.

Table 5.5. Estimates of carbon stock per hectare in 1996 and annual fluxes during the 1984–96 period in the French forest biomass according to different sets of values for branch volume, wood density and carbon concentration (first value: 0.475%, second value: 0.5%). Values retained in the current study are shown in bold.

| | **Branch volume** | | | |
| | **Branch expansion factors from the literature** | | **Allometric equations (Vallet *et al.*, 2006)** | |
Wood specific gravity	C stock 1996 (tC·ha⁻¹)	Annual C flux (MtC·year⁻¹)	C stock 1996 (tC·ha⁻¹)	Annual C flux (MtC·year⁻¹)
Low values	59.7–62.8	13.5–14.2	**71.2–74.9**	**17.9–18.8**
High values	64.5–67.9	14.5–15.3	77.0–81.1	19.3–20.3

We extended these calculations of fluxes for a limited number of different values of the input variables to a more general Monte-Carlo simulation approach, following general guidelines given by Heath and Smith (2000) and Smith and Heath (2001).

According to the following simplified formula, carbon stock S_i at a given date t_i depends on five input variables:

$$S_i = V \times \mathrm{EF}_b \times \mathrm{EF}_r \times \mathrm{WSG} \times C$$

where V is the total merchantable volume, EF_b and EF_r the average branch and root expansion factors, respectively, WSG the average wood specific gravity and C the average carbon concentration. S_1 and S_2 are the stocks at two points in time, t_1 and t_2, the calculated flux is $\left(S_2 - S_1\right)/\left(t_2 - t_1\right)$.

The ranges of possible values for the 10 input variables (the five previous variables on the two study dates) were represented by beta distributions (Figure 5.9(a)–(e)). The beta distribution allows a bounded and asymmetric representation of uncertainty. The four parameters of each beta distribution (i.e. minimum, maximum, skewness and narrowness of the distribution) were determined by expert judgement. Symmetric distributions were chosen except for the branch expansion factor, which was skewed towards negative values, and wood density, slightly skewed towards positive values, according to the previous discussion. A correlation was introduced for each of the five pairs of corresponding variables between the two sampling dates (e.g. between carbon concentrations on the first and second sampling dates): 0.5 for merchantable volume, 0.7 for branch and root expansion factors and 0.8 for wood density and carbon concentration.

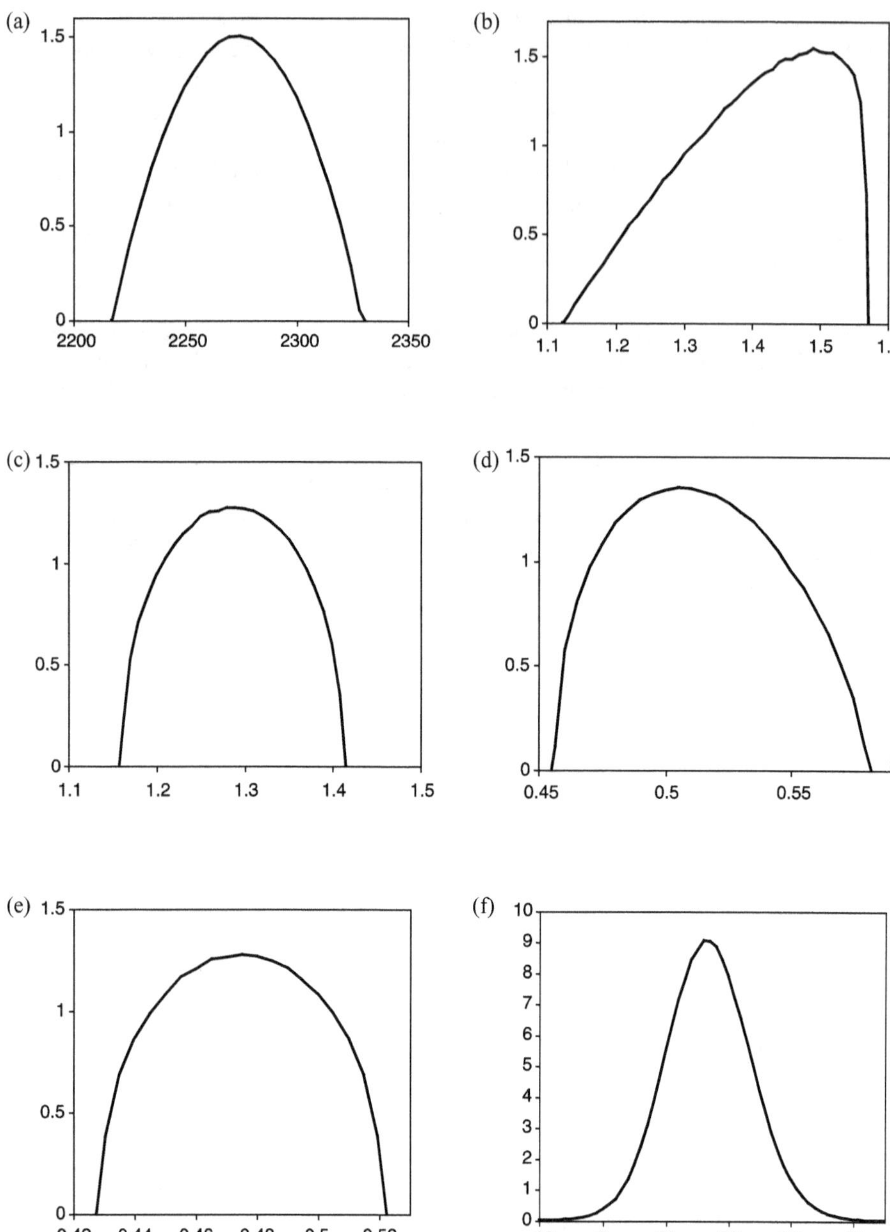

Figure 5.9. Simulated distributions of the input variables based on expert assessment and resulting distribution for the annual carbon flux: (a) merchantable volume ($M \cdot m^{-3}$); (b) branch expansion factor; (c) root expansion factor; (d) wood density; (e) carbon concentration; (f) resulting distribution of values for the calculated annual carbon flux ($MtC \cdot year^{-1}$). Distributions (a)–(d) are given for the second sampling date only. They are similar on the first sampling date. Distribution of carbon concentration is identical on the two sampling dates. Vertical axes are scaled to a mean of 1.

The resulting distribution of values for the annual carbon flux (Figure 5.9(f)) was calculated from 100,000 random sets of the 10 previous variables. Its 95% confidence interval was 4.1–30.3 MtC·year^{-1}. Only 0.5% of the values fell below 0. The coefficient of variation of the distribution, an estimate of the relative error on the carbon flux calculation, was 40%.

Another way to evaluate possible bias in our calculations is to estimate the carbon flux using another, independent, method.

Calculation of carbon flux with the input/output budget method

Wood fluxes in French forests can be estimated by using the equation linking annual production, P, removal, R (wood volumes exported out of the forest), stock variation, S, and tree mortality, M:

$$P = R + S + M$$

Production and mortality are estimated by the French NFI at each inventory for a period including the last 5 years preceding the inventory. We calculated the uptake from the annual survey of forest industry activities (including all forest activities but non-commercial fuelwood collection), conducted by the French Ministry of Agriculture, and from estimates of fuelwood uptake from French forests (Dupouey *et al.*, 1999). From these data, we estimated S, the resulting flux of carbon sequestration. The different terms of this balance are given in Table 5.6 in merchantable volumes.

Table 5.6. Production, wood uptake, stock increase and mortality in French productive forests (merchantable volumes). Poplar plantations and protection forests were excluded.

	Annual production per unit area	(m³·ha⁻¹·year⁻¹)	1984	1996
			5.7	6.5
P	Total annual production from 1984 to 1996	(millions m³·year⁻¹)	87.2	
R	Annual removal between 1984 and 1996:	(millions m³·year⁻¹)		
	– construction timber		19.3	
	– fuelwood		18.4	
	– pulpwood and industrial timber		11.4	
	Total		52.4	
S	Variation of stock between 1984 and 1996	(millions m³·year⁻¹)	30.7	
M	Annual mortality between 1984 and 1996	(millions m³·year⁻¹)	4.2	

P: production; R: wood removal; S: stock increase; M: mortality.

With this budget method, we obtained a value of 30.7 millions m^{-3}·year^{1} of average annual increase in the growing stock between 1984 and 1996 in French productive forests (excluding poplar plantations). The lag with value (31.4 millions m^{-3}·year^{1}), calculated by the difference in merchantable volume stocks between these two same dates, is very low (2%), although the two methods are based on very different data. This agreement suggests that the errors in the estimation of merchantable volume changes in French forests are probably small. However, this small lag could also partly be fortuitous considering the large uncertainty about the estimation of non-commercial fuelwood uptake. Moreover, because of the self-declarative nature of the statistics concerning the volume of wood used in the industry, there could be an unknown underestimation of the annual uptake of wood from forests, leading to an overestimation of the increase in forest stock.

The three preceding error analyses give different views on the uncertainty of carbon flux. The crude Monte-Carlo simulation provides a large confidence interval from 3.7 MtC·year[-1] to 30.1 MtC·year[-1], from which we can conclude that the flux of carbon to French forests is clearly positive. A more reasoned choice of a limited set of possible values gives a more limited range for the carbon flux, between 13.5 MtC·year[-1] and 20.3 MtC·year[-1]. This latter interval does not include errors on merchantable volumes, which are most probably the smallest among all the sources of errors. Finally, a new and partly independent method of evaluation provides, perhaps fortuitously, an exactly identical value for the change in stock volume.

Conclusion: recommendations and improvement of methods for French reporting under the framework of international conventions

The new volume equations used in the present work have helped to improve significantly the estimate of total above-ground volume of trees and enabled a reassessment of expansion factors previously used in France. A comparison of the last two inventories showed that the branch expansion factors have increased for all species, sometimes quite substantially (see Table 5.4). Systematic use of these allometric equations, therefore, seems preferable to a simplistic approach based on constant expansion factors. Annual inventories of greenhouse gas budgets for the whole of France, conducted under the responsibility of CITEPA, should take these new results into account. It must be pointed out that, although switching to the use of adapted allometric equations is obviously an improvement on the previous method, it would be useful to continue calculating and publishing global, averaged, expansion factors aggregated at the national scale for all countries reporting under the carbon conventions. If this is not the case, it will become difficult to compare the different national protocols for carbon assessment and reporting, because even large differences between countries in their conversion parameters (i.e. branch and root expansion factors, wood density and carbon concentration values) will be "hidden" and scattered throughout a multitude of allometric equations.

French forests have certainly been sequestrating carbon in their biomass during the studied period between 1984 and 1996, according to our uncertainty analysis. We estimate this flux at 17.9 MtC·year[-1]. Interestingly, the recent reassessment of carbon flux into forest biomass in Germany gave a close value of 14.9 MtC·year[-1] for a total forest area of 10.4 m ha only (Dieter and Elsasser, 2002). However, large uncertainties remain in our calculations of carbon stocks and fluxes in forest biomass. Several perspectives could be explored in the future to improve or confirm results presented here:

– architectural models of trees offer obvious opportunities for the evaluation and partitioning of above-ground biomass; such models, which already exist for pines (*Pinus pinaster* Ait., *P. sylvestris* L., *P. nigra* J.F. Arnold subsp. *austriaca* var. *austriaca*) (see Chapter 4) should be built for other forest species;

– developing a better knowledge of wood density variations seems also of high priority. Density values could be improved by two actions: measuring wood specific gravity on a subset of the thousands of cores sampled during NFIs and developing models of the internal structure of stems (Leban *et al.*, 1996), which allow the calculation of an average wood specific gravity for each tree of the NFI sample. Such models are already available for oak, spruce, Douglas fir and Corsican pine and should be developed for more species.

The Kyoto Protocol, which recently came into force, but whose forestry issues are still under discussion, will take into account only a limited fraction of previously calculated fluxes. How much of the previous fluxes will be accounted for is the subject of many debates because it could have a large impact on national carbon accounts. For France, the annual 17.9 MtC carbon sink in forest biomass represents more than 18% of current carbon emissions from fossil fuels (98 MtC·year[1] on average between 1984 and 1996, according to Marland *et al.*, 2003). Only deforestation and afforestation or reforestation directly resulting from human activities since 1990 will be completely accounted for. Is natural recolonization, which is the dominant way of reforestation today, directly resulting from land-use management decisions, for example, as a consequence of European common agricultural policy rules, or is it a process out of our control? After a first phase of rejection of this carbon sink from national carbon accounts in the framework of the Kyoto Protocol, the trend is now towards its inclusion. The current increase of standing wood volume in existing forests will only be marginally accounted for (Pignard and Dupouey, 2002). One of the reasons for this is the difficulty in separating the respective roles of voluntary actions (e.g. silvicultural decisions) and uncontrolled environmental changes (e.g. atmospheric CO_2 increase, nitrogen deposition, climatic change, etc.) in the increasing productivity of forest stands. NFIs which, in a first step, provided a sound basis for international negotiations will now contribute to the establishment of national carbon budgets helping to implement and control the application of the Kyoto Protocol. However, the Kyoto Protocol in its current state needs some data which are difficult to provide with NFI data only. Land-use exchanges between forest and non-forest states are not always finely monitored. Joint effort with agricultural statistics (Teruti-Lucas survey) has been undertaken to better meet the requirements of the Kyoto Protocol. The French NRI also recently switched to a new method of sampling, which needs to be evaluated in the context of carbon calculations. This new method should allow reporting fluxes over a shorter time window of integration, that is, 5 years instead of 12 years as in the present study. First results, covering the period 2004–09 should soon be available. This will be particularly welcome in a context of increasing demand for forest products, both for economic and environmental purposes, which could lead to an inversion of the historical trend of increasing carbon stock in forest biomass.

Acknowledgements

We thank the French Ministry of Ecology, Energy and Sustainable Development, the French Ministry of Agriculture and Fisheries and the GIP-ECOFOR for financial support within the framework of the Management and Impacts of Climate Change (GICC) programme. We also thank Vladimir Usoltsev for helpful comments on this manuscript.

References

Dieter M., Elsasser P., 2002. Carbon stocks and carbon stock changes in the tree biomass of Germany's forests. *Forstwissenschaftliches Centralblatt*, 121, 195–210.

Dupouey J.-L., Pignard G., 2001. Quelques problèmes posés par l'évaluation des stocks et flux de carbone forestiers au niveau national. *Revue Forestière Française*, 53 (3–4), numéro spécial *Les 40 ans de l'IFN*, 294–300.

Dupouey J.-L., Pignard G., Badeau V., Thimonier A., Dhôte J.-F., Nepveu G., Bergès L., Augusto L., Belkacem S., Nys C., 1999. Stocks et flux de carbone dans les forêts françaises. *Comptes-rendus de l'Académie d'agriculture de France*, 85 (6), 293–310.

Fontelle J.P., Beguier S., Allemand N., Chang J.P., Oudart B., Audoux N., Druart A., Gaborit G., Sambat S., Serveau L., Vincent J., 2003. *Inventaire des émissions de gaz à effet de serre en France au titre de la convention cadre des Nations unies sur le changement climatique, rapport d'inventaire national.* UNFCCC/CRF/CITEPA/Ministry of Ecology and Sustainable Development, Paris, 115 p.

Goodale C.L., Apps M.J., Birdsey R.A., Field C.B., Heath L.S., Houghton R.A., Jenkins J.C., Kohlmaier G.H., Kurz W., Liu S., Nabuurs G.J., Nilsson S., Shvidenko A.Z., 2002. Forest carbon sinks in the northern hemisphere. *Ecological Applications*, 12 (3), 891–899.

Heath L.S., Smith J.E., 2000. An assessment of uncertainty in forest carbon budget projections. *Environmental Science and Policy*, 3, 73–82.

IFN, 2000. Analyse de l'évolution de la productivité des forêts françaises au cours des 25 dernières années à partir des données de l'inventaire forestier national. Rapport GIP ECOFOR. Ministry of Agriculture and Fisheries, Paris.

IFN/AFOCEL, 2004. *Disponibilités en bois résineux en France. Réévaluation après les tempêtes de 1999. Rapport final.* Ministry of Agriculture and Fisheries, Paris, 129 p.

IFN/MAP, 2000. *Les indicateurs de gestion durable des forêts françaises. Rapport du GIP ECOFOR.* Ministry of Agriculture and Fisheries, Paris, 129 p.

Ito A., Penner J.E., Prather M.J., de Campos C.P., Houghton R.A., Kato T., Jain A.K., Yang X., Hurtt G.C., Frolking S., Fearon M.G., Chini L.P., Wang A., Price D.T., 2008. Can we reconcile differences in estimates of carbon fluxes from land-use change and forestry for the 1990s? *Atmospheric Chemistry and Physics*, 8, 3291–3310.

Leban J.-M., Saint-André L., Pignard G., Colin F., Daquitaine R., 1996. Une méthode non destructive d'évaluation de la qualité de la ressource en bois d'épicéa commun du Nord-Est de la France – application au département de la Moselle. *Actes du 4e colloque Sciences et industries du bois*, pp. 67–74.

Levy P.E., Hale S.E., Nicoll B.C., 2004. Biomass expansion factors and root: shoot ratios for coniferous tree species in Great Britain. *Forestry*, 77 (5), 421–430.

Löwe H., Seufert G., Raes F., 2000. Comparison of methods used within Member States for estimating CO_2 emissions and sink according to UNFCC and EU Monitoring Mechanism: forest and other wooded land. *Biotechnology, Agronomy, Society and Environment*, 4 (4), 315–319.

Marland G., Boden T.A., Andres R.J., 2003. Global, regional, and national fossil fuel CO_2 emissions. *In: Trends: A Compendium of Data on Global Change*. Carbon Dioxide Information Analysis Center, Oak Ridge National Laboratory, US Department of Energy, Oak Ridge.

Mathieu A., 1877. *Flore forestière*, 3rd edition. Berger-Levrault, Paris, 618 p.

Morrin, G.A., Janz K., 1986. *Les Ressources forestières de la région de la CEE (Europe, URSS et Amérique du Nord)*. ONU/CEE/FAO. Food and Agriculture Organization of the United Nations, Rome, 242 p.

Nabuurs G.J., Schelhaas M.J., Mohren G.M.J., Field C.B., 2003. Temporal evolution of the European forest sector carbon sink from 1950 to 1999. *Global Change Biology*, 9, 152–160.

Pignard G., Dupouey J.-L., 2002. Les flux de carbone dans les forêts françaises et européennes : apport des inventaires forestiers. *Comptes-rendus de l'Académie d'agriculture de France*, 88, 5, 7–17.

Pignard G., Dupouey J.-L., Arrouays D., Loustau D., 2000. Carbon stocks estimates for French forests. *Biotechnology, Agronomy, Society and Environment*, 4 (4), 285–289.

Pignard G., Dupouey J.-L., Granier A., Morel M., 2009. Impact des tempêtes de 1999 sur le bilan de carbone des forêts françaises. *In: Forêt, vent et risque: des connaissances enrichies pour une meilleure gestion forestière*. Éditions Quae/ECOFOR, Versailles, in press.

Rivière E., 1999. *Évaluation des puits de CO_2 suivant la nouvelle méthode préconisée par le GIEC. Rapport au ministère de l'Environnement*. Interprofessional Technical Centre for Studies on Atmospheric Pollution, Paris.

Smith J.E., Heath L.S., 2001. Identifying influences on model uncertainty: an application using forest carbon budget model. *Environmental Management*, 27 (2), 253–267.

Somogyi Z., Cienciala E., Mäkipää R., Muukkonen P., Lehtonen A., Weiss P., 2007. Indirect methods of large-scale forest biomass estimation. *European Journal of Forest Research*, 126, 197–207.

Tobin B., Nieuwenhuis M., 2007. Biomass expansion factors for Sitka spruce (*Picea sitchensis* (Bong.) Carr.) in Ireland. *European Journal of Forest Research*, 126, 189–196.

Vallet P., Dhôte J.-F., Le Moguédec D., Ravart M., Pignard G., 2006. Development of total aboveground volume equations for 7 important forest tree species in France. *Forest Ecology and Management*, 229, 98–110.

Appendix

Table A1. Forest biomass carbon stocks per unit area in the French metropolitan area and mean annual stock change per region and department.

Region	Dep.	Year	Last inventory				Previous inventory					Stock change
			Forest area	Produc-tive forest area	NFI volume	Carbon mass	Year	Forest area	Produc-tive forest area	NFI volume	Carbon mass	
			(1000 ha)	(1000 ha)	(m³·ha⁻¹)	(tC·ha⁻¹)		(1000 ha)	(1000 ha)	(m³·ha⁻¹)	(tC·ha⁻¹)	(1000 tC·year⁻¹)
Île-de-France	75	1994	143	123	173	86	1979	134	116	136	65	237
	77	1993	134	127	168	81	1978	126	120	131	62	192
Total region		1994	277	250	170	83	1979	260	237	134	64	429
Champagne-Ardennes	08	1998	149	146	185	81	1987	147	146	157	71	154
	10	1994	135	129	164	82	1983	140	135	132	64	194
	51	1997	136	120	197	88	1986	131	117	160	75	195
	52	1997	245	245	164	83	1985	241	239	141	70	292
Total region		1997	666	640	175	83	1985	659	637	146	70	836
Picardie	02	1991	125	120	157	78	1977	122	118	139	69	105
	60	2001	122	117	183	91	1990	120	116	166	84	98
	80	1989	53	51	152	79	1976	52	50	127	66	56
Total region		1995	300	288	166	84	1982	294	284	148	75	258
Haute-Normandie	27	1988	124	121	144	74	1975	120	117	129	65	108
	76	1989	100	97	176	88	1976	94	90	177	86	59
Total region		1988	224	218	158	81	1975	214	207	150	74	167 … —

Table A1. (continued)

Region	Dep.	Year	Last inventory				Year	Previous inventory				Stock change
			Forest area	Produc-tive forest area	NFI volume	Carbon mass		Forest area	Produc-tive forest area	NFI volume	Carbon mass	
			(1000 ha)	(1000 ha)	(m³·ha⁻¹)	(tC·ha⁻¹)		(1000 ha)	(1000 ha)	(m³·ha⁻¹)	(tC·ha⁻¹)	(1000 tC·year⁻¹)
Centre	18	1999	172	169	152	76	1986	162	158	123	61	238
	28	1992	72	70	153	75	1977	67	66	120	58	103
	36	1997	118	115	141	75	1988	108	106	134	68	167
	37	1999	150	145	151	75	1985	135	130	131	62	213
	41	1998	206	202	137	64	1982	186	180	120	53	200
	45	1992	168	163	134	64	1979	164	158	112	52	171
Total region		1997	886	864	144	71	1983	822	799	123	58	1 093
Basse-Normandie	14	2001	46	43	137	70	1987	40	39	136	68	33
	50	2001	26	25	118	60	1987	21	21	113	58	23
	61	2001	96	93	173	80	1988	88	87	175	78	63
Total region		2001	168	161	155	74	1988	150	146	156	72	119
Bourgogne	21	1990	317	315	123	63	1980	312	310	101	51	407
	58	1996	226	224	176	86	1985	228	224	127	64	443
	71	1989	200	199	136	66	1980	185	182	110	54	374
	89	1999	226	222	148	73	1986	221	218	128	64	198
Total region		1993	969	960	144	72	1983	945	935	115	58	1 422
Nord-Pas-de-Calais	59	2000	44	41	179	91	1986	42	40	161	79	48
	62	2000	46	42	148	79	1986	39	36	127	68	71
Total region		2000	91	82	163	85	1986	82	76	145	74	120 …

Table A1. (continued)

Region	Dep.	Year	Last inventory				Previous inventory					Stock change
			Forest area	Produc-tive forest area	NFI volume	Carbon mass	Year	Forest area	Produc-tive forest area	NFI volume	Carbon mass	
			(1000 ha)	(1000 ha)	($m^3 \cdot ha^{-1}$)	($tC \cdot ha^{-1}$)		(1000 ha)	(1000 ha)	($m^3 \cdot ha^{-1}$)	($tC \cdot ha^{-1}$)	($1000\ tC \cdot year^{-1}$)
Lorraine	54	1990	163	161	193	92	1980	169	166	178	80	152
	55	1991	224	222	163	82	1980	227	227	141	68	262
	57	1993	172	166	242	109	1982	171	168	242	102	120
	88	1992	281	279	252	99	1981	282	281	242	90	210
Total region		1992	840	828	215	95	1981	849	841	202	85	745
Alsace	67	2002	174	170	249	98	1989	170	168	259	104	-44
	68	1999	143	137	256	103	1988	141	139	223	89	196
Total region		2001	316	307	252	100	1989	311	308	243	97	153
Franche-Comté	25	1994	218	213	241	96	1982	215	210	226	84	238
	39	1992	223	216	185	79	1980	227	220	162	65	234
	70	1996	224	223	207	105	1984	220	219	178	87	367
	90	1996	26	26	227	106	1984	25	25	194	91	38
Total region		1994	690	677	211	94	1982	686	673	188	79	877
Pays-de-Loire	44	2000	56	47	135	67	1985	44	40	124	59	75
	49	1997	84	79	131	62	1983	76	70	110	51	95
	53	1999	36	34	162	78	1983	34	32	126	61	45
	72	1999	108	104	160	75	1984	104	100	143	65	86
	85	1994	35	32	107	52	1984	34	31	104	47	20
Total region		1998	318	297	143	68	1984	292	273	125	58	322

...

Table A1. (continued)

Region	Dep.	Year	Last inventory				Previous inventory					Stock change
			Forest area	Produc-tive forest area	NFI volume	Carbon mass	Year	Forest area	Produc-tive forest area	NFI volume	Carbon mass	
			(1000 ha)	(1000 ha)	(m^3·ha^{-1})	(tC·ha^{-1})		(1000 ha)	(1000 ha)	(m^3·ha^{-1})	(tC·ha^{-1})	(1000 tC·year^{-1})
Bretagne	22	1995	83	81	133	65	1981	66	64	116	54	125
	29	1996	75	72	115	54	1981	54	51	114	53	75
	35	1995	57	55	160	76	1980	56	54	131	60	64
	56	1998	113	110	129	60	1980	89	82	109	48	139
Total region		1996	327	317	132	62	1980	266	251	117	53	402
Poitou-Charentes	16	1993	118	116	116	59	1983	111	109	101	49	150
	17	1993	103	98	127	59	1984	99	95	118	52	100
	79	1995	49	47	113	60	1985	46	45	106	53	48
	86	1996	113	109	124	65	1986	103	101	124	60	113
Total region		1994	382	370	121	61	1984	360	349	113	53	410
Aquitaine	24	1992	397	391	129	63	1982	372	364	114	53	533
	33	1998	483	466	147	60	1987	473	457	138	52	389
	40	1999	577	563	166	64	1988	563	554	149	54	615
	47	2000	129	123	149	68	1989	119	116	144	63	116
	64	1995	211	194	166	93	1985	192	179	149	80	421
Total region		1997	1 797	1 737	152	66	1986	1 719	1 670	138	57	2 073 ...—

Table A1. (continued)

| Region | Dep. | Year | Last inventory | | | | Previous inventory | | | | | Stock change |
| | | | Forest area | Produc-tive forest area | NFI volume | Carbon mass | Year | Forest area | Produc-tive forest area | NFI volume | Carbon mass | |
			(1000 ha)	(1000 ha)	(m³·ha⁻¹)	(tC·ha⁻¹)		(1000 ha)	(1000 ha)	(m³·ha⁻¹)	(tC·ha⁻¹)	(1000 tC·year⁻¹)
Midi-Pyrénées	09	1990	201	161	139	70	1978	192	184	114	56	285
	12	1994	246	234	112	62	1981	225	221	97	53	259
	31	2000	125	116	156	80	1987	125	116	127	63	170
	32	2001	85	82	144	78	1989	77	74	123	68	119
	46	2002	215	208	95	56	1990	202	199	82	50	161
	65	1997	132	120	186	90	1986	121	106	160	76	239
	81	1992	164	158	126	62	1979	157	153	106	52	144
	82	2001	69	65	110	62	1989	58	56	87	51	111
Total region		1996	1 237	1 145	129	68	1984	1 156	1 109	109	57	1 489
Limousin	19	2003	267	263	162	75	1990	267	263	141	67	166
	23	1991	155	153	158	74	1981	147	145	130	60	265
	87	1991	141	138	153	74	1981	135	132	129	62	202
Total region		1997	563	554	159	75	1985	549	540	135	64	633
Rhône-Alpes	01	1995	179	167	179	77	1983	177	168	148	60	258
	07	1995	253	205	126	61	1981	224	219	93	46	366
	26	1996	285	237	93	47	1982	263	252	81	38	232
	38	1997	254	210	201	82	1984	243	207	168	65	386
	42	1993	126	124	197	76	1981	123	120	154	58	203
	69	1994	69	66	197	78	1982	71	67	156	60	99
	73	2000	194	135	242	92	1985	179	132	206	75	289
	74	1998	178	140	321	113	1987	171	142	250	86	486
Total region		1996	1 537	1 284	183	75	1983	1 451	1 307	147	58	2 319

Table A1. (continued)

Region	Dep.	Year	Last inventory				Previous inventory					Stock change
			Forest area	Produc-tive forest area	NFI volume	Carbon mass	Year	Forest area	Produc-tive forest area	NFI volume	Carbon mass	
			(1000 ha)	(1000 ha)	(m³·ha⁻¹)	(tC·ha⁻¹)		(1000 ha)	(1000 ha)	(m³·ha⁻¹)	(tC·ha⁻¹)	(1000 tC·year⁻¹)
	03	2001	123	118	200	94	1987	122	121	155	75	171
	15	1989	159	151	146	71	1977	146	137	126	59	224
Auvergne	43	2002	186	181	228	81	1991	182	177	184	67	259
	63	1988	236	229	179	73	1976	236	229	147	59	277
Total region		1994	704	679	188	78	1982	686	664	154	64	932
	11	1989	180	145	108	52	1978	151	146	87	40	294
	30	1993	217	124	91	48	1982	171	165	57	31	463
Languedoc-Roussillon	34	1997	203	191	60	34	1983	162	155	46	27	176
	48	1992	232	222	109	48	1979	206	204	84	38	266
	66	1991	141	120	94	49	1980	114	109	87	42	190
Total region		1993	974	802	92	46	1980	805	779	72	35	1 389
	04	1999	344	317	91	42	1984	298	285	74	34	298
	05	1997	194	167	125	53	1983	161	151	115	46	198
Provence-Alpes-Côte d'azur	06	2002	225	184	129	55	1985	191	155	107	43	241
	13	1988	97	83	41	21	1977	81	73	35	17	63
	83	1999	352	318	57	35	1986	341	315	44	27	232
	84	2001	132	119	55	30	1986	123	117	46	24	63
Total region		1999	1 342	1 189	85	41	1984	1 195	1 096	70	33	1 095
Corse	2A	1988	134	82	136	86	1977	131	124	84	55	398
	2B	1988	118	72	160	92	1977	97	89	133	84	248
Total region		1988	252	155	147	89	1977	228	213	105	67	646
Total France		1996	14 863	13 808	154	71	1984	13 979	13 393	133	60	17 926

Part II

Forest under a changing climate

Chapter 6
Possible future climates in France

MICHEL DÉQUÉ, EMMANUEL CLOPPET

Introduction

The assumption is often made that climate is constant and only conditioned by latitude, topography, altitude or proximity to the sea. According to another point of view, climate seems to change each year: a cold winter may be followed by a warm winter. Modern climatologists consider climate does not correspond to these extreme approaches. According to the World Meteorological Organization's definition, a 30-year period is used to define climate as the ensemble of statistical parameters of the distribution of weather variables. This definition does not exclude the fact that climate can change over a century, but rejects the assertion that inter-annual variability is an expression of climate change. Observations made during the 20th century showed without doubt that climate changed in many regions of the world including France, resulting essentially in a clear warming trend.

What can we expect for future climate? Global warming is likely caused primarily by an increase in the concentration of greenhouse gases (GHGs). Their concentration will probably increase further and the negative feedback due to CO_2 uptake by oceans will play a lesser role as sea surface temperatures (SSTs) increase. Indeed, part of the emitted CO_2 is absorbed by the ocean, but the warmer the ocean, the less the absorption. However, many parameters remain unknown. The projection of future climate change depends partly on the assumptions made about future population or economic growth, which cannot be predicted beyond 50 years. To forecast after 2050 alternative scenarios have to be used, which describe future development paths in various sectors such as energy, and estimate the emissions of GHGs. The use of various models allows us to study the impact of these scenarios on forestry. In this chain of models, the regional climate models give some insights into the effect of the evolution of GHG concentration and SST on a country scale.

This chapter focuses on responses of the ARPEGE climate model, the regional model of Météo-France, in which the simulations are driven by SRES-B2 and SRES-A2 scenario radiative forcing. A study of the uncertainties due to the choice of a particular model in analysing climate response over Europe may be found in Déqué *et al.* (2007). A description of the model and the experiment can be found in the section hereafter. Generally climate change studies are based on simulations performed on short periods (20- or 30-year periods) in order to save computing time while maximizing the statistical significance of the result. However, a project like CARBOFOR requires a long period of data. A long-term continuous forcing is needed to model tree growth (an adult tree at the end of the 21st century will grow during the whole of the next century). The numerical simulations are described in "The climate model". In order to summarize the results, we will first show the evolution of mean temperature over France during the whole simulations (see "Mean temperature evolution along the simulations"). Then we will focus in "Spatial and seasonal distribution of the climate reponse at the end of 21st century" on the geographical distribution of the differences between the last 30 years of the 21st century and the reference period (1960–89).

The climate model

Three types of regionalization techniques have been developed by modellers to provide regional and local climate information. They are high-resolution general circulation models (GCM) (Cubasch *et al.*, 1995), regional models nested in a GCM (Giorgi and Mearns, 1999) or statistical downscaling (Wilby *et al.*, 1998). In the present study, we used the first method, which offers advantages in providing globally consistent simulations. The second method shows computational limitations and the statistical methods are based on the assumption that statistical relationships found for the present climate will also be valid for the future climate under different forcings. The first method is obviously the most expensive of the three methods. This is why high-resolution was restricted to Europe and the Mediterranean basin (Déqué and Piedelievre, 1995). SST data are provided by a coarser resolution simulation of ARPEGE coupled to an ocean model (Royer *et al.*, 2002).

Machenhauer *et al.* (1998) showed that a variable resolution climate model realistically reproduces seasonal and geographical variations of the main climatological parameters over Europe. Déqué *et al.* (1998) focused on responses of the first version of the ARPEGE climate model to a doubling of CO_2 concentration.

Simulations used here were performed with the third version of the ARPEGE climate model. These two versions are very different. Version 3 is described in Gibelin and Déqué (2003), so only model details are repeated here. The model uses semi-lagrangian advection and a two time-level discretization. The variable resolution configuration has a T106 spectral truncation. There are 31 vertical levels (the same as ECMWF ERA 15 reanalysis data, see Gibson *et al.*, 1997). The time step is 30 minutes. The pole of stretching is in the Tyrrhenian Sea (40°N, 12°E) and the factor of stretching is 3. The grid has 120 pseudo-latitudes and 240 pseudo-longitudes with a reduction near the pseudo-poles to maintain isotropy (the so-called reduced Gaussian grid). The resolution thus varies from 0.5° in the Mediterranean to 4.5° in the southern Pacific.

The convection scheme is a mass flux approach with closure by moisture convergence (Bougeault, 1985). The Morcrette (1990) scheme is used to calculate radiative fluxes; it includes the effect of four GHGs (CO_2, CH_4, N_2O and CFC) in addition to water vapour and ozone, and five aerosol types (i.e. land, sea, urban, desert and sulphate). All aerosols are prescribed. Sulphate aerosols may change from month to month and from decade to decade. The indirect effects of sulphate aerosols are parameterized as an empirical function for the cloud drop effective radius (Hu *et al.*, 2001). The cloud-precipitation-vertical diffusion scheme uses the statistical approach of Ricard and Royer (1993). The soil vegetation hydrology scheme is the Interactions Soil-Biosphere-Atmosphere (ISBA) scheme (Douville *et al.*, 2000). Representation of orographic gravity wave drag is described in Déqué *et al.* (1994), modified by Lott and Miller (1997) and Lott (1999).

The climate scenario

The experimental conditions are described in Gibelin and Déqué (2003). Radiative forcings (GHG and sulphate aerosol concentrations) are those of the SRES-B2 and SRES-A2 scenarios defined by the Intergovernmental Panel on Climate Change (IPCC, 2001). These scenarios are calculated by a model describing the evolution of society during the 21st century. In the A-family of scenarios, priority is set for economics, while in the B-family the priority is set for the environment. The number of the scenario evaluates the level of international cooperation. In A2 and B2, decisions are taken at a regional level, whereas in A1 and B1 a global optimum is searched. As a result of the demography, economic development and carbon surface absorption, the A2 scenario leads to a triple concentration of CO_2 by 2100 with respect to the pre-industrial value, whereas a B2 scenario leads to a doubling in concentration. In neither scenario are measures against GHG increase taken.

SSTs are precalculated. From 1960 to 2000, monthly mean observed SSTs are used (Reynolds and Smith, 1994). This allows validation of the model by comparing the simulation with observations. The simulated climate is compared successfully with climatology data from the Climate Research Unit (New *et al.*, 1999). The main difference over France is an overestimate of winter rainfall amounts due to a too zonal circulation in ARPEGE bringing too much mild moist oceanic air across the country (Gibelin and Déqué, 2003). The temperature bias is less than 1 K in all seasons except spring, where it reaches this value. A detailed validation of rainfall amounts on the Alps can be found in Frei *et al.* (2003).

Artificial monthly SSTs were created from 2000 through to 2099 by adding 30-year monthly anomalies updated every 10 years to the above described observed SSTs. These anomalies are calculated with two coarser resolution coupled simulations (Royer *et al.*, 2002). The same version (V3) of the ARPEGE climate model is used with a uniform T63 horizontal resolution and a higher vertical resolution with 45 levels. It is coupled with the global sea ice-ocean model OPA8.0-GELATO (Salas-Mélia, 2002). A 150-year control simulation uses radiative forcings of 1950. Two additional simulations use the above described radiative forcings (observations then B2 or A2 forcings). The use of these two simulations makes it possible to correct the 1 K/century drift, which is obtained in global mean SSTs due to a slight imbalance in the energy fluxes at the ocean surface. The correction is applied pointwise and for each calendar month separately. We also paid attention not to take into account twice ocean warming during the 1960–2000 period. Details on the procedure of SSTs and sea-ice extent pre-computation are given in Gibelin and Déqué (2003). This procedure relies on the assumption that the SST inter-annual variability does not vary during the 21st century.

Mean temperature evolution along the simulations

Figure 6.1(a) shows winter temperature averaged over France during the two simulations. In this chapter we consider climatological seasons, winter means DJF average, spring means MAM average, etc. During the 20th century, as the forcing comes from observations, the two series are identical and labelled as B2 (black diamonds). Note that the chronology along the *x*-axis is just a reference to the external forcing (SST and GHG concentration), not to the observed meteorological events. For example, the coldest winter over France, which occurred in 1963 (0.2°C on average), is not particularly cold in the simulation. One can see that cold winter (less than 3°C) can still occur at the end of the 21st century in the B2 simulation: the warming is not just a linear increase. As expected, the A2 simulation (grey diamonds) produces warmer temperatures, but the year-to-year variability is still large. The 0.05 and 0.95 quantiles of the model temperature distribution during the 1960–99 period (thick horizontal black lines) indicate that beyond 2020 warm winters are significantly more frequent than during the reference period.

Figure 6.1(b) shows the same variables in summer (JJA). In 2003, the mean observed temperature was 22.5°C, more than 3°C above the 20th century average. This high value becomes usual in both scenarios beyond 2070. Beyond 2090, in the A2 simulation, it even becomes a lower boundary; the last 10 summers are warmer than 22.5°C.

Comparing (a) with (b) in Figure 6.1 shows an asymmetry in the distributions: in winter extreme values are on the cold side whereas in summer they are on the warm side. Due to this asymmetry and to the large inter-annual variability, a simple 30-year mean model response is not sufficient to describe the climate change over France in the first part of the 21st century. However, in the last 30 years of the simulations, the signal (response to GHG increase) to noise (natural inter-annual variability) ratio is small enough to prove that the climate is warmer (most points are outside the 90% interval bounded by the horizontal black lines).

It should be remembered that the year-to-year variability at the end of the scenario is possibly biased by the assumption of stability of SST variability. In the forthcoming regional scenarios, this constraint is removed.

Spatial and seasonal distribution
of climate response at the end of 21st century

In order to get a more stable and accurate idea of possible climate change over France, the differences between two climate normals at the beginning and end of the simulations have been computed. We have followed the traditional use of 30-year means to define normals. We thus consider here the periods 1960–89 and 2070–99. Although the model covers the whole globe, we concentrate on model points covering France, excluding sea points. Indeed, in CARBOFOR, only the 360 grid-points of this area were distributed to partners. Figure 6.2 (see Plate 1) indicates the location of these points over France, together with geographical information about the model. A colour code gives additional information about elevation (Figure 6.2(a)), maximum soil water content (Figure 6.2(b)), field capacity (Figure 6.2(c)) and soil water content at wilting point (Figure 6.2(d)). The last three fields are discussed in "Results" (chapter 5).

(a)

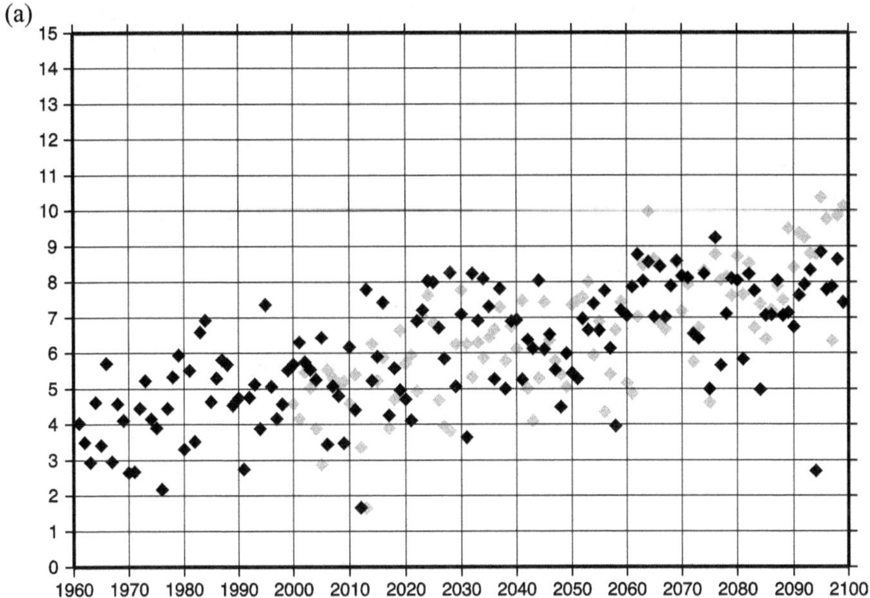

(b)

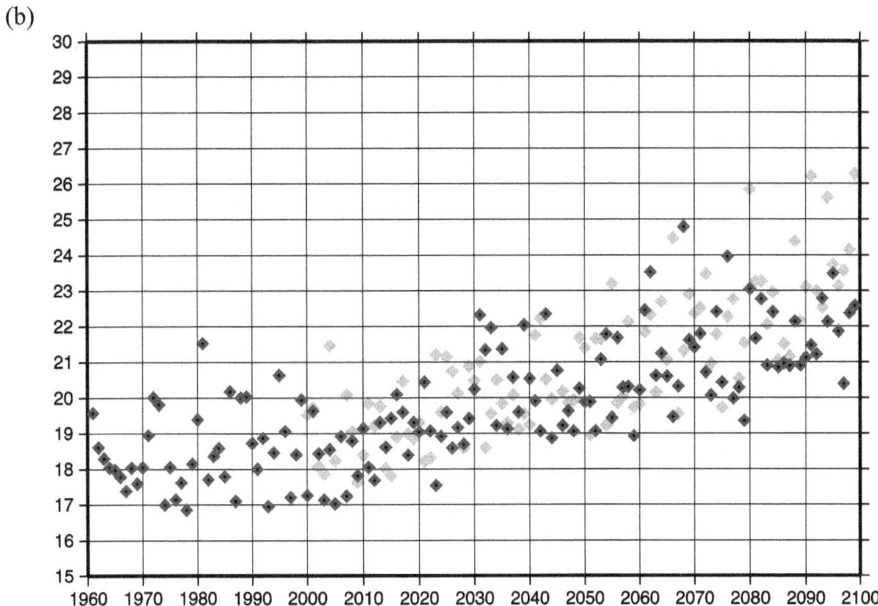

Figure 6.1. Mean simulated temperature over France year by year for (a) winter and (b) summer. Black diamonds correspond to the B2 simulation. Grey diamonds (beyond year 2000) correspond to the A2 simulation.

In the PRUDENCE European project (Christensen and Christensen, 2007), the data have been analysed over Europe. As far as temperature and precipitation are concerned, the quality of the control simulation is discussed in Gibelin and Déqué (2003). A comparison with observations over France can also be found in Déqué (2007).

2 m temperature

Figure 6.2 shows temperature at screen level (referred to as 2 m temperature in the following) difference maps for the four seasons (a: DJF, b: MAM, c: JJA, d: SON) in the B2 scenario. The direct consequence of an increase in GHG concentration is a warming. During the 20th century, France has warmed up by almost 1 K, according to homogenized observations (Moisselin *et al.*, 2002). The simulated warming is maximum in summer and higher in the southern part of the country. Autumn has the second highest simulated warming. In winter and spring, mountain regions exhibit the maximum warming, due to the snow-albedo feedback. The highest warming is 3 K and is seen in the south and west in summer.

With scenario A2 (see Figure 6.3, see Plate 1), the same spatial patterns are obtained, as expected, with higher temperatures. In summer and autumn the warming is 1 K more than in scenario B2; in winter and spring, it is a little less. It can be noted that warming is generally attenuated in the vicinity of the sea, due to the inertia of the oceans in responding to the GHG forcing.

Precipitation

The expected impact of GHG concentration increase on precipitation is an increase, as a warmer atmosphere can contain, and thus transport more water vapour from oceans to continents. This is true at a global scale but not at local scale as modifications to the circulation are at least as important as changes in atmospheric water content. Figure 6.4 (see Plate 2) shows the response for precipitation for the four seasons in scenario B2. There is an increase of about 0.5 mm·day^{-1} in winter, which corresponds approximately to 20% of the present amount. This increase is maximal along the Atlantic and English Channel coasts, as well as over mountainous areas. These regions have the most rain in the present climate. During the other three seasons, precipitation decreases by about 0.25 mm·day^{-1}, which represents about 10% of the present amount. The largest decrease is found north of the Pyrenees chain and is maximal in spring. The response is minimal in autumn.

The spatial patterns in the A2 scenario (see Figure 6.5, Plate 2) are similar to those in the B2 scenario. However, contrary to temperature, the response is not systematically enhanced. In winter, the increase in precipitation is less than in the B2 scenario. In summer, the decrease is larger. In spring and autumn, A2 and B2 responses are similar, the B2 decrease being slightly larger in spring, and the A2 decrease being slightly larger in autumn. This mechanism can be explained as follows and is confirmed by an examination of soil moisture response in "Results" (chapter 5). The soil moisture depletion is more intense in A2 because of higher temperatures and enhanced evaporation. As a result, recycling precipitation is less in A2 than in B2, which explains the winter and summer behaviours. The reason why this precipitation decrease is little changed between B2 and A2 in the other two seasons results from a compensation between the depletion of soil moisture (which reduces the response of A2) and the direct effect through precipitable water increase in a warmer atmosphere (which increases the response of A2 because the soil is not dried out as in summer). A slight precipitation increase in northern France in autumn can also be noted in both scenarios. This mechanism is not exclusive and other contributions have been examined by Rowell and Jones (2006). In particular, the fact that the warming of the ocean is slower than that of the atmosphere increases the land-sea temperature contrast in summer and reduces evaporation from the sea: in the present climate, a relatively cold sea often implies a dry coastal climate (e.g. Peru, California, Angola, etc.).

2 m atmosphere moisture

As mentioned in the last section, atmospheric specific humidity may increase in a warmer atmosphere due to the Clausius-Clapeyron relation. As France is surrounded by seas on more than 50% of its boundaries, water availability for the atmosphere is ensured and, in addition, the main advection streams come from the west (i.e. a wide ocean). Indeed, there is an increase in 2 m specific humidity in all seasons and scenarios (not shown). However, as far as the hydrological cycle is concerned, the important parameter is relative humidity. In its vapour phase, water is just a GHG. Interaction with the surface occurs only when saturation is reached.

Figure 6.6 (see Plate 3) shows the response in relative humidity at screen level (ratio of water vapour pressure over saturation pressure at 2 m above the ground, expressed in percent) in the B2 scenario. In winter the response is almost zero. Vapour brought by the general circulation is enough to fill the additional capacity due to the warmer temperature. The strongest effect is seen in summer, with a strong decrease in the western part of the country (except the Atlantic coast). In spring, the drying effect is moderate and concentrated in the south. Autumn is similar to spring, but with a slightly larger response. The reason why the reduction is smaller in intermediate seasons than in summer is due to the fact that summer in experiencing the largest warming, the Clausius-Clapeyron relation plays the major role.

Figure 6.7 (see Plate 3) indicates that, as for temperature, the A2 scenario produces a stronger response, but keeps the same spatial patterns and maintains the chronology of the seasonal cycle of the response.

Soil moisture

As we have mentioned above, soil moisture may be a key parameter in explaining the details of climate change over France. Indeed, analysis of seasonal mean precipitation gives a partial insight into the impact of climate change on hydrology. Over a year, compensating effects of precipitation responses may occur. An increase in winter precipitation may produce heavy rainfall; in addition, warmer temperatures reduce the solid part of precipitation. This contributes to an increase in river runoff in winter. Warmer temperatures favour enhanced evaporation in all seasons. The latter parameter is not easy to analyse over a continent and may lead to erroneous conclusions due to a "chicken and egg" problem: if the soil becomes drier, evaporation decreases and the surface water budget appears little modified. This may suggest that the impact on hydrology is weak. An alternative method is to analyse potential evapotranspiration (PET). PET is the maximum water flux from the surface (including vegetation) obtained if there was no constraint on soil moisture availability (over seas or lakes it is equivalent to evaporation). Empirical formulae have been calibrated from *in situ* data but they are not suitable for atmospheric model output. Indeed in a GCM, all parameters are linked with each other by physical feedbacks, which is not the case with PET. In fact ARPEGE, as for the other GCMs, includes a simple soil hydrology scheme named ISBA (Douville *et al.*, 2000), which interacts at each time step with low level atmospheric parameters.

Figure 6.8 (see Plate 4) shows the response in soil water content. The reservoir corresponds to the root zone and is able to evaporate through the canopy. The depth of this reservoir varies between 2 and 4 m over France in the model, according to the type of vegetation and the constraint of physiography. The content of this reservoir, which depends on precipitation, evapotranspiration, runoff and soil texture, is expressed in $kg \cdot m^{-2}$. This parameter is the best suited to evaluation of the impact of climate change on water availability for vegetation. In the CARBOFOR project, it is of utmost importance, although local 1D models with more refined soil discretization and better description of biological processes of the vegetation have been used for particular regions and species (see further chapters in this volume). In particular ARPEGE neglects the fact that higher CO_2 rates may reduce evaporation through increased stomatal resistance. Nevertheless, Figure 6.9 (see Plate 4) provides a synthetic view over France to a first order of accuracy. In winter, soil wetness increases, except in the plains of south-west France, in the eastern Pyrenees and in parts of northern France. This increase is consistent with enhanced precipitation. It is maximum in mountain regions because in cold regions soil water may freeze and is no longer accounted for as water available for vegetation. In these regions warming produces soil ice (not to be confused with snow cover) melting. A part of the solid precipitation becomes liquid and thus directly feeds the soil reservoir, which explains the maximum response in soil moisture. However, this increase is just a shift in the seasonal cycle, not a net effect. Indeed snow and ice would have melted later in spring in a colder climate. As expected, soil moisture decreases in spring, partly due to this shift, but also in regions of low elevation due to a decrease in precipitation. In higher elevation regions (core of the Alps), the lagging effect of ice and snow melt is still observed. Maximum drying is obtained in summer in the west and north (30 $kg \cdot m^{-2}$ lost for vegetation), and this strong depletion persists in the north in autumn. This is probably the major threat of climate change for France.

In the A2 scenario the reduction in soil water content is even stronger, particularly in summer and autumn. The average loss in summer is 33 $kg \cdot m^{-2}$ for France. It corresponds to a halving of the relative available soil moisture (0.41 in the reference to 0.19 in the scenario) where this index is defined as the ratio:

$$(W - W\text{wilt})/(W\text{fc} - W\text{wilt}) \tag{1}$$

where W is the actual soil moisture, Wwilt soil moisture at wilting point and Wfc soil moisture at field capacity. Wfc and Wwilt correspond to the two extremes of possible water extraction by the roots (see Figure 6.2(c) and (d)). They are less than the maximum water that the soil can contain (see Figure 6.2(b)). This phenomenon explains why the highest warming is found in summer and in the south-west of France. The absence of thermal regulation of the surface by soil and canopy evaporation may lead, as observed in summer 2003, to heat waves.

Global radiation

Changes in the temperature of the atmosphere and the hydrological cycle produce changes in cloud cover and cloud water content in the scenario. From the tree point of view, this results in a modification of the incoming solar radiation at the surface. In the B2 scenario, a weak decrease is obtained in the Mediterranean area in winter. In summer, radiation increases everywhere with a minimum along the coasts. In the intermediate seasons, a weak increase is obtained over most of the country. Similar behaviour is obtained with the A2 scenario, with a larger response in summer. These features can be linked to the atmosphere drying, which leads to fewer clouds.

Conclusion

The numerical experiment carried out and which provided input to our partners is characterized by a temperature rise in all seasons, particularly in summer, by an increase in precipitation in winter in the coastal regions (except the Mediterranean Sea) and the mountains (except the Pyrenees) and a decrease in precipitation during the other seasons. From a hydrological point of view, the availability of soil water for vegetation is significantly reduced, except in winter.

The basic scenario of this study is SRES-B2. The more pessimistic SRES-A2 scenario, used in the French project IMFREX (Déqué, 2007) or in the European project PRUDENCE, is about 1°K warmer over France in spatial and annual mean. The precipitation increase in winter is reduced and the summer decrease is enhanced. The soil moisture depletion in summer is enforced. The conclusions drawn from the CARBOFOR project should, therefore, be considered as a moderate or even optimistic projection of climatic change for France at the end of the current century, if GHG emissions continue their exponential growth as in the last century.

Acknowledgements

This work has been partly supported by the French Ministry of Ecology and Sustainable Development (GICC-CARBOFOR) and by the European Commission (PRUDENCE, EVK2-2001-00156).

References

Bougeault P., 1985. A simple parameterization of the large-scale effects of cumulus convection. *Monthly Weather Review*, 113, 2108–2121.

Christensen J.H., Christensen O.B., 2007. A summary of the PRUDENCE model projections of changes in European climate by the end of this century. *Climatic Change*, 81 (1), 7–30.

Cubasch U., Waszkewitz J., Hegerl G., Perlwitz J., 1995. Regional climate changes as simulated in time-slice experiments. *Climatic Change*, 31, 273–304.

Déqué M., 2007. Frequency of precipitation and temperature extremes over France in an anthropogenic scenario: model results and statistical correction according to observed values. *Global and Planetary Change*, 57 (1–2), 16–26.

Déqué M., Dreveton C., Braun A., Cariolle D., 1994. The ARPEGE-IFS atmosphere model: a contribution to the French community climate modelling. *Climate Dynamics*, 10, 249–266.

Déqué M., Marquet P., Jones R.G., 1998. Simulation of climate change over Europe using a global variable resolution general circulation model. *Climate Dynamics*, 14, 173-189.

Déqué M., Piedelievre J.-P., 1995. High-resolution climate simulation over Europe. *Climate Dynamics*, 11, 321–339.

Déqué M., Rowell D.P., Lüthi D., Giorgi F., Christensen J.H., Rockel B., Jacob D., Kjellström E., Castro M., van den Hurk B., 2007. An intercomparison of regional climate simulations for Europe: assessing uncertainties in model projections. *Climatic Change*, 81 (1), 53–70.

Douville H., Planton S., Royer J.-F., Stephenson D.B., Tyteca S, Kergoat L., Lafont S., Betts R.A., 2000. The importance of vegetation feedbacks in doubled-CO_2 time-slice experiments. *Journal of Geophysical Research*, 105, 14841–14861.

Frei C., Christensen J.H., Déqué M., Jacob D., Jones R.G., Vidale P.L., 2003. Daily precipitation statistics in regional climate models: evaluation and intercomparison for the European Alps. *Journal of Geophysical Research*, 108, 91–19.

Gibelin A.L., Déqué M., 2003. Anthropogenic climate change over the Mediterranean region simulated by a global variable resolution model. *Climate Dynamics*, 20, 327–339.

Gibson J.K., Kållberg P., Uppala S., Hernandez A., Serano E., 1997. *ERA Description. ECMWF Re-analysis Project Report Series.* European Centre for Medium-range Weather Forecasts, Reading, 77 p.

Giorgi F., Mearns L.O., 1999. Regional climate modeling revisited. An introduction to the special issue. *Journal of Geophysical Research*, 104, 6335–6352.

Hu R.M., Planton S., Déqué M., Marquet P., Braun A., 2001. Why is the climate forcing of sulfate aerosols so uncertain? *Advances in Atmospheric Sciences*, 18, 1103–1120.

IPCC, 2001. *Contribution of Working Group I to the Third Assessment Report of the IPCC.* (J.T. Houghton, Y. Ding, D.J. Griggs, M. Noguer, P.J. Van der Linden, X. Dai, K. Maskell, C.A. Johnson, eds). Cambridge University Press, Cambridge, 881 p.

Lott F., 1999. Alleviation of stationary biases in a GCM through a mountain drag parameterization scheme and a simple representation of mountain lift forces. *Monthly Weather Review*, 125, 788–801.

Lott F., Miller M.J., 1997. A new subgrid-scale orographic drag parametrization: its formulation and testing. *Quarterly Journal of the Royal Meteorological Society*, 123, 101–127.

Machenhauer B., Windelband M., Botzet M., Christensen J.H., Déqué M., Jones J., Ruti P.M., Visconti G., 1998. *Validation and Analysis of Regional Present-day Climate and Climate Change Simulations over Europe. MPI Report*, No 275, Max-Planck-Institute for Meteorology, Hamburg, 87 p.

Moisselin J.-M., Schneider M., Canellas C., Mestre O., 2002. Changements climatiques en France au XXᵉ siècle. Étude des longues séries de données homogénéisées françaises de précipitations et températures. *La Météorologie*, 38, 45–56.

Morcrette J.J., 1990. Impact of changes to the radiation transfer parameterizations plus cloud optical properties in the ECMWF model. *Monthly Weather Review*, 118, 847–873.

New M., Hulme M., Jones P.D., 1999. Representing twentieth century space-time climate variability. Part 1: development of a 1961–90 mean monthly terrestrial climatology. *Journal of Climate*, 12, 829–856.

Reynolds R.W., Smith T.M., 1994. Improved global sea surface temperature analyses using Optimum Interpolation. *Journal of Climate*, 7, 929–948.

Ricard J.-L., Royer J.-F., 1993. A statistical cloud scheme for use in an AGCM. *Annales Geophysicae*, 11, 1095–1115.

Rowell D.P., Jones R.G., 2006. Causes and uncertainty of future summer drying over Europe. *Climate Dynamics*, 27, 281–299.

Royer J.-F., Cariolle D., Chauvin F., Déqué M., Douville H., Planton S., Rascol A., Ricard J.-L., Salas-Mélia D., Sevault F., Simon P., Somot S., Tyteca S., Terray L., Valcke S., 2002. Simulation of climate changes during the 21st century including stratospheric ozone. *Geosciences*, 334 (3), 147–154.

Salas-Mélia D., 2002. A global coupled sea ice-ocean model. *Ocean Modelling*, 4, 137–172.

Wilby R.L., Wigley T.M.L., Conway D., Jones P.D., Hewitson B.C., Main J., Wilks D.S., 1998. Statistical downscaling of general circulation model output: A comparison of methods. *Water Resources Research*, 34, 2995–3008.

Chapter 7
Forest tree phenology and climate change

Isabelle Chuine

Introduction

Phenology is the study of periodic biological events in animals and plants triggered by the environment (Schwartz, 2003). The major phenological events in trees are leaf unfolding (or budburst), flowering, fruit maturation and leaf senescence (leaf colouring). In forest trees, leaf unfolding is by far the most widely monitored phenological trait, followed by leaf colouring or fall. Few investigations have been undertaken on flowering and fruit maturation. While leaf phenology is a major determinant of tree primary productivity and net gas fluxes, flowering and fruit maturation may take on more importance in the context of global climate change in the problem of natural regeneration of forest stands. In this chapter, I review first, the phenological changes documented on tree species during the 20th century with a special focus on forest tree species; second, the different process-based models currently used to simulate and predict the phenology of trees; third, projections of phenology change for the 21st century provided by these models.

Observed changes in forest tree phenology

Phenological traits are among the traits that are most sensitive to climate change. Although under strong genetic control (Howe *et al.*, 2003), they also are under strong environmental control. Studies which have searched for fingerprints of climate change in living organisms all concluded that the two main fingerprints were phenological traits and species distributions (Walther *et al.*, 2002; Parmesan and Yohe 2003; Root *et al.*, 2003). These studies show that spring phenological events now occur earlier, while autumn phenological events occur later.

Parmesan and Yohe (2003) showed, using a meta-analysis, that 62% of 678 species have had significant forward shifts in spring phenological events with a mean rate of

2.3 days·decade[-1], while 9% of the species have had significant backward shifts in these events. Root *et al.* (2003) reviewed trends in spring phenological events in 694 plant and animal species and concluded that spring events had shifted significantly earlier since the 1950s with a mean rate of advancement of 5.1 days·decade[-1]. This meta-analysis also showed that flowering and leaf unfolding of tree species have moved forward at a mean rate of 3.0 days·decade[-1], which is lower than herb species (5.1 days·decade[-1]). Walther *et al.* (2002) found a mean rate of advancement of flowering and leaf unfolding in trees of 1.4–3.1 days·decade[-1] for Europe depending on the species and 1.2–3.8 days·decade[-1] in North America. Focusing on flowering, Fitter and Fitter (2002) have also shown differences in the response to climate change among plant types; annual herb species have a higher rate of advancement than perennial herb species.

In reviewing studies, which looked specifically at trends in European and North American tree species phenophases during the 20th century (Tables 7.1 and 7.2), I estimated a mean advancement since the 1950s of 2.9 days·decade[-1] in leaf unfolding (17 species), and 3.4 days·decade[-1] in flowering (46 species). Observations on fruit maturation and leaf colouring are much rarer. From Penuelas *et al.* (2002), I estimated that since 1974 fruit ripening of Mediterranean woody species ($N = 20$) has been significantly delayed in 35.3% of the species at a mean rate of 8.6 days·decade[-1], and significantly advanced in 52.9% of the species at a mean rate of 8.0 days·decade[-1]. When considering only forest Mediterranean tree species ($N = 6$), I estimated from the same study that since 1974 fruit ripening has been significantly delayed in 16.7% of the species at a mean rate of 7.1 days·decade[-1], and significantly advanced in 50.0% of the species at a mean rate of 9.7 days·decade[-1]. From Menzel *et al.* (2001), I estimated that leaf colouring has been delayed since 1951 at a mean rate of 0.7 day·decade[-1] based on observations for *Betula pendula* Roth, *Quercus robur* L. and *Fagus sylvatica* L. These observations are consistent with satellite data and atmospheric CO_2 data showing that between 1982 and 1990 the start of the growing season has advanced by 8 ± 3 days (7 days between the 1960s and 1990s) (Keeling *et al.*, 1996), while the end has been delayed by 4 ± 2 days (Myneni *et al.*, 1997).

In summary, there is now much evidence that spring phenological events, such as flowering, leaf unfolding and sometimes fruit maturation, have shifted significantly forward in a majority of tree species, while autumn events such as fruit maturation and leaf colouring have been significantly delayed. The rate at which these events have moved forward or backward is species-specific, however (Table 7.2).

Phenological models

Predicting species phenology for the 21st century has become a major issue for several fundamental as well as applied scientific questions. Accurate phenology predictions for the 21st century would: (i) increase the accuracy of predictions of ecosystems productivity, gas exchanges between the vegetation and the atmosphere, and thereby climate; (ii) allow a better understanding of populations dynamics in multi-species interactions (especially between plants and animals) systems; (iii) help farmers and foresters in selecting the most adapted provenances or varieties for the new climatic conditions. Therefore, development of phenological models has been especially intense in the last few decades.

Table 7.1. Review of changes in phenophases between 1950 and 2000 in woody species. Mean change is expressed in number of days per decade. Negative numbers indicate advanced events; positive numbers indicate delayed events.

Region	Phenophase	Number of species	% sign. adv.*	Mean change	Period	Reference
Europe	Spring events	14	–	– 2.0	1959–93	Menzel and Fabian (1999)
Europe, North America, Asia	Spring events	33	–	– 3.9	1950–2000	Root *et al.* (2003)
Germany	Leaf colouring	3	–	+ 0.7	1951–96	Menzel *et al.* (2001)
Germany	Leaf unfolding	5	–	– 3.2	1951–96	Menzel *et al,* (2001)
Estonia	Leaf unfolding	2	–	– 2.3	1948–96	Ahas *et al,* (2000)
Europe, North America	Leaf unfolding	9	–	– 3.5	1950–2000	Root *et al.* (2003)
Spain	Flowering	38	47.4	– 4.2	1952–2000	Penuelas *et al.* (2002)
North America	Flowering	14	78.6	– 3.0	1970–99	Abu-Asab *et al.* (2001)
Estonia	Flowering	3	–	– 2.5	1948–96	Ahas *et al.* (2000)
Europe, North America	Flowering	12	–	– 3.6	1950–2000	Root *et al.* (2003)
Spain	Fruiting	17	52.9	– 8.0	1974–2000	Penuelas *et al.* (2002)
Europe	Autumn events	14	–	+ 1.6	1959–93	Menzel and Fabian (1999)

* % sign. adv.: percentage of species showing significant advancement of the phenophase.

Table 7.2. Review of changes in leaf unfolding dates in major forest tree species. Mean change is expressed in number of days per decade.

Species	Region	Mean change	Period	Reference
Quercus robur	UK	– 4.3 to – 5.8	1950–96	Cannell *et al.* (1999)
Quercus robur	Germany	– 3.1	1951–96	Menzel *et al.* (2001)
Quercus robur	Estonia	– 1.7	1948–96	Ahas *et al.* (2000)
Betula pendula	Germany	– 3.7	1951–96	Menzel *et al.* (2001)
Betula pendula	Estonia	– 2.9	1948–96	Ahas *et al.* (2000)
Betula pendula	Northern Europe	– 2.7	1951–98	Ahas *et al.* (2002)
Fagus sylvatica	Germany	– 2.3	1951–96	Menzel *et al.* (2001)
Picea abies	Germany	– 3.1	1951–96	Menzel *et al.* (2001)
Populus tremuloides	Canada	– 2.6	1900–97	Beaubien and Freeland (2000)

Relationship between climate and phenology

The observed advancement or delay in tree phenophases is related to climate change (Parmesan and Yohe, 2003; Root *et al.*, 2003). The relationship between temperature or degree-days and phenophases, especially flowering and leaf unfolding, is well known and has been widely studied (see for review Schwartz, 2003). The evidence for the role of photoperiod in tree phenology is conflicting, depending on species and location (Heide, 1993; Kramer, 1994b; Falusi and Calamassi, 1996). However, photoperiod alone cannot explain the annual variability of phenology at a given location because photoperiod is the same each year. Further experimental work on the impact of the interaction of photoperiod and temperature on leaf unfolding and flowering dates is required to use photoperiod appropriately in phenological models.

Phenological models used in forestry

Reaumur (1735) first realized that a plant develops quicker at a higher temperature and introduced the concept of degree-day sum, that is, daily average temperatures accumulated between an arbitrary date of onset and the date of an observed phenological event. He proposed that plant development is proportional to the sum of temperature over time rather than to temperature during the phenological event itself. Reaumur's work deeply influenced phenology and degree-day sums have been widely used since because they can provide accurate predictions of many species flowering or leaf unfolding dates.

Three main types of phenology models have been used to predict phenophase occurence: theoretical models, statistical models and process-based models (Chuine *et al.*, 2003). Theoretical models are based on the cost/benefit trade-off of producing leaves to optimize resource acquisition (Kikuzawa, 1991, 1995a, 1995b, 1996; Kikuzawa and Kudo, 1995) and have been designed primarily to understand the evolution of leaf lifespan strategies in trees. Statistical models relate the timing of phenological events to climatic factors with various mathematical linear equations (Boyer, 1973; Spieksma *et al.*, 1995; Emberlin *et al.*, 1997; Schwartz, 1998, 1999). Process-based models formally describe known or assumed cause-effect relationships between biological processes and some driving exogenous factors. Most of the models described in the literature are of this type.

A first category of process-based models, inherited directly from Reaumur's finding, consider that temperature has an effect only during the quiescence phase, that is, the period during which the rate of ontogenetic development increases with increasing temperature (Lamb, 1948; Campbell and Sorensen, 1973; Landsberg, 1974; Cannell and Smith, 1983). The thermal time model (Robertson, 1968; Cannell and Smith, 1983) or spring warming model (Hunter and Lechowicz, 1992) has been used for many species and in many locations. It is one of the simplest phenology models, requiring three parameters (F^*, t_0, T_{b1}):

$$y \text{ such that } S_f\left(x_t\right) = \sum_{t_0}^{y} R_f\left(x_t\right) = F^* \tag{1}$$

$$R_f\left(x_t\right) = \begin{cases} 0 & \text{if } x_t \geq T_{b1} \\ x_t - T_{b1} & \text{if } x_t > T_{b1} \end{cases} \tag{2}$$

where y is the date of the phenological event; x_t is the daily mean temperature; $R_f(x_t)$ is the rate of development during quiescence; S_f is the state of development; T_{b1} is the threshold temperature and t_0 is the day when rate of development starts to accumulate. R_f is also commonly called "degree-days" or "forcing units", and S_f is the sum of degree-days or state of forcing.

However, experimental evidence shows that bud dormancy (endodormancy) needs to be broken before plants enter the quiescence phase. Lower temperatures are required to break dormancy (Sarvas, 1974; Hänninen, 1990). However, the dormancy release process remains a black box in process-based phenology models because there is no method for direct measurement of the state of development during dormancy that could be used to assess the accuracy of the model parameter estimates.

The first type of model described here does not take dormancy into account. These models thus assume either that at t_0 the chilling requirements of the species to break dormancy have been met, or that the species does not have chilling requirements. Five process-based phenological models take dormancy into account: (i) the sequential model (Richardson *et al.*, 1974; Sarvas, 1974; Hänninen 1987, 1990); (ii) the parallel model (Landsberg, 1974; Sarvas, 1974; Hänninen, 1990); (iii) the alternating model (Cannell and Smith, 1983; Murray *et al.*, 1989; Kramer, 1994b); (iv) the deepening rest model (Kobayashi *et al.*, 1982); and (v) the four phases model (Vegis, 1964; Hänninen, 1990). The sequential model assumes that forcing (warm) temperatures are not effective unless chilling requirements are fulfilled because plants have no competence to respond to forcing temperatures as long as chilling requirements are not met. The parallel model assumes that forcing temperatures are active simultaneously with chilling temperatures. The alternating model assumes a negative exponential relationship between the sum of forcing units required for completion of quiescence and the sum of chilling units received. The four phases model assumes three phases of dormancy (i.e. pre-rest, true-rest and post-rest) before the quiescence phase. This is formalized by an increasing temperature threshold for forcing during pre-rest and a decreasing temperature threshold for forcing during post-rest, and buds cannot respond to forcing temperature at all during true-rest.

In the different models, the developmental responses to temperature during dormancy and during quiescence have been described by various types of functions (Figure 7.1).

Forcing units have commonly been formulated either as degree-days (Equation 2) or as follows (Figure 7.1(b)–(d)):

$$R_f\left(x_t\right) = \begin{cases} 0 & \text{if } x_t < 0 \\[2mm] \dfrac{a}{1 + e^{b\left(x_t - c\right)}} & \text{if } x_t \geq 0 \end{cases} \tag{3}$$

with $a = 28.4$, $b = -0.185$, $c = 18.4$ (Sarvas, 1974; Hänninen, 1990).

Chilling units have commonly been formulated as chilling days:

$$R_c\left(x_t\right) = \begin{cases} 1 & \text{if } x_t < T_{b2} \\[2mm] 0 & \text{if } x_t \geq T_{b2} \end{cases} \tag{4}$$

where T_{b2} is the base temperature (Figure 7.1(a)), or as a more complex function:

$$R_c\left(x_t\right) = \begin{cases} 0 & \text{or } x_t \geq T_M \\[2mm] \dfrac{x_t - T_m}{T_o - T_m} & \text{if } x_t > T_m \\[2mm] \dfrac{x_t - T_m}{T_o - T_m} & \text{if } x_t < T_M \end{cases} \tag{5}$$

where T_o is the optimal temperature, $T_m = -3.4$ and $T_M = 10.4$ (Sarvas, 1974; Hänninen, 1990) (Figure 7.1(c)).

The variety of model assumptions and formulations led some authors to propose model comparison, as well as formulation and notation homogenization (Hänninen, 1990; Kramer, 1994a, 1994b; Hänninen, 1995; Chuine, 2000). For example, Chuine (2000) developed a unified model based on two general functions that describe the relationships between temperature and the rates of chilling and forcing development:

$$R_c\left(x_t\right) = \frac{1}{1 + e^{a\left(x_t - c\right)^2 + b\left(x_t - c\right)}} \tag{6}$$

$$R_f\left(x_t\right) = \frac{1}{1 + e^{d\left(x_t - e\right)}} \tag{7}$$

where $R_c(x_t)$ is the rate of development during dormancy and $R_f(x_t)$ is the rate of development during quiescence. Most of the mechanistic models described above are special cases of the unified model, depending on the choice of parameter values in Equations (6) and (7), and a few additional parameters (for details see Chuine, 2000).

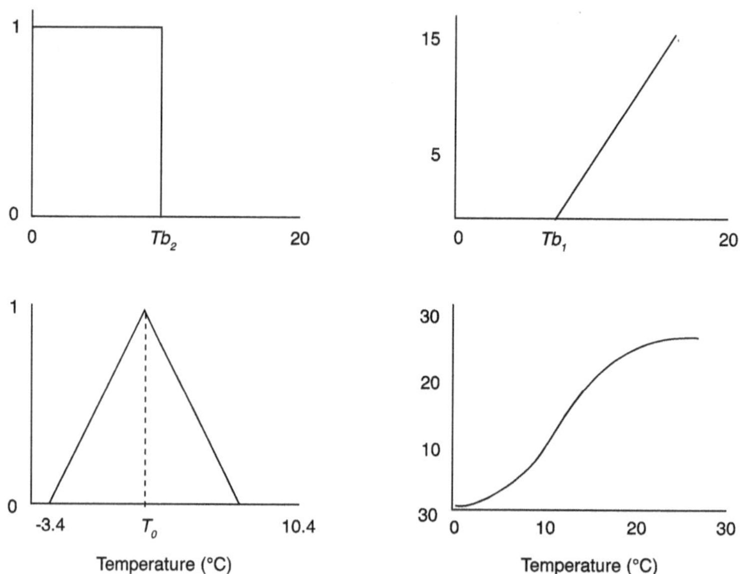

Figure 7.1. Examples of responses to temperature used to calculate chilling: (a) Equation (4); (c) Equation (5); and forcing units (b) Equation (2); (d) Equation (3). T_{b1}, T_{b2}, base temperatures, T_o, optimal temperature.

Methodological problems

Process-based model parameter estimation involves least squared residuals methods. The classically used method, consisting of fixing all but one parameter to a given value and finding the value of the free parameter that minimizes the sum of squared residuals, usually does not provide accurate estimates for the following reasons. First, parameter values are estimated independently from each other although they are usually not independent. Second, the least squares function usually has several local optima and it is almost impossible to find the global optimum without a more systematic search.

More efficient methods consist in estimating all parameters simultaneously using optimization algorithms. Traditional optimization algorithms, for example, downhill simplex, Newton methods (Press *et al.*, 1989), however, rarely converge towards the global optimum (Kramer, 1994b). Simulated annealing algorithms, which are especially designed for functions with multiple optima, have proven to be more efficient in fitting phenology models (Chuine *et al.*, 1998).

Accurate parameter estimation is a first step towards accurate predictions of phenology. A second step consists of an adequate testing method of the model and parameter estimates. Cross-validation (Chatfield, 1988) is an adequate method, but is not always possible to use when observations are scarce so that it is impossible to split the dataset into two parts, one to fit the model, and the other to test its prediction accuracy.

Forest tree phenology under climate change

The observed changes in tree phenophases during the last decades suggest that leaf unfolding, flowering, fruit ripening and leaf colouring will occur in the next few decades at much different dates than currently. The review of changes in leaf unfolding dates in major forest trees (Table 7.2) shows that on average leaf unfolding has advanced at a mean rate of 2.7 days per decade since 1950. Considering that the global surface temperature has increased at a rate of 0.1°C per decade since 1950 (IPCC, 2001), leaf unfolding has thus advanced on average at a mean rate of 27 days per °C. Considering that the global surface temperature is projected to increase by 1–2°C over the period 2000–50, that is a mean increasing rate of 0.2°C to 0.4°C per decade, leaf unfolding should thus advance on average at a mean rate of 5.4–10.8 days·decade^{-1} over the period 2000–50. Thus, by 2050 leaf unfolding of forest trees could occur on average 27 to 54 days earlier than currently. This tremendous change should have major consequences on forests productivity and species composition.

Few studies have projected leaf unfolding dates of forest tree species under climate change scenarios (Kramer, 1994a; Kellomäki *et al.*, 1995; Kramer *et al.*, 1996; Leinonen, 1996; Linkosalo *et al.*, 2000). These studies were aimed primarily at projecting frost damage risk for particular populations. Some authors have indeed argued that frost damage risk could increase because of climate change (Hänninen, 1991). This paradoxical situation would arise because leaf unfolding would occur much earlier due to warming, but at a period when frost occurrence is still possible. Morin *et al.* (2006) have calibrated and validated leaf unfolding phenological models for 22 North American temperate tree species, and projected their frost damage under climate scenarios A2 (+2.6°C by 2080) and B2 (+1.6°C by 2080) run with a HadCM3 climate model. This study showed that leaf unfolding of the 22 species studied should move forward, but the amount of change depends on the species (up to 26 days for *Acer rubrum* L. by 2100), and only three species showed an increase of frost damage risk over their distribution.

Global simulation of forest tree species phenology and frost risk have never been undertaken to our knowledge, while several recent studies have modelled ecosystem phenology at a global scale (Botta *et al.*, 2000; Cook *et al.*, 2004; Arora and Boer, 2005; Jolly *et al.*, 2005). Projections of global ecosystem phenology under climate change have, however, rarely been undertaken (Cook *et al.*, 2005). One reason why global simulations of tree species phenology remain difficult is the lack of large phenological datasets allowing the calibration of phenological models for several populations of tree species. Chuine *et al.* (unpublished data) have undertaken such a study for four major forest tree species: *Betula pendula*, *Fagus sylvatica*, *Larix decidua* L. and *Quercus robur*. In this study the unified phenological model (Chuine, 2000) was used to simulate leaf unfolding of 8–21 populations of these species evenly distributed over Europe. These models were used to project leaf unfolding dates under two climate change scenarios A2 (+2.6°C by 2080) and B2 (+1.6°C by 2080) run with the climate model HadCM3. These projections show that leaf unfolding could advance by 10 to 20 days by 2050 and 30 to 50 days in some regions by the end of the century. These estimates are slightly lower than what can be extrapolated from current observed changes for tree species, but this study concerned only four species.

One major problem ecologists face in projecting the impact of climate change on species and ecosystems is their diversity; each species and each ecosystem has a particular response to climate. Biodiversity is extremely difficult to handle in global vegetation models, even in species-specific models, because climate exerts a strong selective pressure on populations, which adapt genetically locally. Indeed, forest tree species exhibit steep genetic clines along climatic gradients for many traits, such as phenology and cold hardiness (Rehfeldt *et al.*, 1999; Howe *et al.*, 2003; Savolainen *et al.*, 2004). In contrast, genetic differentiation among populations in neutral genetic markers such as allozymes is low, showing that gene flow among populations is high (Hamrick and Godt, 1990). Thus, natural selection exerted by climate within species distributions appears to be strong enough to counterbalance gene flow which hampers local adaptation (Lenormand, 2002). Selection by climate of ecotypes does not prevent, however, a high genetic diversity within populations, which can be the result of either strong gene flow or a differential selection regime both in time and space, which is the case with climate. Genetic local adaptation of forest tree species can be taken into account in process-based models by using different parameter estimates for each provenance. This requires gathering observations for several populations evenly distributed within the species distribution and testing for differentiation (Chuine *et al.*, 2000). When local adaptation is strong, using homogeneous parameter estimates over all the species distribution may flaw the model predictions. The phenology of the vegetation significantly affects the productivity of the ecosystems (Rotzer *et al.*, 2004; Loustau *et al.*, 2005) and the exchanges of heat and gas between the Earth's surface and the atmosphere. Although current geographical patterns in phenology changes can be estimated from satellites, we still lack the ability to predict accurately and widely future trends in the response of phenology to climate change. While progress has been made in phenology modelling, the physiological bases of the control of environment on leaf phenology are far from being completely understood. Further progress in phenology modelling requires new experimental investigations on the determinism of tree phenology.

References

Abu-Asab M.S., Peterson P.M., Shetler S.G., Orli S.S., 2001. Earlier plant flowering in spring as a response to global warming in the Washington, DC, area. *Biodiversity and Conservation*, 10, 597–612.

Ahas R., Aasa A., Menzel A., Fedotova V.G., Scheifinger H., 2002. Changes in European spring phenology. *International Journal of Climatology*, 22, 1727–1738.

Ahas R., Jaagus J., Aasa A., 2000. The phenological calendar of Estonia and its correlation with mean air temperature. *International Journal of Biometeorology*, 44, 159–166.

Arora V.K., Boer G.J., 2005. A parameterization of leaf phenology for the terrestrial ecosystem component of climate models. *Global Change Biology*, 11, 39–59.

Beaubien E.G., Freeland H.J., 2000. Spring phenology trends in Alberta, Canada: links to ocean temperature. *International Journal of Biometeorology*, 44, 53–59.

Botta A., Viovy N., Ciais P., Friedlingstein P., Monfray P., 2000. A global prognostic scheme of vegetation growth onset using satellite data. *Global Change Biology*, 6, 709–725.

Boyer W.D., 1973. Air temperature, heat sums, and pollen shedding phenology of longleaf pine. *Ecology*, 54, 421–425.

Campbell R.K., Sorensen F., 1973. Cold-acclimation in seedling Douglas-fir related to phenology and provenance. *Ecology*, 54, 1148–1151.

Cannell M.G.R., Palutikof J.P., Sparks T., 1999. *Indicators of Climate Change in the UK*. Department of the Environment, Transport and the Regions, London.

Cannell M.G.R., Smith R.I., 1983. Thermal time, chill days and prediction of budburst in *Picea sitchensis*. *Journal of Applied Ecology*, 20, 951–963.

Chatfield C., 1988. What is the best method of forecasting? *Journal of Applied Statistics*, 27, 264–279.

Chuine I., 2000. A unified model for the budburst of trees. *Journal of Theoretical Biology*, 207, 337–347.

Chuine I., Cour P., Rousseau D.D., 1998. Fitting models predicting dates of flowering of temperate-zone trees using simulated annealing. *Plant, Cell and Environment*, 21, 455–466.

Chuine I., Kramer K., Hänninen H., 2003. Plant development models. *In: Phenology: An Integrative Environmental Science* (M.D. Schwarz, ed.). Kluwer, Dordrecht, pp. 217–235.

Chuine I., Mignot A., Belmonte J., 2000. A modelling analysis of the genetic variation of phenology between tree populations. *Journal of Ecology*, 88, 561–570.

Cook B.I., Mann M.E., D'Odorico P., Smith T.M., 2004. Statistical simulation of the influence of the NAO on European winter surface temperatures: Applications to phenological modelling. *Journal of Geophysical Research-Atmosphere*, 109, D16106.

Cook B.I., Smith T.M., Mann M.E., 2005. The North Atlantic Oscillation and regional phenology prediction over Europe. *Global Change Biology*, 11, 919–926.

Emberlin J., Mullins J., Corden J., Millington W., Brooke M., Savage M., Jones S., 1997. The trend to earlier Birch pollen season in the UK: a biotic response to changes in weather conditions? *Grana*, 36, 29–33.

Falusi M., Calamassi R., 1996. Geographic variation and bud dormancy in beech seedlings (*Fagus sylvatica* L). *Annales des Sciences Forestières*, 53, 967–979.

Fitter A.H., Fitter R.S.R., 2002. Rapid changes in flowering time in British plants. *Science*, 296, 1689–1691.

Hamrick J.L., Godt M.J., 1990. Allozyme diversity in plant species. *In: Plant Population Genetics, Breeding, and Genetic Resources* (A.H.D. Brown, M.T. Clegg, A.L. Kahler, B.S. Weir, eds). Sinauer Associates, Sunderland, pp. 43–63.

Hänninen H., 1987. Effects of temperature on dormancy release in woody plants: implications of prevailing models. *Silva Fennica*, 21, 279–299.

Hänninen H., 1990. Modelling bud dormancy release in trees from cool and temperate regions. *Acta Forestalia Fennica*, 213, 1–47.

Hänninen H., 1991. Does climatic warming increase the risk of frost damage in northern trees? *Plant, Cell and Environment*, 14, 449–454.

Hänninen H., 1995. Effects of climatic change on trees from cool and temperate regions: an ecophysiological approach to modelling of budburst phenology. *Canadian Journal of Botany*, 73, 183–199.

Heide O.M., 1993. Dormancy release in beech buds (*Fagus sylvatica*) requires both chilling and long days. *Physiologia Plantarum*, 89, 187–191.

Howe G.T., Aitken S.N., Neale D.B., Jermstad K.D., Wheeler N.C., Chen T.H.H., 2003. From genotype to phenotype: unraveling the complexities of cold adaptation in forest trees. *Canadian Journal of Botany*, 81, 1247–1266.

Hunter A.F., Lechowicz M.J., 1992. Predicting the timing of budburst in temperate trees. *Journal of Applied Ecology*, 29 (3), 597–604.

IPCC, 2001. *Climate Change 2001: Impacts, Adaptation and Vulnerability. Contribution of the Working Group II to the Third Assessment Report of IPCC*. Intergovernmental Panel on Climate Change, Geneva.

Jolly W.M., Nemani R., Running S.W., 2005. A generalized, bioclimatic index to predict foliar phenology in response to climate. *Global Change Biology*, 11, 619–632.

Keeling C.D., Chin J.F.S., Whorf T.P., 1996. Increased activity of northern vegetation inferred from atmospheric CO_2 measurements. *Nature*, 382, 146–149.

Kellomäki S., Hänninnen H., Kolström M., 1995. Computations on frost damage to Scots pine under climatic warming in boreal conditions. *Ecological Applications*, 5, 42–52.

Kikuzawa K., 1991. A cost-benefit analysis of leaf habit and leaf longevity of trees and their geographical pattern. *The American Naturalist*, 138, 1250–1263.

Kikuzawa K., 1995a. The basis for variation in leaf longevity of plants. *Vegetatio*, 121, 89–100.

Kikuzawa K., 1995b. Leaf phenology as an optimal strategy for carbon gain in plants. *Canadian Journal of Botany*, 73, 158–163.

Kikuzawa K., 1996. Geographical distribution of leaf life span and species diversity of trees simulated by a leaf-longevity model. *Vegetatio*, 122, 61–67.

Kikuzawa K., Kudo G., 1995. Effects of the length of the snow-free period on leaf longevity in alpine shrubs: a cost-benefit model. *Oikos*, 73, 214–220.

Kobayashi K.D., Fuchigami L.H., English M.J., 1982. Modelling temperature requirements for rest development in *Cornus sericea*. *Journal of American Society of Horticultural Science*, 107, 914–918.

Kramer K., 1994a. A modelling analysis of the effects of climatic warming on the probability of spring frost damage to tree species in The Netherlands and Germany. *Plant, Cell and Environment*, 17, 367–377.

Kramer K., 1994b. Selecting a model to predict the onset of growth of *Fagus sylvatica*. *Journal of Applied Ecology*, 31, 172–181.

Kramer K., Friend A., Leinonen I., 1996. Modelling comparison to evaluate the importance of phenology and spring frost damage for the effects of climate change on growth of mixed temperate-zone deciduous forests. *Climate Research*, 7, 31–41.

Lamb R.C., 1948. Effects of temperature above and below freezing on the breaking of rest in the Latham raspberry. *Journal of the American Society of Horticultural Science*, 51, 313–315.

Landsberg J.J., 1974. Apple fruit bud development and growth; analysis and an empirical model. *Annals of Botany*, 38, 1013–1023.

Leinonen I., 1996. A simulation model for the annual frost hardiness and freeze damage of Scots Pine. *Annals of Botany*, 78, 687–693.

Lenormand T., 2002. Gene flow and the limits to natural selection. *Trends in Ecology and Evolution*, 17, 183–189.

Linkosalo T., Carter T.R., Hakkinen R., Hari P., 2000. Predicting spring phenology and frost damage risk of *Betula* spp. under climatic warming: a comparison of two models. *Tree Physiology*, 20, 1175–1182.

Loustau D., Bosc A., Colin A., Ogée J., Davi H., François C., Dufrêne E., Déqué M., Cloppet E., Arrouays D., Le Bas C., Saby N., Pignard G., Hamza N., Granier A., Bréda N., Ciais P., Viovy N., Delage F., 2005. Modeling climate change effects on the potential production of French plains forests at the sub-regional level. *Tree Physiology*, 25, 813–823.

Menzel A., Estrella N., Fabian P., 2001. Spatial and temporal variability of the phenological seasons in Germany from 1951 to 1996. *Global Change Biology*, 7, 657–666.

Menzel A., Fabian P., 1999. Growing season extended in Europe. *Nature*, 397, 659.

Morin X., Chuine I., Poirier M., Améglio T., 2006. Impacts du changement climatique sur la phénologie des essences forestières et les risques de gel précoce. *In : La Forêt face aux changements climatiques : acquis et incertitudes* (G. Landmann, ed.). GIP/ECOFOR, Paris, in press.

Murray M.B., Cannell M.G.R., Smith R.I., 1989. Date of budburst of fifteen tree species in Britain following climatic warming. *Journal of Applied Ecology*, 26, 693–700.

Myneni R.B., Keeling C.D., Tucker C.J., Asrar G., Nemani R.R., 1997. Increasing plant growth in the northern high latitudes from 1981 to 1991. *Nature*, 386, 698–702.

Parmesan C., Yohe G., 2003. A globally coherent fingerprint of climate change impacts across natural systems. *Nature*, 421, 37–42.

Penuelas J., Filella I., Comas P., 2002. Changed plant and animal life cycles from 1952 to 2000 in the Mediterranean region. *Global Change Biology*, 8, 531–544.

Press W.H., Flannery B.P., Teukolsky S.A., Vetterling W.T., 1989. *Numerical Recipes in Pascal.* Cambridge University Press, Cambridge, 759 p.

Rehfeldt G.E., Ying C.C., Spittlehouse D.L., Hamilton D.A., 1999. Genetic responses to climate for *Pinus contorta* in British Columbia: niche breadth, climate change and reforestation. *Ecological Monographs*, 69, 375–407.

Richardson E.A., Seeley S.D., Walker D.R., 1974. A model for estimating the completion of rest for "Redhaven" and "Elberta" peach trees. *HortScience*, 9, 331–332.

Robertson G.W., 1968. A biometeorological time scale for a cereal crop involving day and night temperatures and photoperiod. *International Journal of Biometeorology*, 12, 191–223.

Root T.L., Price J.T., Hall K.R., Schneider S.H., Rosenzweig C., Pounds J.A., 2003. Fingerprints of global warming on wild animals and plants. *Nature*, 421, 57–60.

Rotzer T., Grote R., Pretzsch H., 2004. The timing of bud burst and its effect on tree growth. *International Journal of Biometeorology*, 48, 109–118.

Sarvas R., 1974. Investigations on the annual cycle of development of forest trees. Autumn dormancy and winter dormancy. *Communicationes Instituti Forestalis Fenniae*, 84, 1–101.

Savolainen O., Bokma F., Garcia-Gil R., Komulainen P., Repo T., 2004. Genetic variation in cessation of growth and frost hardiness and consequences for adaptation of *Pinus sylvestris* to climatic changes. *Forest Ecology and Management*, 197, 79–90.

Schwartz M.D., 1998. Green-wave phenology. *Nature*, 394, 839–840.

Schwartz M.D., 1999. Advancing to full bloom: planning phenological research for the 21st century. *International Journal of Biometerology*, 42, 113–118.

Schwartz M.D., 2003. *Phenology: An integrative Environmental Science.* Kluwer, Dordrecht, 564 p.

Spieksma F.T.H., Emberlin J., Hjelmroos M., Jäger S., Leuschner R.M., 1995. Atmospheric birch (*Betula*) pollen in Europe: trends and fluctuations in annual quantities and the starting dates of the seasons. *Grana*, 34, 51–57.

Vegis A., 1964. Dormancy in higher plants. *Annual Review of Plant Physiology*, 15, 185–224.

Walther G.-R., Post E., Convey P., Menzel A., Parmesan C., Beebee T.J.C., Fromentin J.-M., Hoegh-Guldberg O., Bairlein F., 2002. Ecological responses to recent climate change. *Nature*, 416, 389–395.

Chapter 8
Forest tree phenology in the French Permanent Plot Network (RENECOFOR, ICP forest network)

FRANÇOIS LEBOURGEOIS, ERWIN ULRICH

Characteristics of the French Permanent Plot Network (RENECOFOR)

In 1992, a forest network was created by the French National Forest Office to complete the existing French forest health monitoring activities. The main objective of this French Permanent Plot Network for the Monitoring of Forest Ecosystems (RENECOFOR), covering the whole of France, is to detect possible long-term changes in the functioning of a great variety of forest ecosystems, selected as regionally representative stands, and to determine the reasons for such changes. It consists of 102 permanent plots and 26,000 trees (10 different species), which are to be monitored for at least 30 years. Each plot has an area of about 2 ha, the central 0.5 ha of which is fenced (Ulrich, 1995). A total of 36 trees were numbered trees per plot for observation purposes. Stands were sampled in the different French bioclimatic areas (i.e. oceanic, continental, mountain and Mediterranean) and covered a wide range of ecological conditions (Table 8.1). The general topography is gently rolling (mean slope = 10%) with elevation ranging from 18 to 1850 m a.s.l. (mean = 501 m). The examined forests are composed of pure stands from 30 to 180 years, naturally regenerated or derived from plantations. The stands are structurally uniform with a closed canopy. Trees are mostly managed in high forests with regular thinnings (Lebourgeois, 1997; Cluzeau et al., 1998).

Phenological observations (i.e. leaf unfolding and leaf yellowing) have been performed annually for 90 stands since 1997 (48 broadleaved stands and 42 coniferous forests) (Figure 8.1). Oak stands were mainly sampled in plain areas of northern France with a homogeneous distribution from west to east (20–350 m a.s.l.; mean: 200 m). Beech stands are located in the north-eastern plain (250–570 m a.s.l.; mean: 413 m) and in mountain areas in the south and south-east of France (400–1400 m a.s.l.; mean: 1031 m). Norway spruce, silver fir and larch stands correspond to high altitude forests in the east of France (400–1850 m a.s.l.; mean: 1000 m). For pine and Douglas fir stands, geographical locations and ecological conditions are more heterogeneous (15–1670 m a.s.l; mean: 450 m). From 1997 to 2003, 81 stands were observed continuously for 5 years and 46 stands during the whole period (7 years). Thus, from 1997 to 2003, a total of 503 and 278 observations, respectively, for leaf unfolding and leaf yellowing are available (Table 8.1). The course of phenological events was observed by local foresters at a weekly time step with binoculars (in most cases) between March and June for leaf unfolding (all species) and September and November for leaf yellowing (broadleaved trees and larch). Each year, the same local forester noted the date in julian days (jd) at which a given percentage of the 36 numbered trees presented a given stage of phenological development (Lebourgeois *et al.*, 2001). For budburst, the first date (bb1) corresponds to the day of year on which 10% of the trees present at least 20% of open buds in the crown: new green leaves are clearly visible but not completely developed. The second date (bb9) corresponds to the day of year on which 90% of the trees have reached the same stage of crown development. The notation principle for leaf yellowing is similar. The dates ly1 and ly9 correspond to the days of year on which 10% or 90% of the trees present at least 20% of crown yellowing. For oak, beech and larch stands, the length of the growing season (lgs) has been evaluated in four different ways according to the stage of development taken into account. In all cases, the duration is the number of days between a date for budburst and a date for leaf yellowing. The four lengths are the number of days between bb1 and ly1 (lgs11), bb1 and ly9 (lgs19), bb9 and ly1 (lgs91) and bb9 and ly9 (lgs99), respectively.

Table 8.1. Characteristics of the phenological data available in the French Permanent Plot Network (RENECOFOR) for the period 1997–2003. The two values in square brackets indicate the number of broadleaved and coniferous stands, respectively.

Species	Number of stands with phenological data only (number of observed trees)	Number of stands with phenological and daily climatic data	Number of phenological observations	
			Budburst	Leaf yellowing
Abies alba (AA)	10 (360)	10 (360)	66	
Fagus sylvatica (FS)	19 (684)	18 (648)	117	121
Larix decidua (LD)	1 (36)	1 (36)	7	7
Picea abies (PA)	8 (288)	8 (288)	53	
Pinus nigra ssp. *laricio (PN)*	2 (72)	2 (72)	10	
Pinus pinaster (PP)	5 (180)	2 (72)	13	
Pinus sylvestris (PS)	11 (396)	10 (360)	65	
Pseudostuga menziesii (PM)	5 (180)	4 (144)	27	
Quercus petraea (QP)	19 (684)	15 (540)	89	95
Quercus robur (QR)	8 (324)	8 (324)	50	49
Mixed *QR-QP*	2 (72)	1 (36)	6	6
Total	90 [48, 42] (3276)	79 [42, 37] (2880)	503	278

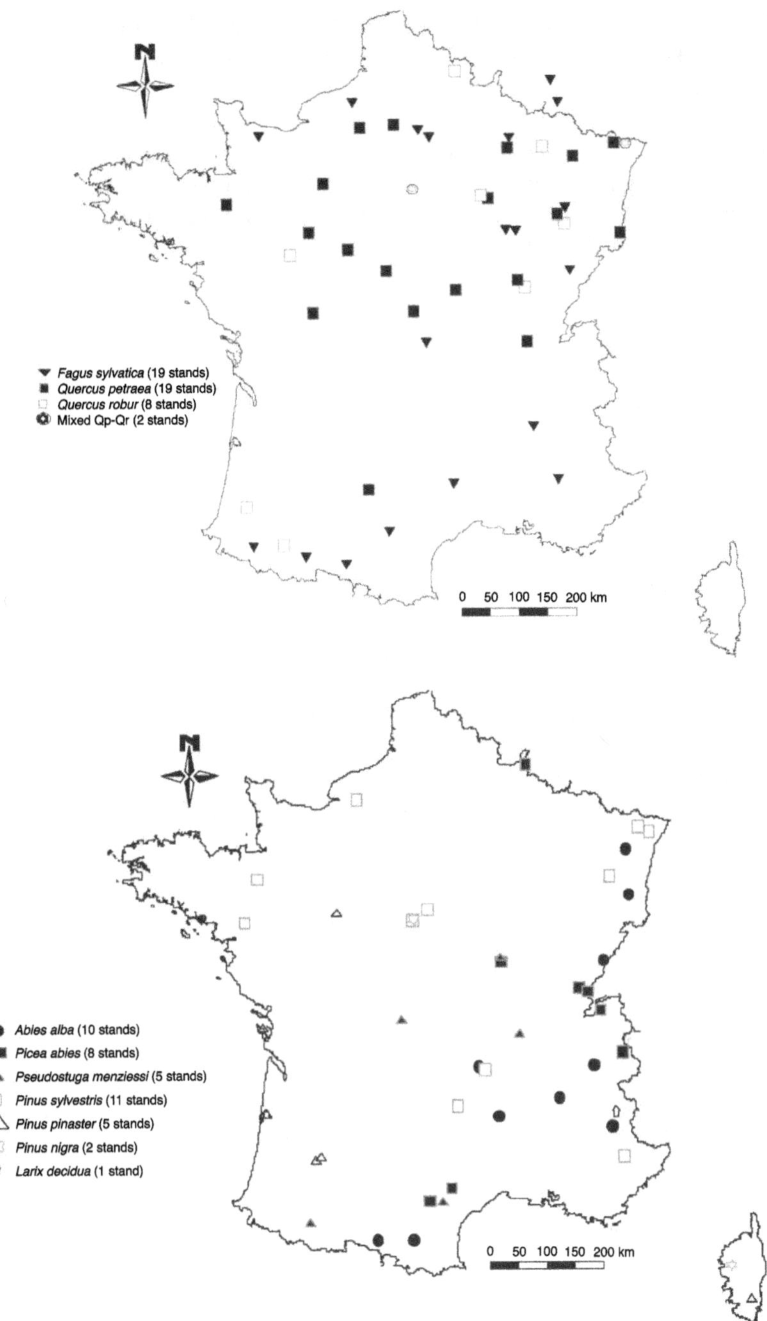

Figure 8.1. Geographical location of the 90 stands (48 broadleaved stands and 42 coniferous stands) sampled in the French Permanent Plot Network (RENECOFOR). For beech stands, two plots were sampled in the Grand-Duchy of Luxembourg.

Daily meteorological data from January 1997 to December 2003 were obtained from the RENECOFOR meteorological sub-network (20 stations) (Ponette *et al.*, 1996) and from the French National Climatic Network "Météo-France" (58 stations). Monthly minimum, maximum, mean temperatures and monthly sunshine durations were calculated for each year and for the whole period 1997–2003. Climatic data were complemented with global radiation estimations (in MJ·m^{-2}). A GIS-based solar radiation model was used, taking both local parameters (topography) and global parameters (cloudiness and latitude) into account. The latter was run for the whole extent of French territory with a 50-m digital elevation model and for each month of the year.

The relationships between the annual timing of phenological phases and monthly climatic parameters and local ecological data (i.e. latitude, longitude, altitude, aspect and slope) were established by using simple or multiple regression analysis. The main objective of this first analysis of the French forest database is to provide simple models usable at a large regional scale. The results of multiple regression analysis are derived from the means of the period 1997–2003. Multiple regression procedure leads to an estimate of a linear equation of the form:

$$Y = a + b1 \times X1 + b2 \times X2 + \ldots + bp \times Xp \tag{1}$$

The regression coefficient b represents the contributions of each independent variable X (e.g. climatic data, geographical variables, etc.) to the prediction of the dependent variable Y (phenological phase). In each case, the selected model was the simplest and easiest to understand from a biological point of view. Each model is adjusted according to least squares criterion. Thus, the model maximizes the percentage (r^2) of variance explained. Relative model quality has been evaluated by the mean absolute error (MAE) and the root mean square error (RMSE). RMSE is representative of the size of a "typical" error but is more sensitive than other measures to the occasional large error: the squaring process gives disproportionate weight to very large errors. The RMSE is evaluated by the equation:

$$RMSE = \sqrt{\frac{1}{n} \sum_{j=1}^{n} \left(P_j - O_j\right)^2} \tag{2}$$

where P_j is the value predicted by the model, O_j the actual value and n the total number of observations. By means of the resulting equation, phenological phases can be calculated for any location or species within the area covered by the network.

Average phenological events in forest stands

On average, the beginning of the growing season starts on 24 April (bb1, jd 114), but greatly depends on species (Table 8.2 and Figures 8.2 and 8.3). In oak stands, budburst starts at the first week of April (jd 96), whereas the onset for beech stands occurs on 21 April (jd 111). The difference between both broadleaved species averages 15 days. The latest onset of spring can be observed at high altitude forests of silver fir, spruce and larch trees. For these stands, budburst starts on 10 May. For all species, the difference between the onset (bb1) and the end (bb9) of budburst averages 10 days. For broadleaved trees, the end of the growing season begins on 5 October (jd 278). The crown appears totally yellow on 22 October (jd 295). The average length of the growing season lasts between 153 and

204 days according to the species and phenological stages taken into account (Table 8.2). Compared with oak stands, a shorter duration of about 22 days can be observed in beech stands as a consequence of later budding and earlier autumn leaf colouring.

Table 8.2. Average characteristics of the growing season for each species in the French Permanent Plot Network (RENECOFOR). Means have been calculated from the period 1997–2003. The different dates are expressed in julian days (jd). For spring and autumn phases and for each species, the first line corresponds to the stage 10% (bb1; ly1) and the second line to the stage 90% (bb9; ly9).

Species	Date of budburst (bb1/bb9)					Date of leaf yellowing (ly1/ly9)				
(Number of stands)	Mean	STD	Min.	Max.	No.	Mean	STD	Min.	Max.	No.
Quercus petraea	94	9.5	78	124	95	282	14.1	248	330	101
(19)	106	10	88	131		299	14.1	154	330	
Quercus robur	97	14.3	48	125	50	281	14.4	251	326	49
(8)	111	17.2	55	137		298	14	147	326	
Fagus sylvatica	111	11.2	71	138	117	274	16.5	223	326	121
(19)	120	8.6	98	143		291	16.5	105	326	
Pinus pinaster	113	9.5	99	131	13					
(2)	124	7.7	111	136						
Pinus nigra laricio	120	17	104	159	10					
(2)	131	16.4	114	166						
Pseudostuga menz	120	11.8	97	141	27					
(5)	133	12.9	111	161						
Pinus sylvestris	123	12.3	90	154	65					
(11)	135	13.2	104	164						
Larix decidua	129	4.6	122	134	7	272	3.8	264	293	7
(1)	139	6.2	129	146		287	3.8	133	293	
Picea abies	130	11.5	97	152	53					
(8)	140	10.7	114	162						
Abies alba	131	11.7	102	172	66					
(10)	140	11.1	116	179						
Mean	114					278				
	124					295				

Species	Lgs11		Lgs19		Lgs91		Lgs99	
	Mean	STD	Mean	STD	Mean	STD	Mean	STD
Quercus petraea	187	16	204	14	176	17	192	15
Quercus robur	184	25	201	25	170	28	187	26
Fagus sylvatica	162	22	179	23	153	20	170	21
Mean	**175**	24	**192**	24	**164**	23	**181**	22

Lgs: length of the growing season in days (see text for details); no.: number of phenological observations from 1997 to 2003.

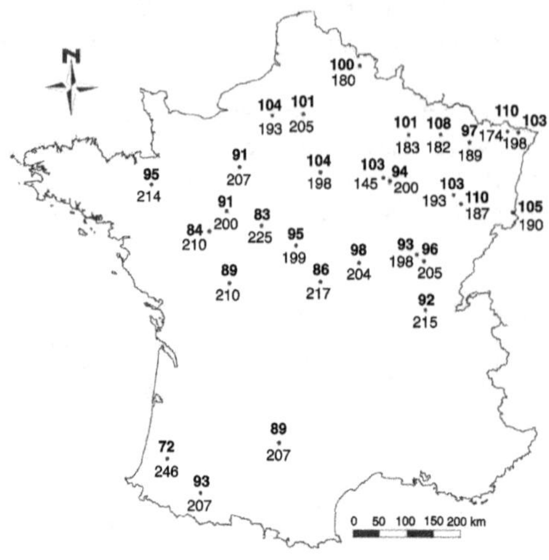

Figure 8.2. Average dates (in julian days) of budburst (in bold) and length of growing season (lgs 19) for broadleaved species (means for the period 1997–2003).

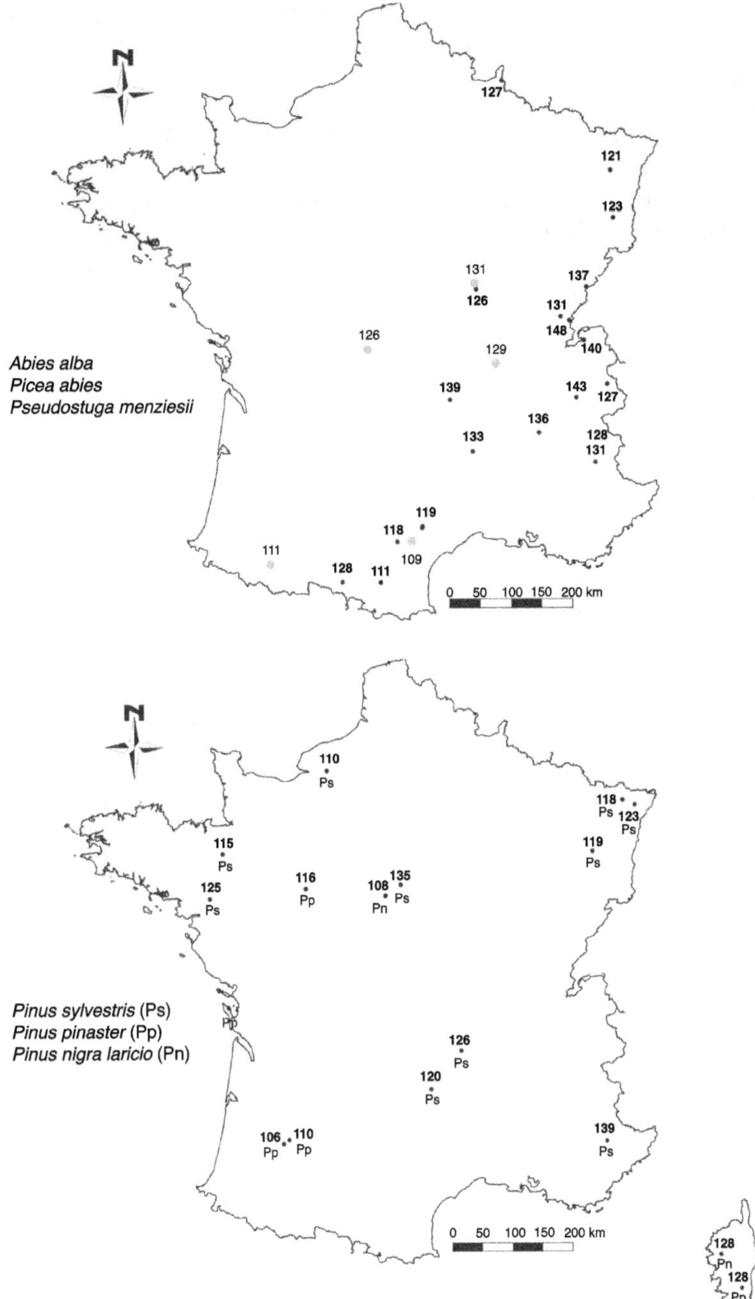

Figure 8.3. Average dates (in julian days) of budburst (bb1) for coniferous species (means for the period 1997–2003).

The timing of spring and autumn events shows strong regional variations and depends highly on the local ecological parameters such as altitude or mean air temperature. On average the green wave in France shifts annually by 3.6 days per degree of longitude from west to east (Figure 8.4). For oak stands, an early onset of the growing season between 10 and 15 days can be observed in oceanic and south regions (at the end of March) (Figure 8.2). In these warmer regions, the length of the growing season is often above 210 days. A later budburst occurs in the continental part of France (in the middle of April) with a growing season of 180–190 days. For oak trees, the regression equation indicates a mean delay of leaf unfolding and duration of growing season by 2.4 days and 4.6 days per degree of longitude from west to east, respectively.

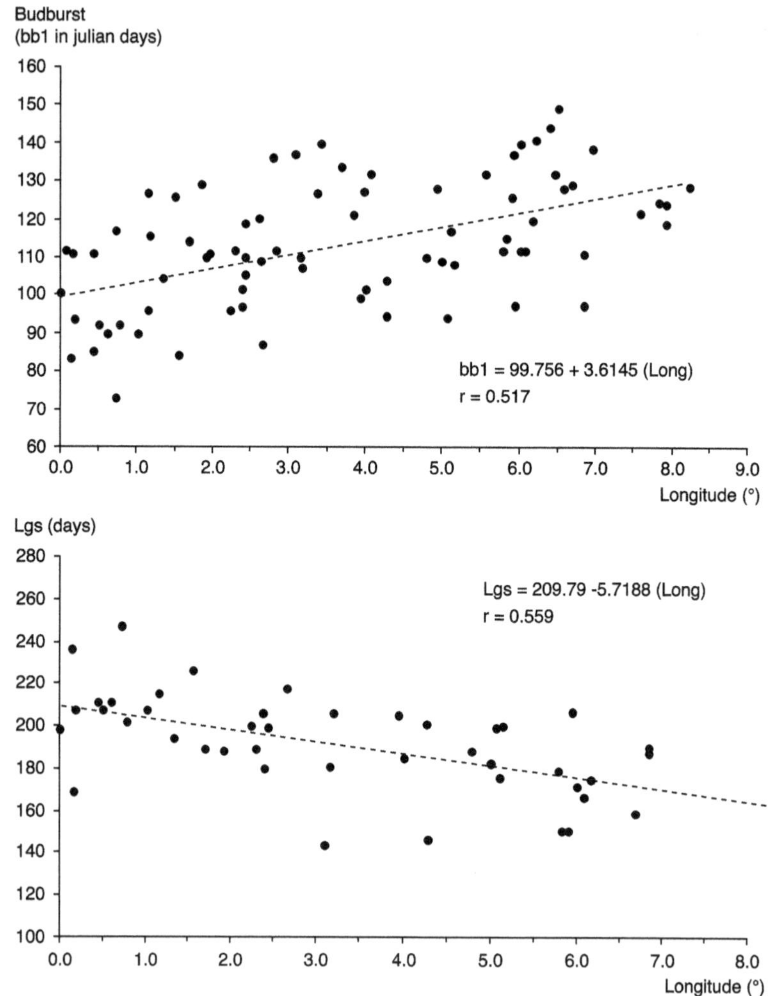

Figure 8.4. Relationship between longitude and average date of budburst (bb1: 90 stands) and length of growing season (lgs19: 48 broadleaved stands) in the French Permanent Plot Network (RENECOFOR). Means have been calculated for the period 1997–2003.

Concerning the effect of altitude, the regression equation indicates that a 100 m increase in altitude is associated with a delay in budburst and length of growing season of about 2 and 3 days, respectively (Figure 8.5). For all phenological parameters, temperature is the most important driving force for the timing of the different events. Correlations with the mean annual temperature appear highly significant (r values superior to 0.6), but the highest correlations are observed with monthly spring conditions. Thus, an increase of 1°C in March–April mean temperature corresponds with an advance of budburst by about 6 days and with an extension of the growing season by about 10 days (Figure 8.6). Analyses of autumn phases show that high maximum September–October temperatures delay the onset of leaf colouring. The response to temperature is about +3.8 days per °C (r = 0.63; 42 broadleaved stands).

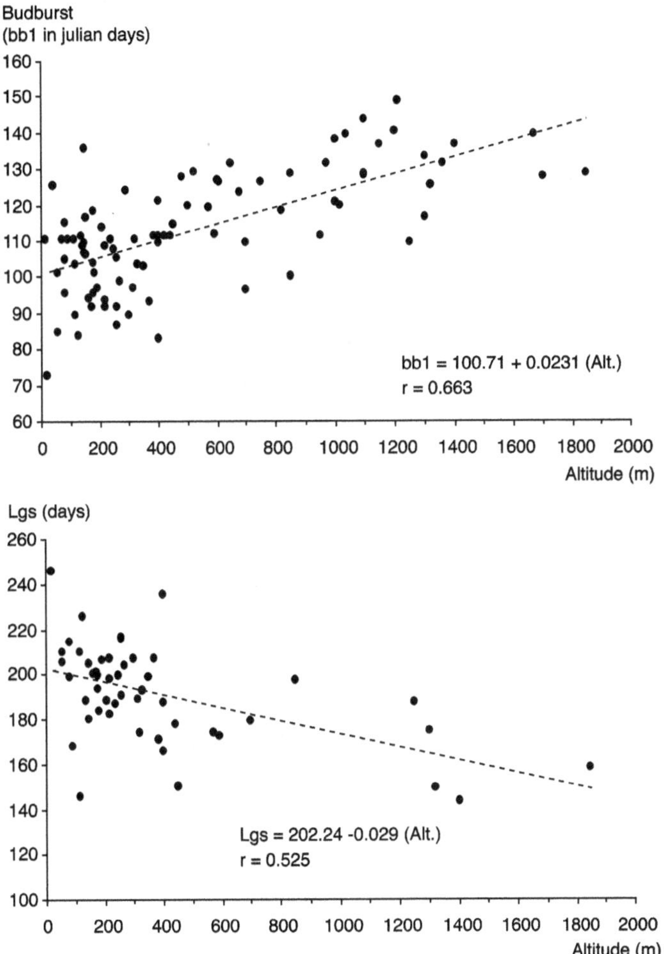

Figure 8.5. Relationship between altitude (m) and average date of budburst (bb1: 90 stands) and length of growing season (lgs19: 48 broadleaved stands) in the French Permanent Plot Network (RENECOFOR). Means have been calculated from the period 1997–2003.

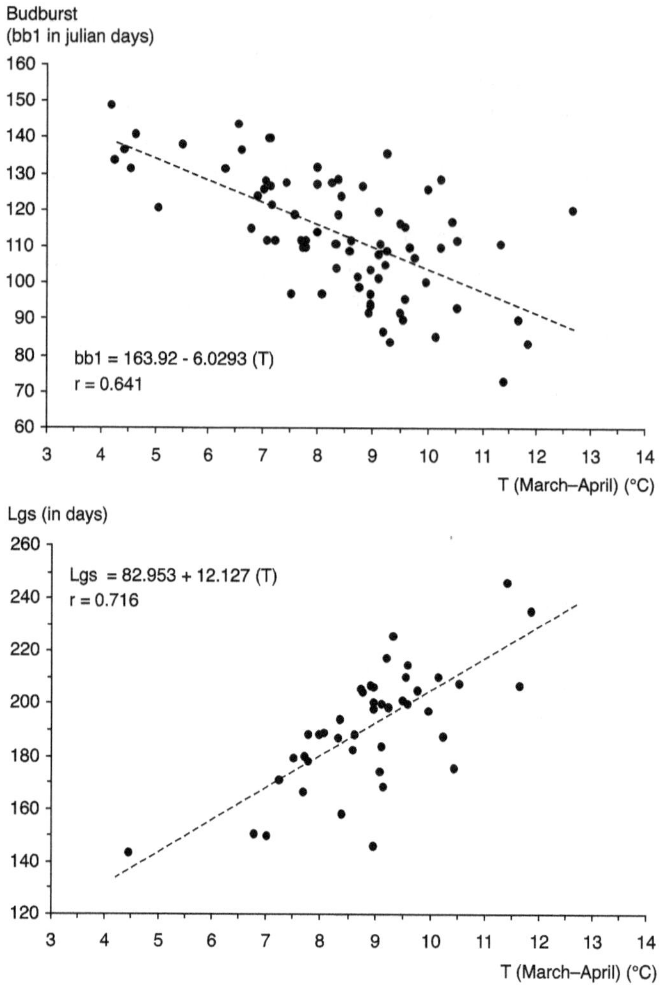

Figure 8.6. Relationship between early spring temperature and average date of budburst (bb1: 90 stands) and length of growing season (lgs19: 48 broadleaved stands) in the French Permanent Plot Network (RENECOFOR). Means have been calculated from the period 1997–2003.

Predictive models of phenological events

The different models explain between 47% and 92% of the variance depending on species and phenophases (Table 8.3 and Figure 8.7). Altitude, latitude and temperature play a major role in the timing of the growing season for the studied forests. For leaf unfolding, leaf yellowing and length of the growing season, the best adjustments have been observed for bb1, ly9 and lgs19, respectively. In all cases, the goodness of fit r^2 for leaf yellowing was always smaller. The accuracy of the models varies from 3 to 10 days according to the species and phenological parameters taken into account. For budburst (bb1), the global model explains 81% of the variance with a mean absolute error of 5.7 days. The goodness of fit is higher for broadleaved trees than for coniferous stands with values of 79.3% and 60.5%, respectively. The mean absolute error of the regressions is 5.6 days and 6.6 days, respectively. The onset of spring events is predicted with an accuracy of less than 3 days for beech and silver fir stands.

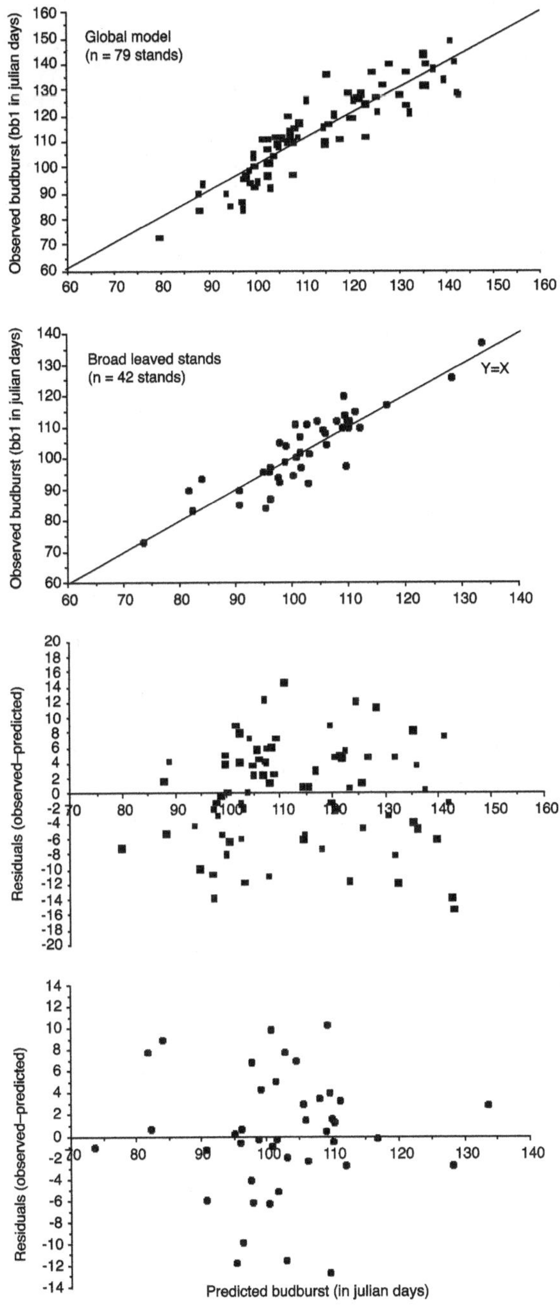

Figure 8.7. Observed and predicted dates of budburst (bb1 in julian days) (see Table 8.3 for details).

Table 8.3. Results of the multiple regression analysis of the different phenophases for each group of species. The results are derived from the means of the period 1997–2003.

Species	Phase	Intercept	Sp.	Alt.	Lat.	Long.	Annual mean	Max March	Max Sept.–Oct.	Jan.	Feb.	Jan.–Feb.
							Temperature			Rg		Sunshine duration
				(m)	°	°	(°C)			(MJ·m^{-2})		(hours)
All species (79)	bb1	− 9.87	16.071	0.021	2.638		− 1.734					
	bb9	44.09	15.544	0.017	1.728		1.574					
Broadleaved species (42)	bb1	− 60.68		0.032			− 2.394					
	ly9	235.44		− 0.011					3.464			
	lgs 19	255.25		− 0.038	− 2.595			5.707				
Fagus sylvatica (18)	bb1	− 111.77		0.032	4.881		− 2.164					
	ly9	228.44		− 0.010					3.875			
	lgs 19	312.55		0.037	4.303		8.273					
Quercus sp. (24)	bb1	− 491.24			10.779	2.466				0.020		
	ly9	423.11		− 0.054	− 2.415							
	lgs 19	1094.33			− 16.366	− 4.139				− 0.029		
Coniferous (37)	bb1	142.50		0.022							− 0.319	
Abies alba (10)	bb1	1956.89			31.977							3.68
Pinus sylvestris (10)	bb1(*)	55.81					8.238					0.937

Table 8.3: continued.

Species	Phase	r²	F	MAE	RMSE
All species (79)	bb1	80.6	77.0	5.7	7.1
	bb9	75.3	56.4	5.9	7.6
Broadleaved species (42)	bb1	79.3	48.6	4.3	5.6
	ly9	50.4	19.7	6.1	7.8
	lgs 19	78	45	7.5	9.8
Fagus sylvatica (18)	bb1	92.7	59.1	2.3	2.9
	ly9	47.3	6.7	7.7	9.7
	lgs 19	76.6	15.3	8.5	10.4
Quercus sp. (24)	bb1	76.8	22.01	3.9	4.7
	ly9	50.7	10.81	4.4	5.3
	lgs 19	72.4	17.49	6.2	7.6
Coniferous (37)	bb1	60.5	25.9	5.3	6.6
Abies alba (10)	bb1	91.9	39.46	1.9	2.6
Pinus sylvestris (10)	bb1(*)	55.6	4.382	4.1	6.3

bb1 and bb9: budburst; ly9: leaf yellowing; lgs19: length of the growing season (see text for details). Sp: species (code 0 for broadleaved trees; code 1 for coniferous species); m: mean temperature; x: maximum temperature; Rg: monthly sum of global solar radiation. Insol: sunshine duration. r²: coefficient of determination. MAE and RMSE: mean absolute error and root mean square error. The number in brackets corresponds to the number of stands.
All the equations are significant at the 99.9% level excepted (**) $P < 0.01$ and (*) $P < 0.05$.

Conclusions

On average, the spring phase starts on 24 April and the growing season lasts 181 days in the French Permanent Plot Network. Our observations are in agreement with the average growing season observed in national phenological networks in Central Europe, which starts on 23 April and lasts 188 days (Chmielewski and Rötzer, 2001). Moreover, phenological events show high species and spatial variability. The growing season for oak stands in western and southern France starts at the beginning of April and lasts at least 210 days whereas the budburst of high altitude coniferous stands located in eastern France occurs in the middle of May. Compared with oak stands, the length of the growing season is reduced by about 20 days in beech forests (about 170 days). The difference of spring events averages 15 days (later leaf unfolding for beech trees at the beginning of May).

The regression models for the calculation of the different phenophases of forest stands, which are presented in this chapter, are intended to provide a simple improvement to phenological models at a large regional scale. The equations described are particularly suitable because of their low input requirements, that is, altitude, latitude and mean monthly temperature. As previously observed for Europe (Kramer, 1995a; Chmielewski and Rötzer, 2001, 2002; Rötzer and Chmielewski, 2001; Menzel *et al.*, 2003), the annual timing of spring and autumn phenophases is to a great extent a response to temperature. In the forest network, phenological events highly reflect the thermal regime in France. The mean annual temperature is a decisive parameter, but higher correlations are also often observed with the spring thermal regime (March–April for leaf unfolding) or early autumn conditions (September–October for leaf yellowing). Thus, a temperature increase of 1°C leads to an advance in leaf unfolding of 6 days, a delay in leaf colouring of 4 days and a lengthening of the growing season by 10 days. Our observations agree with observations made throughout Europe. On average, an increase by 1°C in the mean February to April temperature, T_{24}, corresponds to an advanced beginning of growing season by 6.7 days (period 1969–1998; budburst = 149.4 − 6.68 × T_{24}; r = − 0.83) (Chmielewski and Rötzer, 2001). In Germany, such an increase promotes earlier leaf unfolding by 6.4 days and 2.7 days for beech and pedunculate oak, respectively (Menzel *et al.*, 2003). In England, thermal conditions in early spring explain 70% and 50% of the leaf unfolding of pedunculate oak and beech, respectively (Sparks and Carey, 1995; Sparks and Yates, 1997). Leaf colouring in autumn seems to be a more complex process and less dependent on air temperature. The goodness of fit is largely lower than for spring phases (about 50% against 80% or more) and no significant relationship to air temperature has been observed for oak stands. Further more detailed studies are necessary. The goodness of fit and the accuracy of our models appear to be very similar to those obtained with more sophisticated models that have a complex structure or need several hourly or daily climatic input data. The MAE and the explained variance average 4.6 days and 76.7% for leaf unfolding and 7.5 days and 78% for duration of the growing season. These values are close to the MAE of 4.4–5.0 days found by Rötzer *et al.* (2004) for the daily temperature leaf unfolding models of beech, oak, pine and spruce in southern Germany. In the Netherlands, the sequential thermal model of Sarvas (1974) predicts leaf unfolding of beech with accuracy between 4 and 6 days (Kramer, 1994, 1995b).

Timing of phenological events play a major role in calibrating forest growth models (Chuine *et al.*, 1999; Kramer and Mohren, 1996; Kramer *et al.*, 1996, 2000). The impact of global warming on the extension of the growing season will depend on the extent of the change in leaf unfolding and leaf fall in the future (Menzel, 2000; Ahas *et al.*, 2002). Thus, further investigations have to be carried out to elaborate predictive models at a lower temporal scale and to analyse the interrelations on phenology and tree growth (Rötzer *et al.*, 2004). In this first step of the work, we intended only to study spatial and species variability for French forest ecosystems and to define simple phenological models usable by foresters or researchers at a large-scale with simple local variables.

Acknowledgements

J.C. Pierrat, S. Cecchini, L. Croisé and M. Lanier were involved in data collection and synthesis and helped to interpret the data used in this chapter.

References

Ahas R., Aasa A., Menzel A., Fedotova V.G., Scheifinger H., 2002. Changes in European spring phenology. *International Journal of Climatology*, 22, 1727–1738.

Chmielewski F.M., Rotzer T., 2001. Response of tree phenology to climate change across Europe. *Agricultural and Forest Meteorology*, 108 (2), 101–112.

Chmielewski F.M., Rotzer T., 2002. Annual and spatial variability of the beginning of growing season in Europe in relation to air temperature changes. *Climate Research*, 19 (3), 257–264.

Chuine I., Cour P., Rousseau D.D., 1999. Selecting models to predict the timing of flowering of temperate trees: implications for tree phenology modelling. *Plant, Cell and Environment*, 22 (1), 1–13.

Cluzeau C., Ulrich E., Lanier M., Garnier F., 1998. *RENECOFOR : interprétation des mesures dendrométriques de 1991 à 1995 des 102 peuplements du réseau*. National Forests Office, Paris, 309 p.

Kramer K., 1994. Selecting a model to predict the onset of growth of *Fagus sylvatica*. *Journal of Applied Ecology*, 31, 172–181.

Kramer K., 1995a. Modeling comparison to evaluate the importance of phenology for the effects of climate-change on growth of temperate-zone deciduous trees. *Climate Research*, 5 (2), 119–130.

Kramer K., 1995b. Phenotypic plasticity of the phenology of 7 European tree species in relation to climatic warming. *Plant, Cell and Environment*, 18 (2), 93–104.

Kramer K., Friend A., Leinonen I., 1996. Modelling comparison to evaluate the importance of phenology and spring frost damage for the effects of climate change on growth of mixed temperate-zone deciduous forests. *Climate Research*, 7 (1), 31–41.

Kramer K., Leinonen I., Loustau D., 2000. The importance of phenology for the evaluation of impact of climate change on growth of boreal, temperate and Mediterranean forests ecosystems: an overview. *International Journal of Biometeorology*, 44 (2), 67–75.

Kramer K., Mohren G.M.J., 1996. Sensitivity of FORGRO to climatic change scenarios: a case study on *Betula pubescens*, *Fagus sylvatica* and *Quercus robur* in the Netherlands. *Climatic Change*, 34 (2), 231–237.

Lebourgeois F., 1997. *RENECOFOR – étude dendrochronologique des 102 peuplements du réseau*. National Forests Office, Paris, 307 p.

Lebourgeois F., Granier A., Breda N., 2001. An analysis of regional climate change in France between 1956 and 1997. *Annals of Forest Science*, 58 (7), 733–754.

Menzel A., 2000. Trends in phenological phases in Europe between 1951 and 1996. *International Journal of Biometeorology*, 44 (2), 76–81.

Menzel A., Jakobi G., Ahas R., Scheifinger H., Estrella N., 2003. Variations of the climatological growing season (1951–2000) in Germany compared with other countries. *International Journal of Climatology*, 23 (7), 793–812.

Ponette Q., Belkacem S., Nys C., 1996. Ion dynamics in acid forest soils as affected by addition of Ca fertilizers. *Geoderma*, 71 (1/2), 53–76.

Rotzer T., Chmielewski F.M., 2001. Phenological maps of Europe. *Climate Research*, 18 (3), 249–257.

Rotzer T., Grote R., Pretzsch H., 2004. The timing of bud burst and its effect on tree growth. *International Journal of Biometeorology*, 48 (3), 109–118.

Sarvas R., 1974, Investigations on the annual cycle of development of forest trees. Autumn dormancy and winter dormancy. *Communicationes Instituti Forestalis Fenniae*, 84, 1–101.

Sparks T.H., Carey P.D., 1995. The responses of species to climate over 2 centuries – an analysis of the Marsham phenological record, 1736–1947. *Journal of Ecology*, 83 (2), 321–329.

Sparks T.H., Yates T.J., 1997. The effect of spring temperature on the appearance dates of British butterflies 1883–1993. *Ecography*, 20 (4), 368–374.

Ulrich E., 1995. The RENECOFOR network: objectives and realization. *Revue Forestière Française*, 47 (2), 107–124.

Chapter 9

Modelling tools for predicting the carbon cycle

HENDRIK DAVI, CHRISTOPHE FRANÇOIS, JÉRÔME OGÉE, ÉRIC DUFRÊNE,
PHILIPPE CIAIS, ALEXANDRE BOSC, DENIS LOUSTAU, GUERRIC LE MAIRE

Introduction

Predicting the effect of climatic change on the carbon cycle requires the use of models
to represent adequately the complexity and connectivity of the various processes. The aim
of this chapter is to review the various existing models for predicting the carbon cycle and
to evaluate some of these models. This chapter is composed of three parts ranging from
generalities to a case study: (i) the general state-of-the-art regarding existing models;
(ii) an evaluation of three models against eddy-flux data; and (iii) a detailed investigation
of one model using an uncertainty analysis.

The first part discusses the existing models and areas of active research. Schematically,
the models can be classified using two criteria: accuracy in representation of the
processes (forestry models versus process-based models) and the spatial scale over which
the models are used (stand level versus regional level). In the second part, we test three
process-based models (from the more to the less specific) and the discrepancies between
simulations and measurements are analysed. This part by combining simulations and
eddy-flux measurements can help to improve understanding of carbon cycling in forests.
It also allows the testing of the error level when a regional model is used in comparison
with specific models. Finally, the third part gives a precise description of model sensitivity.
This is obviously an essential step in determining the accuracy of the model. However, it
is a preliminary study for scaling-up of these processes because it provides an exploration
of the sensitivity of the spatially variable input parameters of each model.

Diversity of models at various scales

Model type: growth and yield models versus process-based models

During the last century, several types of forest models were developed with different objectives. Among the oldest, the empirical "growth and yield" models predict stem growth with a time step of one year or more (Schober, 1975; Dhôte, 1990, 1991). These models use empirical rules based on large datasets from field plots. They are able to reproduce tree growth over a century, according to forest management and age, but they do not consider seasonal and inter-annual climatic changes. Climate change, however, probably explains the increase in radial growth (Becker *et al.*, 1994) and tree height (Dhôte and Hervé, 2000).

During the last two decades, several biophysical models have been developed to simulate carbon and water fluxes. These models may consider the canopy as a single layer (i.e. "big-leaf'), a multi-layer or a three-dimensional volume (Wang and Jarvis, 1990; Aber and Federer, 1992; Amthor, 1994; Baldocchi and Harley, 1995; de Pury and Farquhar, 1997). They assess mass and energy exchange between the canopy and the atmosphere by coupling the fluxes of CO_2 and water vapour. The physiologically based photosynthesis model of Farquhar *et al.* (1980), linked with a stomatal conductance model (Jarvis, 1976; Ball *et al.*, 1987; Collatz *et al.*, 1991) provides a representation of leaf fluxes that can be spatially integrated to canopy level. These process-based models integrate canopy functioning over time from minutes to days (Caldwell *et al.*, 1986; Baldocchi, 1992; Leuning *et al.*, 1995; Williams *et al.*, 1996; Wang and Leuning, 1998). Some daily time step models also use a physiologically based canopy photosynthesis model, for example, Sands (1995).

These models are not designed, however, to predict the seasonal and inter-annual variations of tree growth and stand biomass increment. On the other hand, forest growth models based on ecological and biogeochemical principles focus on how carbon and water fluxes vary from daily to annual and decadal time scales. These models are able to predict changes in plant carbon pools (i.e. organs) taking into account their respective size and turnover time (Mohren, 1987; McMurtrie *et al.*, 1990). Some of these models consider litter and soil mineralization processes to predict soil organic matter dynamics (Running and Coughlan, 1988; Korol *et al.*, 1991; Running and Hunt, 1993; Hoffmann, 1995; Bossel, 1996). They tend to ignore, however, the effects of microclimate spatial variability within the plant canopy. Moreover, as they generally use a daily time step, they often use empirical (i.e. not physiologically based) leaf and canopy photosynthesis sub-models. Since hydrological processes control the drought effect on photosynthesis (Schulze, 1986), soil carbon and nitrogen dynamics (Parton *et al.*, 1987), some of these models also couple the carbon budget with a simulation of the water cycle. The rainfall reaching the ground is partitioned between soil evaporation, transpiration, interception, infiltration and runoff. Depending on the application domain, the hydrology models are more or less sophisticated. For example, the evapotranspiration (ETR) can be calculated as a function of potential evaporation or estimated following Monteith (1965) or Shuttleworth and Wallace (1985). The soil can be divided into numerous layers (Braud *et al.*, 1995) or parameterized into one or several buckets (Eagleson, 1978). Detailed models are difficult to use for the investigation of spatial and temporal variability of land surface fluxes. The large number of parameters involved requires detailed field studies and experimentation to derive parameter estimates (Boulet *et al.*, 2000).

In conclusion, various models have been developed in order to respond to different scientific questions. What is forest production? How does the environment influence vegetation functioning? Both questions converge today: the current need to understand the relative impacts of climate and forest management on wood production and carbon sink size makes it necessary to couple mechanistic approaches based on ecophysiological rules with empirical forestry knowledge (Mäkelä *et al.*, 2000; Johnsen *et al.*, 2001). This should help to improve predictions of forest production under global change. Such couplings between process and empirical models have been carried out for a few years and evaluated using stand-level forestry data such as basal area, volume or dominant height (Mäkelä, 1988; Sievänen and Burk, 1993; Landsberg and Waring 1997; Valentine *et al.*, 1997; Bartelink, 2000).

Spatial scale: stand-level models versus global models

Numerous studies using process-based models have assessed the effects of climate change on the net ecosystem exchange (NEE) at stand (Kirschbaum, 1999; Grant and Nalder, 2000), regional (Häger *et al.*, 1999; Coops and Waring, 2001; Joos *et al.*, 2002; Minkkinen *et al.*, 2002; Nabuurs *et al.*, 2003), and global scales (Sellers *et al.*, 1997; Cramer *et al.*, 1999; Kicklighter *et al.*, 1999; White *et al.*, 2000; Cramer *et al.*, 2001). Modelling the exchanges of energy, water and carbon between continents and the atmosphere has been performed through three generations of global models (Sellers *et al.*, 1997). The first generation, developed in the late 1960s and 1970s, was based on simple aerodynamic bulk transfer formulae and often on uniform prescriptions of surface parameters over the continents. In the early 1980s, a second generation of models explicitly recognized the effects of vegetation in the calculation of the surface energy balance. At the same time, global, spatially varying data of land surface properties were assembled from ecological and geographical surveys published in the scientific literature. The latest models use modern theories relating photosynthesis and plant water relations to provide a consistent description of energy exchange, ETR and carbon exchange by plants (Krinner *et al.*, 2005).

Now, regional and global models often estimate the carbon balance using the same algorithms as process-based models developed at stand level, with a broader resolution (typically at least 50 km × 50 km) in order to be coupled with a global circulation model (GCM) and with a simplification of the input parameter description. To date, model parameterization is based on studies performed at the stand level or using remotely sensed data, and is homogeneously applied at broader scales using spatial aggregation of the input parameters. Carbon and water fluxes, however, are controlled by a combination of several biophysical processes (e.g. photosynthesis, respiration, transpiration, evaporation, drainage, etc.), which generally do not respond linearly to the biochemical and structural properties of ecosystems. Consequently, the use of an arithmetic average of spatially variable parameters could be inappropriate when processes exhibit strong non-linear responses to this parameter. This can lead to a significant bias in output variables (Kicklighter *et al.*, 1994; Wollenweber, 1995; Arain *et al.*, 1999). In spite of non-linear responses of simulated NEE to the leaf area index (LAI) and wood biomass, Davi *et al.* (2006a) have shown at three different scales that using parameters averaged over the whole area induces only a slight bias in the estimation of carbon fluxes and almost no bias in the estimation of water fluxes. This last result tends to show that the use of a process-based model at larger scales may be justified.

Scaling-up from stands to region and from regions to the globe also requires: (i) the determination of the parameters needed by the model at the different scales; and (ii) the evaluation of the models at the different scales. Scaling-up can be achieved using methodology proposed by Running *et al.* (1999). At stand level, canopy measurements and soil characteristics (e.g. leaf nitrogen content, LAI, wood biomass and wood growth, soil water content, soil carbon biomass, etc.), as well as carbon and water fluxes estimated from correlation methods, allow the parameterization and validation of the models in the complex ground-vegetation-atmosphere (see le Maire *et al.*, 2005). Then at regional level, remote sensing data and databases can be used to parameterize the models on an area of 100 km^2. This can be done using algorithms connecting the main canopy characteristics to reflectance measurements with relationships determined on ancillary sites. Simulations at the regional scale can then be used to validate global models such as products of the Earth Observation System (EOS).

Summary

Many models exist in the scientific community. Each model has been developed to answer a particular question and falls more or less into one type of model. We can draw a classification following the two criteria: process-based model/forestry model and stand-level model/regional model. However, often the models are hybrid and there is a continuous gradient from forestry models to process-based models (Table 9.1).

Table 9.1.   Classification of models.

	Process-based model	**Forestry models**
Stand level	CASTANEA[1]/GRAECO[2]	FAGACEES[4]
Regional level	ORCHIDEE[3]	

[1] Dufrêne *et al.* (2005); [2] Porté (1999); [3] Krinner *et al.* (2005); [4] Dhôte (1991).

Model evaluation using measurements

In recent years several research programmes have been monitoring CO_2 and water fluxes above temperate, boreal and tropical vegetation canopies (see Chapter 2). There is a global network (FLUXNET), which consists of more than 250 long-term, eddy-flux stations (Running *et al.*, 1999, Baldocchi *et al.*, 2001) and the number of stations is rapidly increasing. With the increase of longer datasets, there is a new challenge for testing coupled carbon-water flux models on multi time and spatial scales. Some studies have been carried out in this direction (Aber *et al.*, 1996; Grant *et al.*, 1999; Kimball *et al.*, 1999; Law *et al.*, 2000a, 2000b), but often using only integrative measurements (i.e. estimating the carbon and water fluxes of the whole ecosystem and consequently summarizing all the carbon and water processes) with the eddy-covariance method. Recent studies have shown that there are some discrepancies between integrative flux measurements and other independent NEE measurements (Granier *et al.*, 2000; Ehman *et al.*, 2002; Davi *et al.*, 2005). Another issue is to evaluate the accuracy of specific models (i.e. a model developed for one type of forest canopy) and global models on several sites representing the diversity of ecosystems.

Presentation of the three evaluated models

In this study, three process-based models were evaluated against flux measurements. The first model, GRAECO, was developed to simulate the growth and functioning of maritime pine stands (Porté, 1999; Loustau *et al.*, 2005). The second model, CASTANEA (Dufrêne *et al.*, 2005), was developed with the aim of bridging the gap between SVAT (soil-vegetation-atmosphere transfer) and growth models. CASTANEA consists of a multi-layer, process-based physiological model coupled with a carbon allocation model and a soil model. CASTANEA was first parameterized and evaluated for beech (*Fagus sylvatica* L.) and has also been adapted to other forest ecosystems (Davi *et al.*, 2006b), either dominated by deciduous species (*Quercus robur* L.), coniferous species (*Pinus pinaster* Ait., *P. sylvestris* L.) or sclerophyllous evergreen species (*Quercus ilex* L.). The third model, ORCHIDEE (Krinner *et al.*, 2005), is a dynamic global vegetation model designed to study the coupled atmosphere-biosphere system.

The three models are conceptually similar. They are all based on ecophysiological processes (e.g. photosynthesis, respiration and transpiration) and physical principles (e.g. soil water transfer and radiative budget). They describe canopy photosynthesis and transpiration, maintenance and growth respiration, seasonal development, partitioning of assimilates, ETR, soil heterotrophic respiration, water and carbon balances of the soil. Net primary productivity (NPP) is calculated as the difference between gross photosynthesis and plant respiration. The net ecosystem exchange (NEE = Reco-GPP (gross primary production), see Chapter 1) between soil-plant system and atmosphere is calculated as the difference between total respiration (soil + plants) and gross photosynthesis. The meteorological driving variables are global radiation, rainfall, wind speed, air humidity and temperature (either half-hourly or daily values). It is beyond the scope of this chapter to describe precisely all the equations implemented in the models (Krinner *et al.*, 2005; Dufrêne *et al.*, 2005) and we only focus here on the main differences between the three models. How is the vegetation represented? Which formulation is used for the computation of photosynthesis and stomata conductance? How is the carbon allocation modelled? What are the differences in the calculation of soil carbon budget? All these differences reflect the limits and uncertainties of knowledge as well as the different goals pursued by each model and the diversity of the study sites where they have been developed and tested.

Representation of the vegetation

In CASTANEA and ORCHIDEE no between-tree variation is taken into account; that is, one "average" tree is considered as representative of the whole stand. In contrast, in GRAECO, the carbon is allocated to each tree. ORCHIDEE is a generic global model that does not consider the fertility and history of each site. CASTANEA and GRAECO take into account the nitrogen (leaf nitrogen content) and water status (soil depth and soil texture) in a more or less empirical way, and the effects of forestry practices can be represented via the initialization of the stand biomass and soil carbon content.

ORCHIDEE and CASTANEA do not represent either the understorey or heterogeneous stands (it is not possible to take into account two species in the same stand). In contrast to these two models, GRAECO explicitly represents the understorey since it was developed for a homogeneous maritime pine forest (*Pinus pinaster*) with an important understorey composed mainly of grass (*Molinia coerulea* L.).

Photosynthesis and stomatal conductance

Owing to the non-linearity of the photosynthetic light response curve, the canopy in CASTANEA and ORCHIDEE is vertically sub-divided into a variable number of layers (i.e. multi-layer canopy model). Leaf gas exchange is represented using the biochemical photosynthesis model of Farquhar *et al.* (1980), coupled with the stomatal conductance model of Ball *et al.* (1987). The CO_2 demand is determined as the minimum of Rubisco carboxylation and RuBP regeneration, while CO_2 supply depends on the difference in CO_2 concentration between the air outside the leaf and the carboxylation sites. The leaf nitrogen effect on photosynthesis is modelled by assuming a linear relationship between the maximal carboxylation rate (V_{cmax}) and leaf nitrogen content per unit area, and a fixed ratio between V_{cmax} and the potential rate of electron flow (V_{jmax}). Consequently, both the leaf nitrogen content and the absorbed light drive the photosynthetic rate. Leaf nitrogen is assumed to be constant in ORCHIDEE for a given plant functional type (generic approach) and is forced in CASTANEA according to the site. In GRAECO, photosynthesis is calculated using a fitted response curve of photosynthesis and conductance to light, vapour pressure deficit, CO_2 concentration, soil water reserve and a fertility index. These response curves were fitted beforehand using simulations of MAESTRO (Wang and Jarvis, 1990), a detailed three-dimensional model simulating water and carbon fluxes in conifer canopies. The optimized parameters are V_{cmax} and V_{jmax} using a constant ratio between the two. The fitting was done "manually" in this study but gave similar results as in a more detailed work (Santaren *et al.*, 2006).

Phenology, maintenance respiration and carbon allocation

The phenology and respiration modules are similar across the three models. Phenological events (i.e. budburst, full leaf area development, full leaf maturity, start of leaf yellowing, complete canopy yellowing) depend on day-degrees and day duration. Only the parameterization changes from one model to another. Maintenance respiration is calculated assuming an exponential relationship to account for temperature dependence (Ryan, 1991). In CASTANEA, the base respiration also depends on the nitrogen content of living biomass.

The allocation schemes differ considerably across the models (Figure 9.1). In all three models, photosynthesis is calculated at the stand level. GRAECO allocates the carbon to each tree in proportion to its leaf area. For each tree, the carbon is allocated either to the above-ground biomass or to the roots according to a water stress index. ORCHIDEE and CASTANEA allocate the carbon to four compartments: reserves, fine roots, leaves and wood (divided into trunk, branches and coarse roots). In ORCHIDEE the allocation is driven by environmental stress indices, while in CASTANEA the carbon is first allocated to leaves (priority sink during leaf growth). The remaining assimilates are then allocated to other compartments using allometric rules.

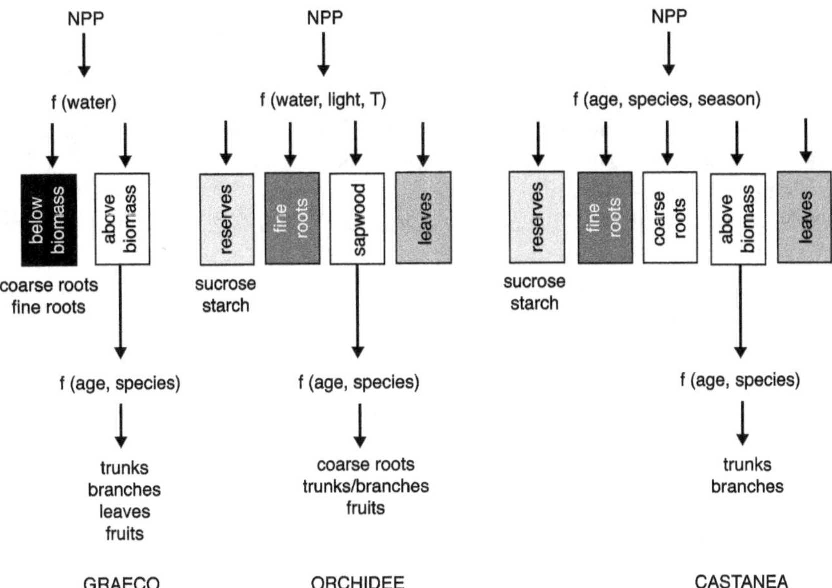

Figure 9.1. Description of allocation module in GRAECO, ORCHIDEE and CASTANEA.

Carbon cycle in the soil

In CASTANEA and ORCHIDEE, the soil organic carbon sub-model is based on the soil organic matter sub-model of CENTURY (Parton *et al.*, 1987). Soil organic carbon is divided into three major components including active, slow and passive soil carbon. The model also includes a surface microbial pool, which is associated with decomposing surface litter (mainly leaf litter). Carbon flows between these pools are controlled by decomposition rate and microbial respiration loss parameters, both of which are a function of soil texture, soil temperature and soil water content. GRAECO uses a unique soil compartment with a decomposition rate depending on soil water content, soil temperature and carbon inputs arriving at the soil.

Evaluation of the models using eddy-flux measurements

The three models were evaluated on three sites (Hesse, le Bray and Puéchabon) belonging to the CARBOEUROFLUX network. The water and carbon fluxes simulated by the model (NEE and ETR) were compared with fluxes estimated by eddy covariance. These three sites are dominated by different species (*Fagus sylvatica*, *Pinus pinaster* and *Quercus ilex*) and have contrasting climates (continental, oceanic and Mediterranean). ORCHIDEE was evaluated on all three sites whereas CASTANEA was evaluated on the *Fagus sylvatica* and *Quercus ilex* sites and GRAECO on the *Pinus pinaster* site. Two versions of ORCHIDEE were evaluated, a "basic" version and an optimized version where the photosynthetic parameters were fitted to minimize the bias between simulated and measured NEE and ETR.

Description of sites

The Hesse site is located in the east of France (48°40' N, 7°05' E, altitude 300 m) and is mainly composed of beech with a very sparse understorey. The studied plot covers 0.63 ha of a pole beech forest (30 years old in 1997) with a density of 3482 trees·ha⁻¹ and a dominant height close to 14 m. The soil type was intermediate between a luvisol and a stagnic luvisol, with a clay content ranging between 25% and 40%. The annual precipitation is 820 mm and average temperature 9.2°C. For more details see Granier *et al.* (2000) and Chapter 2 of this volume.

The Bray site is located in the south-west of France (44°43' N, 0°46' W, altitude 61 m) and is composed of homogeneous maritime pine trees seeded in 1970. The stand density is 520 trees·ha⁻¹ and the understorey consists mainly of grass (*Molinia coerulea*). The soil is a sandy podzol lying over a hard iron pan. The water table never goes deeper than 200 cm and sometimes during the winter its level may reach the soil surface (Ogée *et al.*, 2003). This limitation to water drainage can also increase the water available at the beginning of summer. The climate is temperate oceanic with an annual mean temperature of 13.5°C and 930 mm of precipitation, and is characterized by a strong seasonal contrast in water condition. For more details see Loustau *et al.* (1990) and Berbigier *et al.* (2001).

The Puéchabon site is located in the Puéchabon State Forest in the south of France (43°44' N, 3°35' E, altitude 270 m). This woodland has been managed as a coppice for centuries and the last clear-cut was performed in 1942. The vegetation is largely dominated by holm oak (*Quercus ilex*) with a sparse understorey mainly composed of a shrubby layer with *Buxus sempervirens* L., *Phyllirea latifolia* L., *Pistacia terebinthus*, L. and *Juniperus oxycedrus* L. The stand density is 8500 trees·ha⁻¹. The soil is classified as calcareous fersiallitic with high clay content. The stone and rock fraction is about 90% across the whole-soil profile. The area has a Mediterranean-type climate. Rainfall occurs during autumn and winter with about 75% falling between September and April. Mean annual precipitation is 883 mm and mean annual temperature is 13.6°C over the 1984–2002 period. For more details see Hoff *et al.* (2002).

In the Hesse and Bray sites, the windstorm of December 1999 led to the fall and consequent change of the flux tower and all the measuring instruments. This may consequently bias the conclusions concerning the inter-annual dynamics.

Seasonal variations

At Hesse, CASTANEA and ORCHIDEE correctly reproduce the variations in ETR and NEE, but with a constant positive bias for ETR (Figure 9.2, Plate 5). This bias may be due to an underestimation of measured ETR when the canopy is wet. In previous work (Davi *et al.*, 2005), in Hesse in 1997, CASTANEA simulated an annual transpiration (329 mm) close to that derived from sapflow measurements (313 mm). At the same time, however, the simulated total ETR was much higher in the model (603 mm) than the measured total (351 mm). This difference was attributed to a bias during rainy days when the eddy-covariance measurements underestimate ETR. This hypothesis was confirmed by the analysis of the annual water balance. This phenomenon probably also explains the overestimation of ETR at Hesse observed in this study, particularly in 2001 and 2002, which were rainy years.

In the sites dominated by evergreen species (*Pinus pinaster* and *Quercus ilex*), the seasonal variations of NEE are not as well simulated (see Figure 9.3, Plate 5 and Figure 9.4,

Plate 6). Several studies have demonstrated that the photosynthetic capacity of the canopy shows seasonal variation (Wilson *et al.*, 2000; Frak, 2002; Ogée *et al.*, 2003; Davi *et al.*, 2005). These variations are generally badly taken into account in the models because they do not simulate the mechanisms involved in nitrogen allocation. This study confirms that all the seasonal variations are not reproduced, particularly in evergreen species where two or three cohorts of leaves (with different photosynthetic capacities) coexist. Nevertheless, seasonal variation in respiration rates could also be behind this mismatch.

Lastly, GRAECO and CASTANEA show stronger variation than ORCHIDEE and the latter underestimates the carbon sink in ecosystems dominated by evergreen species.

Inter-annual variations

At Hesse, where we have the most complete long-term dataset, CASTANEA accurately predicts NEE from 1997 to 1999 and underestimates it from 2000 to 2003. It is the opposite for ORCHIDEE and it seems that the windstorm during the winter 1999–2000 separates the two periods.

After the storm, the measured NEE at Hesse decreased (increase of the carbon sink). Since the tower was changed for a taller one, the footprint was modified and it is difficult to know whether the change in NEE is due to a direct effect of the storm on the forest ecosystem or an indirect effect of footprint modification. Nevertheless, neither model was able to predict accurately the inter-annual dynamics before and after the storm. When tree clearing occurs, the remaining trees experience a change in light environment and a modification of nutrient availability. Light regime and soil water balance are taken into account in the models and, therefore, their effect, when tree fall occurred, can be partially reproduced even if the effects of the storm are highly spatially variable and may not be well represented by a homogeneous canopy model. However, this is not the case for nitrogen availability. The inter-annual variations of nitrogen are at best only empirically taken into account in CASTANEA by forcing the measured leaf nitrogen content.

All the models show similar patterns even if the specific model gives better results. The optimized version of ORCHIDEE should be used if one wants to limit bias. An interesting avenue to explore is to use the specific model to parameterize the optimized version of ORCHIDEE. The next model improvement appears to be the introduction of a mechanistic approach simulating nitrogen and carbon allocation between trees and inside a tree. Such an improvement would help to improve the simulation of seasonal and inter-annual dynamics and to provide a link between the fluxes and tree growth simulations.

Uncertainty estimation and model precision

In the preceding sections, simulations were compared with available measurements (mainly fluxes and growth data). These comparisons allowed the assessment of the errors made by the model for a limited number of output variables. As underlined in Knorr (2000), the identification of the relative importance of the various sources of uncertainty is necessary to improve our ability to model carbon cycling and to identify the weak points in the models. We present in this section a systematic uncertainty analysis method to predict and quantify the uncertainty (error ranges) associated with the output variables of an ecosystem functioning model. Such methodology allows us to assess the model precision (or uncertainty) before measurements are available for long-term simulations, for instance. It also allows us to be able to predict the uncertainties on all simulated variables, not only on variables where validation measurements are available. The last objective is to identify which parameter is most responsible for the uncertainties in the simulations.

The method consists of screening the input model parameters (about 200) in order to select only the sensitive parameters. Sensitive parameters are defined with respect to chosen output variables. The chosen output variables are, in this study, NEE and the above-ground woody biomass growth (AWBG). For the subsequent uncertainty analysis, only the 30 most important (i.e. sensitive) input parameters are kept. The second stage is to evaluate the uncertainty of each of these key parameters. The last stage is the uncertainty study itself: it consists of performing a series of simulations (here 1500), where the key parameters are randomly varied around their nominal value through a normal distribution whose standard deviation is given by the parameter uncertainty.

This section is divided into four parts: the site and model presentation, the sensitivity study (parameter screening), the uncertainty determination for each of the key input parameters and the results of the predicted uncertainty of the output variables.

Study site and forest process model presentation

We consider here, as an illustration, a beech stand in Fontainebleau, which was 55 years old in 1970. Thirty years of simulations were performed from 1970 until 1999. The beech stand has a summer LAI plateau ranging between 4 $m^2 \cdot m^{-2}$ and 6 $m^2 \cdot m^{-2}$ for the considered years. The forest process model used is CASTANEA, described above. For the uncertainty analysis, the model is used in its dynamic version, that is, where the maximum LAI is simulated and computed each year at budburst, as a function of the available reserves.

Sensitivity study

First, the 200 input parameters were set to their nominal value and a reference simulation was performed for the 30 years. Then a series of simulations was performed in which each parameter was individually set to ±10% of its reference value. For each parameter, the average variation coefficient for the 30 years $\Delta^{10\%}$ is computed for both NEE and AWBG.

First, the average variations $\Delta^{+10\%}$ and $\Delta^{-10\%}$ over the 30 years are computed for each parameter on each of the studied outputs Y (here NEE and AWBG):

$$\Delta^{+10\%} = \frac{\sum\limits_{a\,=\,1970}^{1999}\left(Y_a^{+10\%} - Y_a^{ref}\right)}{30} \tag{1}$$

$$\Delta^{-10\%} = \frac{\sum\limits_{a\,=\,1970}^{1999}\left(Y_a^{-10\%} - Y_a^{ref}\right)}{30} \tag{2}$$

Then the maximum variation $\Delta^{10\%}$ (in absolute value) between $\Delta^{+10\%}$ and $\Delta^{-10\%}$ is kept and divided by the average over the 30 years of the reference simulation outputs Y^{ref} to obtain a percentage variation coefficient:

$$\Delta^{10\%}(\%) = \frac{\max\left(\Delta^{-10\%},\,\Delta^{+10\%}\right)}{\dfrac{1}{30}\sum\limits_{a\,=\,1970}^{1999} Y_a^{ref}} \tag{3}$$

The results for the 41 most influential variables are given in Table 9.2. The most sensitive parameters (but not necessarily the most uncertain, and thus not necessarily the most important) are: the wood and roots construction cost; the maintenance respiration nitrogen dependency; the allocation to carbohydrate reserves; the dependency between V_{cmax} and leaf nitrogen density; the quantum yield; the leaf mass per area of sun leaves; the ratio between V_{cmax} and V_{jmax}; leaf, branch (and, somewhat less significantly, fine root) nitrogen content; the critical value of state of forcing for budburst; the storage allocation coefficient; the initial ratio between fine roots and leaf biomass; the fine root mortality; the soil extractable water; the clumping factor; and the maximum carbohydrate reserve stock.

Table 9.2. Results of the screening sensitivity study of the CASTANEA model (41 most sensitive parameters over 200). Right columns represent the average variation coefficient over the 30 years for AWBG and NEE when the input parameter is varied by 10%. The positive (resp. negative) sign indicates that an increase in the input parameter tends to increase (resp. decrease) the considered output variable (AWBG or NEE).

No.	Variable	Name	AWBG variation coefficient (%)	NEE variation coefficient (%)
1	Wood construction cost ($gC \cdot gC^{-1}$)	CR_{wood}	− 27.3	16.5
2	Nitrogen dependency for all organs ($gC \cdot gN^{-1} \cdot h^{-1}$)	MRN	− 15.1	15.5
3	Allocation coefficient to carbohydrate reserves	GSS	− 15.2	6.7
4	Dependency between $V_{cmax}{}^1$ and leaf nitrogen density ($\mu mol_{CO2} \cdot gN^{-1} \cdot s^{-1}$)	α_{Na}	13.6	− 14.6
5	Quantum yield	α	12.3	− 13
6	Leaf mass per area of sun leaves ($g \cdot m^{-2}$)	LMA_{sunmax}	− 2.2	− 11.9
7	Ratio between $VC_{max}{}^1$ and $VJ_{max}{}^2$	β	10.6	− 11.3
8	Leaf nitrogen content ($gN \cdot g^{-1}$)	$N_{mleaves}$	10	− 10.9
9	Critical value of state of forcing for budburst (K)	F_{critBB}	− 10.7	10.5
10	Initial ratio between fine roots and leaf biomass	$coef_{ROOTS}$	− 9.7	6.5
11	Fine roots construction cost ($gC \cdot gC^{-1}$)	$CR_{fine\ roots}$	− 9.1	7.3
12	Fine roots mortality ($year^{-1}$)	TMRF	− 7.8	3.8
13	Extractable water (mm)	EW	5.9	− 7.4
14	Clumping factor	Agreg	− 5.3	6.1
15	Maximum carbohydrate reserve stock ($gC \cdot m^{-2}$)	BSS_{max}	− 5.1	3.3
16	Coarse roots construction cost ($gC \cdot gC^{-1}$)	$CR_{coarse\ roots}$	0	4.3
17	Proportion of living biomass in branches	co_live_{BRAN}	− 1.4	4.1
18	Branch nitrogen content ($mg_N \cdot g_{dm}{}^{-1}$)	NBR	− 1.4	4.1
19	Slope of the Ball relationship (maximum value)	gl_{max}	− 2.9	3.9
20	Initial biomass of aerial wood ($gC \cdot m^{-2}$)	$B_{wood}init$	− 1.7	3.7
21	Ratio between coarse root and living biomass	rootshoot	− 3.5	− 0.4
22	Coefficient for LAI computing from carbohydrate reserves	$coef_{LAI}$	3.4	− 2.7
23	Leaf mass per area decrease coefficient	KLMA	2.2	3.1
24	Leaf construction cost ($gC \cdot gC^{-1}$)	CR_{leaf}	0.3	2.9
25	Root percentage in the soil (first 20 cm) (%)	P_{roots}	− 2.4	− 2.7

...

Table 9.2. (continued)

No.	Variable	Name	AWBG variation coefficient (%)	NEE variation coefficient (%)
26	Q10 temperature effect for branch respiration	$Q_{10}BR$	− 2.6	1.5
27	Fine roots nitrogen content ($mg_N \cdot g_{dm}^{-1}$)	NRF	− 0.9	2.5
28	Exponential coefficient for LAI computing from carbohydrate reserves	$coef_{LAI2}$	− 2.5	1.9
29	Lignin fraction in root structural carbon ($g \cdot g_{dm}^{-1}$)	$LIGN_{root}$	0	− 2.5
30	Proportion of living biomass in stem wood	co_live_{STEM}	0.3	2.1
31	Critical value of state of forcing for leaf fall (K)	F_{critLF}	1	− 2.1
32	Q10 temperature effect for root respiration	QDIXrac	− 1.9	0.5
33	Stem wood nitrogen content ($mg_N \cdot g_{dm}^{-1}$)	NTV	0.2	1.8
34	Slow carbon pool initial value of top soil ($gC \cdot m^{-2}$)	$slowC_{top}$	0	1.5
35	Initial carbohydrate reserve biomass ($gC \cdot m^{-2}$)	BSS_{init}	− 0.7	1.5
36	Soil maximal conductivity for evaporation ($m \cdot s^{-1}$)	gsolmax	− 0.9	1
37	Q10 temperature effect for carbohydrate reserve respiration	QDIX	− 0.4	0.9
38	Ratio between branch and wood biomass	co_ratioBR1	− 0.9	0.8
39	Depth of total soil root zone (m)	hsol	0.3	− 0.9
40	Coarse root nitrogen content ($mg_N \cdot g_{dm}^{-1}$)	NRG	0.1	0.8
41	Cuticular conductance ($mol_{H2O} \cdot m^{-2} \cdot s^{-1}$)	g0	− 0.3	0.7

1 VCmax: maximum carboxylation rate.

2 VJmax: maximum electron transport rate.

Parameter uncertainty

The uncertainty of a parameter is defined as the standard deviation of this parameter, σ. This means that the actual value of the parameter has a probability of 0.68 to be comprised between its mean value and +/- σ. This allows us to define the output error as the standard deviation on the output variables. This uncertainty is composed of the variability of the parameter and an uncertainty on the measurement of the parameter. The determination of the uncertainty σ of each parameter is a difficult task since measurement errors are not always available. Table 9.3 sums up the estimated uncertainty expressed as a percentage for each parameter. When available, the reference for the parameter uncertainty has been given in Table 9.3 Assuming linearity of the parameter sensitivity around the 10% shift and combining it with the parameter uncertainty, the effect on the model of each parameter uncertainty is obtained, that is, its influence:

$$E(\%) = U(\%) \times \left(\frac{\Delta^{10\%}}{10\%} \right) \qquad (4)$$

where $\dfrac{\Delta^{10\%}}{10\%}$ is the variation coefficient of the studied parameter for a 1% variation (i.e. the variation of the output for a 1% variation of the parameter), U (column 4 in Table 9.2) is the uncertainty of this parameter (i.e. its expected variation in %), and E (last column in Table 9.3) is the error one expects due both to U and to model sensitivity to this parameter. $\Delta^{10\%}$ was chosen as the absolute maximum between $\Delta^{10\%}$ for AWBG and $\Delta^{10\%}$ for NEE (column 3 in Table 9.3, obtained from columns 4 and 5 in Table 9.2, see also Equation 3). Finally we obtain a sorted list of the most influential parameters, that is, the parameters that will be considered in the random simulations with a noise given by the uncertainty column in Table 9.3. This sorted list differs from the sensitivity list given in the Table 9.2. Some parameters are very sensitive in the model but their uncertainty is low, so that their total effect E is moderate (e.g. wood construction cost, dependency between V_{cmax} and leaf nitrogen density). Conversely, some parameters are not very sensitive in the model but their uncertainty is high, so that their total effect E is moderate or high (e.g. allocation coefficient to carbohydrate reserves, maximum carbohydrate reserve stock).

Table 9.3. First 29 most influential parameters of the CASTANEA model, given the parameter uncertainty (column 4) and model sensitivity (column 3), sorted in decreasing order according to their influence (total error, last column and Equation 1).

Variable	Name	Max (AWBG or NEE) absolute variation coefficient $\Delta_{10\%}$	Parameter uncertainty (U) and reference	Total error (E)
Allocation coefficient to carbohydrate reserves	GSS	15.2%	25% (Barbaroux, pers. comm.)	38.0%
Initial ratio between fine roots and leaf biomass	$coef_{ROOTS}$	9.7%	17% (Davi, pers. comm.)	16.4%
Nitrogen dependency for all organs ($gC \cdot gN^{-1} \cdot h^{-1}$)	MRN	15.5%	10% (Ryan, 1991)	15.5%
Wood construction cost ($gC \cdot gC^{-1}$)	CR_{wood}	27.3%	5% (Damesin *et al.*, 2002)	13.6%
Leaf nitrogen content ($gN \cdot g^{-1}$)	$N_{mleaves}$	10.9%	11% (unpublished)	12.0%
Extractable water (mm)	EW	7.4%	15% (Granier *et al.*, 2000)	11.1%
Leaf mass per area of sun leaves ($g \cdot m^{-2}$)	LMA_{sunmax}	11.9%	8.8% (unpublished)	10.4%
Quantum yield	α	13%	8% (unpublished)	10.4%
Maximum carbohydrate reserve stock ($gC \cdot m^{-2}$)	BSS_{max}	5.1%	20% (Barbaroux, pers. comm.)	10.1%
Fine roots construction cost ($gC \cdot gC^{-1}$)	$CR_{fine\ roots}$	9.1%	10% (unpublished)	9.1%
Dependency between V_{cmax} and leaf nitrogen density ($\mu mol_{CO2} \cdot gN^{-1} \cdot s^{-1}$)	α_{Na}	14.6%	6.1% (Liozon *et al.*, 2000)	8.9%
Clumping factor	Agreg	6.1%	14% (Soudani, pers. comm.)	8.6%
Fine roots mortality (year^{-1})	TMRF	7.8%	10% (Davi, pers. comm.)	7.8%
Initial biomass of aerial wood ($gC \cdot m^{-2}$)	$B_{wood}init$	3.7%	15% (Bouriaud, 2003)	5.5%
Coefficient for LAI computing from carbohydrate reserves	$coef_{LAI}$	3.4%	15% (unpublished)	5.1%
				...

Table 9.3. (continued)

Variable	Name	Max (AWBG or NEE) absolute variation coefficient $\Delta_{10\%}$	Parameter uncertainty (U) and reference	Total error (E)
Critical value of state of forcing for leaf fall (K)	F_{critLF}	2.1%	22% (Bouriaud, 2003)	4.5%
Critical value of state of forcing for budburst (K)	F_{critBB}	10.7%	4.1% (Bouriaud, 2003)	4.4%
Proportion of living biomass in branches	co_live_{BRAN}	4.1%	10% (Davi, pers. comm.)	4.1%
Branch nitrogen content ($mg_N \cdot g_{dm}^{-1}$)	NBR	4.1%	10% (unpublished)	4.1%
Root percentage in the soil (first 30 cm) (%)	P_{roots}	2.7%	15% (unpublished)	4.0%
Slope of the Ball relationship (maximum value)	gl_{max}	3.9%	10% (Davi, pers. comm.)	3.9%
Exponential coefficient for LAI computing from carbohydrate reserves	$coef_{LAI2}$	2.5%	15% (unpublished)	3.7%
Ratio between VC_{max} and VJ_{max}	β	11.3%	3% (Liozon *et al.*, 2000)	3.4%
Initial carbohydrate reserve biomass ($gC \cdot m^{-2}$)	BSS_{init}	1.5%	20% (Barbaroux, pers. comm.)	2.9%
Fine roots nitrogen content ($mg_N \cdot g_{dm}^{-1}$)	NRF	2.5%	10.6% (unpublished)	2.7%
Q10 temperature effect for branch respiration	$Q_{10}BR$	2.6%	10% (Damesin, pers. comm.)	2.6%
Lignin fraction in root structural carbon ($g \cdot g_{dm}^{-1}$)	$LIGN_{root}$	2.5%	10% (unpublished)	2.5%
Coarse roots construction cost ($gC \cdot gC^{-1}$)	$CR_{coarse\ roots}$	4.3%	5% (unpublished)	2.2%
Proportion of living biomass in stem wood	co_live_{ST}	2.1%	10% (Davi, pers. comm.)	2.1%

Results: uncertainty prediction for the output variables

In order to predict the uncertainty of AWBG and NEE, 1500 series of 30 years simulations were performed. In these simulations, the 29 most influential parameters (total error > 2%) given in Table 9.3 are affected by a random noise following a Gaussian distribution whose standard deviation is given by the parameter uncertainty U (column 4 in Table 9.3).

The results of the noisy simulations are given in Figures 9.5–9.7 (respectively AWBG, NEE and simulated LAI). The uncertainty is given as error bars (standard deviation on the 1500 simulations). For comparison, the reference noise-free simulation is plotted on the figures. The reference simulation sometimes significantly differs from the mean of the 1500 simulations, indicating that a single simulation with average parameters is not equivalent to the average of simulations taking into account the variability or uncertainty of the parameter. The available growth measurements are also plotted in Figure 9.5.

The percentage uncertainty values are computed for each year i as follows:

$$CV_i = 100 \frac{\sigma_i}{m_i} \tag{5}$$

where CV_i (gC·m^{-2}·year^{-1}) is the coefficient of variance expressed as a percentage, σ_i (gC·m^{-2}·year^{-1}) the standard deviation of AWBG or NEE over the 1500 runs and m_i the mean of AWBG or NEE over the 1500 runs.

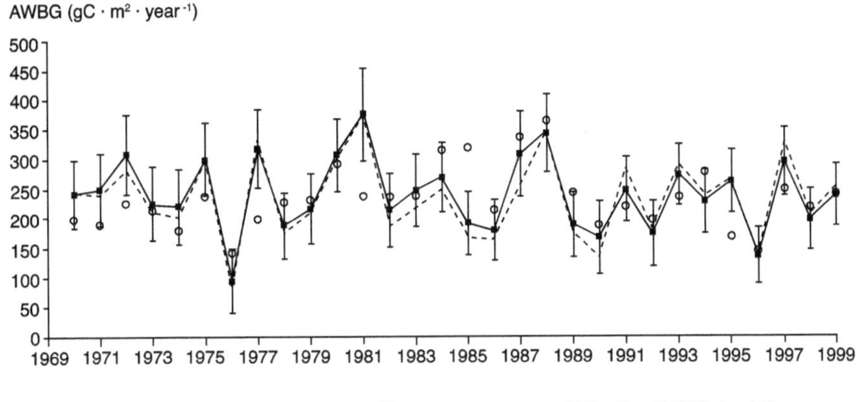

Figure 9.5. Above-ground woody biomass growth (AWBG) simulated by the CASTANEA model at Fontainebleau from 1970 to 1999. Uncertainties due to parameter uncertainty are shown with error bars. Open circles indicate measured values obtained from tree-ring data and the dotted line shows the original noise-free simulation.

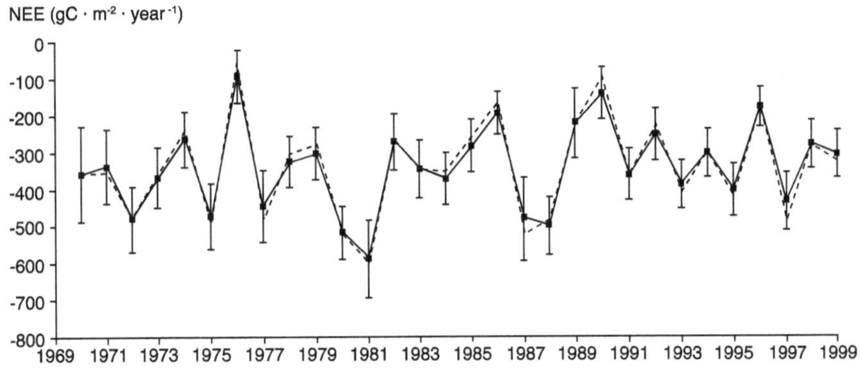

Figure 9.6. Net ecosystem exchange (NEE) simulated by the CASTANEA model at Fontainebleau from 1970 to 1999. Uncertainties due to parameter uncertainty are shown with error bars. The dotted line shows the original noise-free simulation.

The year to year uncertainty variations for AWBG and NEE are plotted in Figures 9.8 and 9.9, both in raw and percentage values. The statistics of the uncertainties in raw and percentage values are given in Tables 9.4 and 9.5 (mean values, minimum and maximum).

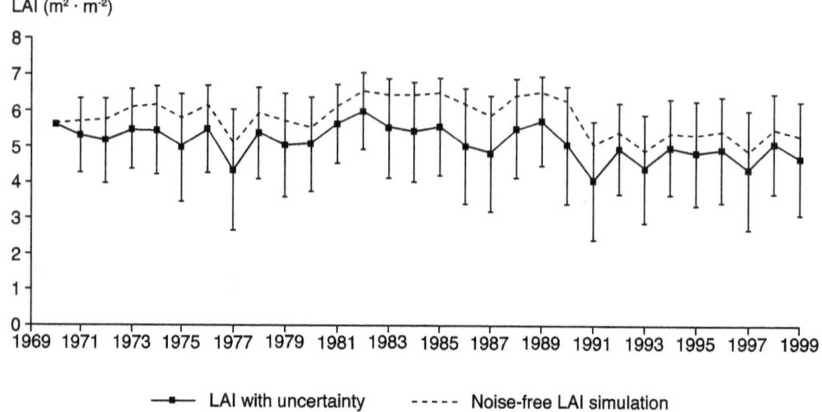

Figure 9.7. Leaf area index (LAI) simulated by the CASTANEA model at Fontainebleau from 1970 to 1999. Uncertainties due to parameter uncertainty are shown with error bars. The dotted line shows the original noise-free simulation.

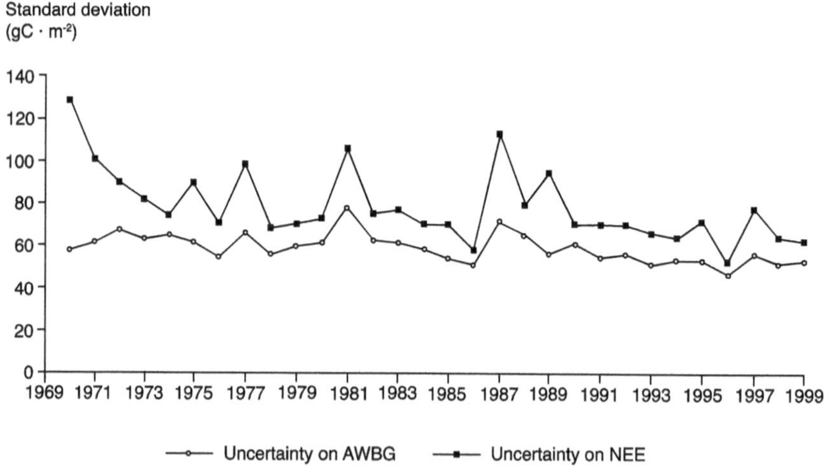

Figure 9.8. Uncertainties in above-ground woody biomass growth (AWBG) and NEE estimated by the CASTANEA model at Fontainebleau for the period 1970–99. Uncertainties are calculated as standard deviations of the 1500 random simulations.

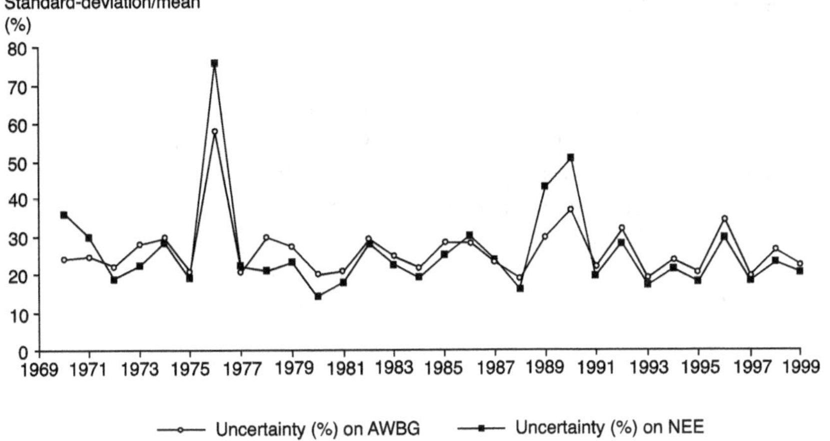

Figure 9.9. Uncertainties in above-ground woody biomass growth (AWBG) and NEE expressed as percentages derived from raw uncertainties given in Figure 9.8.

Table 9.4. Statistics of the uncertainties presented in Figure 9.8 (in raw values).

	AWBG uncertainty $(gC·m^{-2})$	NEE uncertainty $(gC·m^{-2})$	LAI uncertainty $(m^2·m^{-2})$
Minimum	47	53	1.0
Average	59	79	1.3
Maximum	78	128	1.7

Table 9.5. Statistics of the uncertainties presented in Figure 9.9 (in percentages).

	AWBG uncertainty (%)	NEE uncertainty (%)	LAI uncertainty (%)
Minimum	19	14	18
Average	26	26	27
Maximum	58	76	41

Effect of noise on the simulations

The first remark is linked to the difference between noise-free simulations and the average of the noisy simulations (solid and dotted lines in Figures 9.5–9.7): the difference is low on average over the 30 years for AWBG growth and NEE (respectively $- 6$ gC·m^{-2}·year^{1} and 0 gC·m^{-2}·year^{1}), with a moderate standard deviation (respectively, 18 gC·m^{-2}·year^{1} and 24 gC·m^{-2}·year^{1}) (Figures 9.5 and 9.6). On the other hand, for LAI, the average bias between the noise-free and noisy simulations is 0.7 m^2·m^{-2} and the standard deviation 0.2 m^2·m^{-2} (Figure 9.7).

Uncertainty variations

In raw values the uncertainties are very stable for AWBG (around 60 $gC \cdot m^{-2} \cdot year^{-1}$) (Figure 9.8 and Table 9.4). The raw uncertainty is more variable for NEE (around 79 $gC \cdot m^{-2} \cdot year^{-1}$) (Figure 9.8 and Table 9.4), while the percentage uncertainty is more stable (26% in average for NEE, as well as for AWBG) (Figure 9.9 and Table 9.5). Note the uncertainty peaks during dry periods in 1976 and 1990, when NEE raw values are very low. The high uncertainty values during the first 2 years of simulations for NEE (Figure 9.8) are due to the uncertainty of the initial aerial wood biomass, which has a relatively low influence on average during the 30 years (Table 9.3), but is very important during the first years (Figure 9.10).

In order to assess the quality of the uncertainty prediction, we compared the actual error, that is, the average standard deviation over the 30 years defined as the average of the standard deviations of the differences (average simulation-measurement), with the simulated error, that is, the average simulated standard deviation as given in Table 9.4 (59 $gC \cdot m^{-2}$). The actual average standard deviation of the error between the average simulation and the measurements is 56 $gC \cdot m^{-2}$. Since the errors are in agreement with the observed model-measurements errors, one may conclude that the uncertainties are correctly estimated by the model and the parameterization given in Table 9.3 (column 4). Note, however, that the differences between AWBG measurements and simulations are assumed to come only from input parameter uncertainties, which is partly wrong since they also come from the process simulations, from processes that are not simulated and from errors on AWBG or NEE measurements. If taken into account, this would tend to increase the simulated uncertainty.

Inter-annual variation of parameter influence

The inter-annual variation of parameter influence was investigated by changing the initial value (first year) of a parameter, and analysing the effect of this change on the 29 following years. A 15% increase was applied to the four chosen parameters: initial above-ground woody biomass, the allocation coefficient to carbohydrate reserves (GSS), the leaf nitrogen content ($N_{mleaves}$) and the ratio between fine root and leaf biomass ($coef_{ROOTS}$).

The results show that the inter-annual variation of the parameter influence is very variable, either with a clear tendency (a decrease for the aerial wood biomass influence) (Figure 9.10), an important inter-annual variability (GSS and $N_{mleaves}$) (Figures 9.11 and 9.12), or a quasi-constant influence ($coef_{ROOTS}$) (Figure 9.13).

The effect of the initial aerial wood biomass decreases as the years go by because of a negative feedback on NEE: the higher the aerial wood biomass, the higher the maintenance respiration and the lower the NEE (in absolute value), and thus less available carbon for tree growth so that the biomass increases less. Conversely, a lower initial wood biomass will lead to a higher NEE (in absolute value) and more available carbon for growth, and finally a higher biomass. Note, however, that even with this negative feedback, the initial aerial wood biomass still has noticeable effects (variability) throughout the 30 years. For parameters that influence photosynthesis such as $N_{mleaves}$ (Figure 9.12), the influence is variable and more important in the "extreme" years (dry years like 1976 and 1990, or "good" years like 1981 and 1987). When all parameters are taken into account and varied, the effect on NEE (the uncertainty) (Figure 9.8) is smoothed out, but the outstanding years still clearly appear (low uncertainties in 1976 and 1990, and high uncertainties in 1981 and 1987).

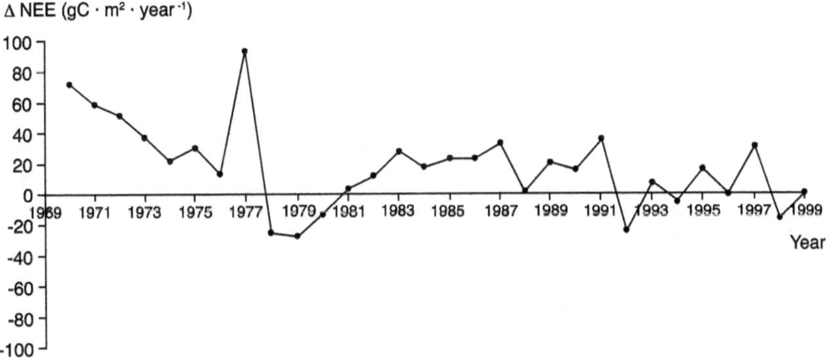

Figure 9.10. Effect on NEE of a 15% increase in initial above-ground woody biomass for a simulation by the CASTANEA model at Fontainebleau for the period 1970–99.

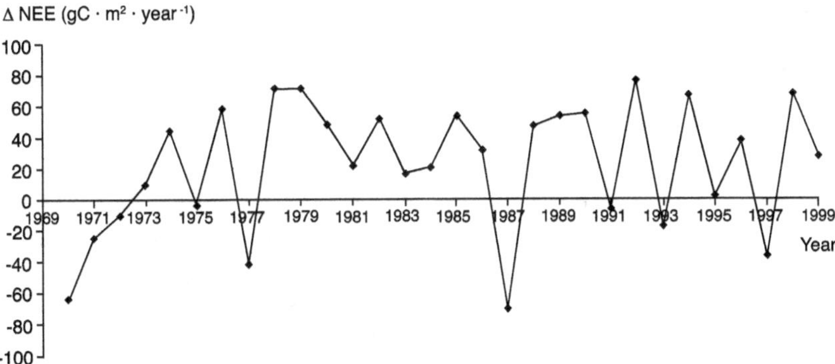

Figure 9.11. Effect on NEE of a 15% increase in the allocation coefficient to carbohydrate reserves (GSS) for a simulation by the CASTANEA model at Fontainebleau for the period 1970–99.

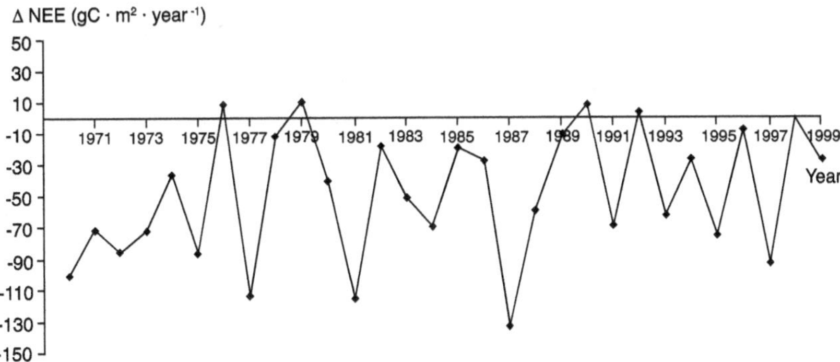

Figure 9.12. Effect on NEE of a 15% increase in leaf nitrogen content ($N_{mleaves}$) for a simulation by the CASTANEA model at Fontainebleau for the period 1970–99.

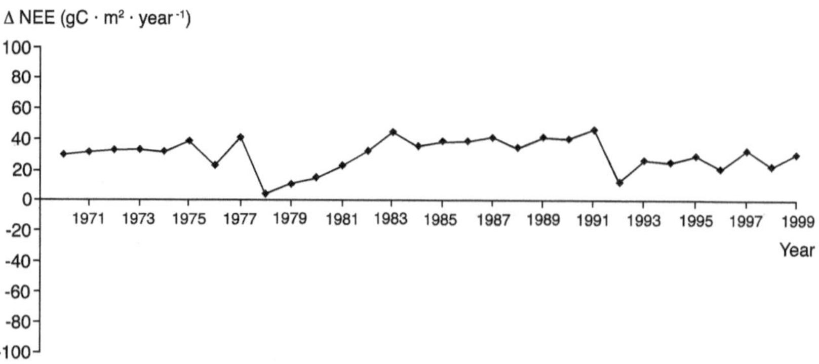

Figure 9.13. Effect on NEE of a 15% increase in the initial ratio of fine roots to leaf biomass (coef$_{ROOTS}$) for a simulation by the CASTANEA model at Fontainebleau for the period 1970–99.

Conclusions

Developing and testing process models simulating carbon and water fluxes at various scales allows the gap between the measurements (see Chapter 2) and the forecast of climate effects (see chapter 10) to be bridged. In this chapter, we presented a review of the various tools for predicting the carbon cycle and the main ways to improve them. We presented an evaluation of three models and an uncertainty analysis for one of them. The evaluation of the models showed that the models give good predictions of carbon and water fluxes if compared with the measurements described in Chapter 2. Moreover, when the global model (ORCHIDEE) is calibrated on each ecosystem, it gives good results. On the contrary, the simulation of above-ground wood biomass is not so good, since the allocation processes are less well understood than the photosynthesis or respiration processes.

The first objective of the uncertainty study was to show the feasibility and applicability of the uncertainty estimation method: we were able to estimate the year to year variations of the simulation uncertainties using ensemble simulations and uncertainties on the key parameters, identified as the most influential parameters on the studied outputs, AWBG and NEE. The uncertainties obtained (around 60 gC·m^{-2}·year^{-1} for wood growth and 80 gC·m^{-2}·year^{-1} for NEE, that is, an uncertainty of 26% on average) give an indication of our capacity to simulate stand carbon fluxes. Concerning NEE, the uncertainty magnitude of 80 gC·m^{-2}·year^{-1} is to be compared with the 300 gC·m^{-2}·year^{-1} NEE inter-annual variations: the ratio is small enough to give confidence in the inter-annual variations of the simulated NEE. The picture is less favourable for AWBG since the AWBG range of variation is about 150 gC·m^{-2}·year^{-1}, while the uncertainty reaches 60 gC·m^{-2}·year^{-1}. This is not surprising since the carbon allocation scheme is the most innovative part of the model and for now, the less robust part. In order to decrease this uncertainty, the identification of sensitive parameters shows where efforts have to be made: allocation, growth and maintenance respiration, nitrogen cycle and soil water stress. Note also that in this study the noise only concerned the parameters and not the processes. This prevents us from identifying the processes that generate most uncertainty. This supplementary step would be important in linking simulated uncertainties and observed errors.

To conclude, this work confirms the necessity of improving the link between models based on ecophysiology and growth models. A recent study seems to point out that taking into account the variations of allocation coefficients between sites and years (Davi, 2005) using functional rules strongly improves the simulation of growth.

References

Aber J.D., Federer A., 1992. A generalized, lumped-parameter model of photosynthesis, evapotranspiration and net primary production in temperate and boreal forest ecosystems. *Oecologia*, 92, 463–474.

Aber J.D., Reich P.B., Goulden M.L., 1996. Extrapolating leaf CO_2 exchange to the canopy: A generalized model of forest photosynthesis compared with measurements by eddy correlation. *Oecologia*, 106, 257–265.

Amthor J.S, 1994. Scaling CO_2-photosynthesis relationships from the leaf to the canopy. *Photosynthesis Research*, 39, 321–350.

Arain M.A., Burke E.J., Yang Z.L., Shuttleworth W.J., 1999. Implementing surface parameter aggregation rules in CCM3 global climate model: regional responses at the land surface. *Hydrology and Earth System Sciences*, 3 (4), 463–476.

Baldocchi D.D., 1992. A lagrangian random walk model for simulating water vapour, CO_2, and sensible heat flux densities and scalar profiles over and within a soybean canopy. *Boundary Layer Meteorology*, 61, 113–144.

Baldocchi D.D., Falge E., Gu L., Olson R., Hollinger D., Running S., Anthoni P., Bernhofer C., Davis K., Evans R., Fuentes J., Goldstein A., Katul G., Law B., Lee X., Mahli Y., Meyers T., Munger W., Oechel W., Paw U.K.T., Pilegaard K., Schmid H.P., Valentini R., Verma S., Vesala T., Wilson K., Wofsy S., 2001. FLUXNET: a new tool to study the temporal and spatial variability of ecosystem-scale carbon dioxide, water vapor and energy flux densities. *Bulletin of the American Meteorological Society*, 82 (11), 2415–2434.

Baldocchi D.D., Harley P.C. 1995. Scaling carbon dioxide and water vapour exchange from leaf to canopy in a deciduous forest. II. Model testing and application. *Plant Cell and Environment*, 18 (10), 1157–1173.

Ball J.T., Woodrow I.E., Berry J.A., 1987. A model predicting stomatal conductance and its contribution to the control of photosynthesis under different environmental conditions. *In: Progress in Photosynthesis Research,* Volume 4 (J. Biggins, ed.). M. Nijhoff Publishers, Dordrecht, pp. 221–224.

Bartelink H.H., 2000. A growth model for mixed forest stands. *Forest Ecology and Management*, 134, 29–43.

Becker M., Nieminen T.M., Gérémia F., 1994. Short-term variations and long term changes in oak productivity in northeastern France. The role of climate and atmospheric CO_2. *Annales des Sciences Forestières*, 51, 477–492.

Berbigier P., Bonnefond J.-M., Mellmann O., 2001. CO_2 and water vapour fluxes for 2 years above Euroflux forest site. *Agricultural and Forest Meteorology*, 108, 183–197.

Bossel H., 1996. TREEDYN3 forest simulation model. *Ecological Modelling*, 90, 187–227.

Boulet G., Chechbouni A., Braud I., Vauclin M., Haverkamp R., Zammint C., 2000. A simple water and energy balance model designed for regionalization and remote sensing data utilization. *Agricultural Forest and Meteorology*, 105, 117–132.

Bouriaud O., 1999. Analyse de la variabilité spatiale de la structure et de l'indice foliaire de quatre parcelles de hêtre en forêt de Hesse (57). Conséquences sur la croissance radiale. PhD Thesis, université de Nancy 1, 221 p.

Braud I., Dantas-Antonino A.C., Vauclin M., Thony J.L., Ruelle P., 1995. A simple soil-plant atmosphere transfer model (SiSPAT) development and field verification. *Journal of Hydrology*, 166, 213–250.

Caldwell M.M., Meister H.P., Tenhunen J.D., Lange O.L., 1986. Canopy structure, light micro-climate and leaf gas exchange of *Quercus coccifera* L. in a Portuguese macchia: measurements in different canopy layers and simulations with a canopy model. *Trees*, 1, 25–41.

Collatz G.J., Ball J.T., Grivet C., Berry J.A., 1991. Physiological and environmental regulation of stomatal conductance, photosynthesis and transpiration: a model that includes a laminar boundary layer. *Agricultural and Forest Meteorology*, 54 (2–4), 107–136.

Coops N.C., Waring R.H., 2001. Assessing forest growth across southwestern Oregon under a range of current and future global change scenarios using a process model, 3-PG. *Global Change Biology*, 7, 15–29.

Cramer W., Bondeau A., Woodward I., Prentice I.C., Betts R.A., Brovkin V., Cox P.M., Fisher V., Foley J., Friend A.D., Kucharik C., Lomas M.R., Ramankutty N., Sitch S., Smith B., White A., Young-Molling C., 2001. Global response of terrestrial ecosystem structure and function to CO_2 and climate change: results from six dynamic global vegetation models. *Global Change Biology*, 7, 357–373.

Cramer W., Kicklighter D.W., Bondeau A., Churkina G., Nemry B., Ruimy A., Schloss A.L., and the Participants of the Postdam NPP Model Intercomparison, 1999. Comparing global models of terrestrial net primary productivity (NPP): overview and key results. *Global Change Biology*, 5, 1–15.

Damesin C., Ceschia E., Goff N.l., Ottorini J.M., Dufrene E., 2002. Stem and branch respiration of beech: from tree measurements to estimations at the stand level. *New Phytologist*, 153 (1), 159–172.

Davi H., Bouriaud O., Dufrêne E., Soudani K., Pontailler J.-Y., Le Maire G., François C., Bréda N., Granier A., Le Dantec V., 2006a. Effect of aggregating spatial parameters on modelling forest carbon and water fluxes. *Agricultural and Forest Meteorology*, 139 (3/4), 269–287.

Davi H., Dufrêne E., François C., Le Maire G., Loustau D., Bosc A., Rambal S., Granier A., Moors E., 2006b. Sensitivity of water and carbon fluxes to climate changes from 1960 to 2100 in European forest ecosystems. *Agricultural and Forest Meteorology*, 141 (1), 35–56.

Davi H., Dufrêne E., Granier A., Le Dantec V., Barbaroux B., François C., Bréda N., Montpied P., 2005. Modelling carbon and water cycles in a beech forest. Part II: Validation for each individual processes from organ to stand scale. *Ecological Modelling*, 185 (2–4), 387–405.

De Pury D.G.G., Farquhar G.D., 1997. Simple scaling of photosynthesis from leaves to canopies without the errors of big-leaf models. *Plant, Cell and Environment*, 20, 537–557.

Dhôte J.-F., 1990. Modèles de la dynamique des peuplements forestiers : articulation entre les niveaux de l'arbre et du peuplement. Application à la sylviculture des hêtraies. Thèse de l'université Claude Bernard Lyon I, 240 p.

Dhôte J.-F., 1991. Modélisation de la croissance des peuplements réguliers de hêtre: dynamique des hiérarchies sociales et facteurs de production. *Annales des Sciences Forestières*, 48, 389–416.

Dhôte J.-F., Hervé J.-C., 2000. Changements de productivité dans quatre forêts de chêne sessiles depuis 1930 : une approche au niveau du peuplement. *Annales des Sciences Forestières*, 57, 651-680.

Dufrêne E., Davi H., Francois C., le Maire G., Le Dantec V., Granier A., 2005. Modelling carbon and water cycles in a beech forest Part I: Model description and uncertainty analysis on modelled NEE. *Ecological Modelling*, 185 (2–4), 407–436.

Eagleson P.S., 1978. Climate, soil and vegetation. 3. A simplified model of soil moisture movement in the liquid phase. *Water Resources Research*, 14 (5), 713–721.

Ehman J.L., Schmid H.P., Grimmond C.S.B., Randolph J.C., Hanson P.J., Wayson C.A., Cropley F.D., 2002. An initial intercomparison of micrometeorological and ecological inventory estimates of carbon exchange in a mid-latitude deciduous forest. *Global Change Biology*, 8, 575–589.

Farquhar G.D., von Caemmerer S., Berry J.A., 1980. A biochemical model of photosynthetic CO_2 assimilation in leaves of C_3 species. *Planta*, 149, 78–80.

Frak E., Le Roux X., Millard P., Adam B., Dreyer E., Escuit C., Sinoquet H., Vandame M., Varlet-Grancher C., 2002. Spatial distribution of leaf nitrogen and photosynthetic capacity within the foliage of an individual tree: disentangling the effects of local light quality, leaf irradiance, and transpiration rate. *Journal of Experimental Botany*, 53, 2207–2216.

Granier A., Ceschia E., Damesin C., Dufrêne E., Epron D., Gross P., Lebaube S., Le Dantec V., Le Goff N., Lemoine D., Lucot E., Ottorini J.M., Pontailler J.-Y., Saugier B., 2000. The carbon balance of a young beech forest. *Functional Ecology*, 14, 312–325.

Grant R.F., Black T.A, den Hartog G., Berry J.A., Neumann H.H., Blanken P.D., Yang P.C., Russel C., Nalde I.A., 1999. Diurnal and annual exchanges of mass and energy between an aspen hazelnut forest and the atmosphere: testing the mathematical model Ecosys with data from the BOREAS experiment. *Journal of Geophysical Research*, 104, 27699–27717.

Grant R.F., Nalder I.A., 2000. Climate change effects on net carbon exchange of a boreal aspen-hazelnut forest: estimates from the ecosystem model ECOSYS. *Global Change Biology*, 6, 183–200.

Häger C., Wurth G., Kohlmaier G.H., 1999. Biomass of forest stands under climatic change: a German case study with the Frankfurt biosphere model (FBM). *Tellus B*, 51, 385–401.

Hoff C., Rambal S., Joffre R., 2002. Simulating carbon and water flows and growth in a Mediterranean evergreen *Quercus ilex* coppice using the FOREST-BGC model. *Forest Ecology and Management*, 164, 121–136.

Hoffmann F., 1995. FAGUS, a model for growth and development of beech. *Ecological Modelling*, 83, 327–348.

Jarvis P.G., 1976. The interpretation of the variations in leaf water potential and stomatal conductance found in canopies in the field. *Philosophical Transactions of the Royal Society London, Series B*, 273, 593–610.

Johnsen K., Samuelson L., Teskey R., McNulty S., Fox T., 2001. Process models as tools in forestry research and management. *Forest Science*, 47 (1), 2–8.

Joos F., Prentice I.C., House J.I. 2002. Growth enhancement due to global atmospheric change as predicted by terrestrial ecosystem models: consistent with US forest inventory data. *Global Change Biology*, 8, 299–303.

Kicklighter D.W., Bruno M., Donges S., Esser G., Heimann M., Helfrich .J, Ift F., Joos F., Kaduk J., Kohlmaier G.H., Mcguire A.D., Melillo J.M., Meyer R., Iii B.M., Nadler A., Prentice I.C., Sauf W, Schloss A.L., Sitch S., Wittenberg U., Wurth G., 1999. A first-order analysis of the potential role of CO_2 fertilization to affect the global carbon budget: a comparison of four terrestrial biosphere models. *Tellus B*, 51, 343–366.

Kicklighter D.W., Melillo J.M., Peterjohn W.T., Rastetter E.B., McGuire A.D., Steudler P.A., Aber J.D., 1994. Aspects of spatial and temporal aggregation in estimating regional carbon dioxide fluxes from temperate forest soils. *Journal of Geophysical Research*, 99, 1303–1315.

Kimball J.S., Running S.W., Saatchi S.S., 1999. Sensitivity of boreal forest regional water flux and net primary productivity simulations to sub grid scale land cover complexity. *Journal of Geophysical Research,* 104, 27789–27801.

Kirschbaum M.U.F., 1999. Modelling forest growth and carbon storage in response to increasing CO_2 and temperature. *Tellus B*, 51, 871–888.

Knorr W., 2000. Annual and interannual CO_2 exchanges of the terrestrial biosphere: process-based simulations and uncertainties. *Global Ecology and Biogeography*, 9, 225–252.

Korol R.L., Running S.W., Milner K.S., Hunt E.R., 1991. Testing a mechanistic carbon balance model against observed tree growth. *Canadian Journal of Forest Research,* 21, 1098–1105.

Krinner G., Viovy N., de Noblet-Ducoudré N., Ogée J., Polcher J., Friedlingstein P., Ciais P., Sitch S., Prentice I.C., 2005. A dynamic global vegetation model for studies of the coupled atmosphere-biosphere system. *Global Biogeochemical Cycles*, 19, article 1015.

Landsberg J.J., Waring R.H., 1997. A generalized model of forest productivity using simplified concepts of radiation-use efficiency, carbon balance and partitioning. *Forest Ecology and Management*, 95, 209–228.

Law B.E., Waring R.H., Anthoni P.M., Aber J.D., 2000a. Measurements of gross and net ecosystem productivity and water vapour exchange of a *Pinus ponderosa* ecosystem, and an evaluation of two generalized models. *Global Change Biology*, 6 (2), 155–168.

Law B.E., Williams M., Anthoni P.M., Baldocchi D.D., Unsworth M.H. 2000b. Measuring and modelling seasonal variation of carbon dioxide and water vapour exchange of a *Pinus ponderosa* forest subject to soil water deficit. *Global Change Biology*, 6 (6), 613–630.

Le Maire G., Davi H., Soudani K., François C., Le Dantec V., Dufrêne E., 2005. Modelling annual production and carbon dioxide fluxes of a large managed temperate forest using forest inventories, satellite data and field measurements. *Tree Physiology*, 25, 859–872.

Leuning R., Kelliher F.M., de Pury D.G.G., Schulze E.D., 1995. Leaf nitrogen, photosynthesis, conductance and transpiration: scaling from leaves to canopy. *Plant Cell and Environment*, 18 (10), 1183–1200.

Liozon R., Badeck F.W., Genty B., Meyer S., Saugier B., 2000. Leaf photosynthetic characteristics of beech (*Fagus sylvatica*) saplings during three years of exposure to elevated CO_2 concentration. *Tree Physiology*, 20 (4), 239–247.

Loustau D., Bosc A., Colin A., Ogée J., Davi H., François C., Dufrêne E., Déqué M., Cloppet E., Arrouays D., Le Bas C., Saby N., Pignard G., Hamza N., Granier A., Bréda N., Ciais P., Viovy N., Delage F., 2005. Modelling climate change effects on the potential production of French plains forests at the sub-regional level. *Tree Physiology*, 25 (7), 813–823.

Loustau D., Granier A., El-Hadj-Moussa F., 1990. Évolution saisonnière du flux de sève dans un peuplement de pins maritimes. *Annales des Sciences Forestières*, 47 (6), 599–618.

Mäkelä A., 1988. Performance analysis of a process-based stand growth model using Monte Carlo techniques. *Scandinavian Journal of Forest Research,* 3, 315–331.

Mäkelä A., Sievänen R., Lindner M., Lasch P., 2000. Application of volume growth and survival graphs in the evaluation of four process-based forest growth models. *Tree Physiology*, 20, 347–355.

McMurtrie R.E., Rook D.A., Kelliher F.M., 1990. Modelling the yield of *Pinus radiata* on a site limited by water and nitrogen. *Forest Ecology and Management*, 30, 381–413.

Minkkinen K., Korhonen R., Savolainen I., Laine J., 2002. Carbon balance and radiative forcing of Finnish peatlands 1900–2100 – the impact of forestry drainage. *Global Change Biology*, 8, 785–799.

Mohren G.M.J., 1987. Simulation of forest growth, applied to Douglas fir stands in the Netherlands. PhD Thesis, University of Wageningen, 184 p.

Monteith J.L., 1965. Evaporation and environment. *In: The state and movement of water in living organisms, 19th symposium of the society for experimental biology.* Cambridge University Press, New York, pp. 205–233.

Nabuurs G.J., Schelhaas M.J., Mohren G.M.J., Field C.B., 2003. Temporal evolution of the European forest sector carbon sink from 1950 to 1999. *Global Change Biology*, 9, 152–160.

Ogée J., Brunet Y., Loustau D., Berbigier P., Delzon S., 2003. MuSICA, a CO_2, water and energy multilayer, multileaf pine forest model: evaluation from hourly to yearly time scales and sensitivity analysis. *Global Change Biology*, 9, 697–717.

Parton W.J., Schimel D.S., Cole C.V., Ojima D.S., 1987. Analysis of factors controlling soil organic matters levels in great plains grasslands. *Soil Science Society of America Journal*, 51, 1173–1179.

Porté A., 1999. Modélisation des effets du bilan hydrique sur la production primaire et la croissance d'un couvert de pin maritime (*Pinus pinaster* Ait.) en lande humide. Thèse de doctorat, université de Paris XI, Orsay, 160 p.

Running S.W., Baldocchi D.D., Turner D., Gower S.T., Bakwin O., Hibbard K., 1999. A global terrestrial monitoring network, scaling tower fluxes with ecosystem modelling and EOS satellite data. *Remote Sensing and Environment*, 70, 108–127.

Running S.W., Coughlan J.C., 1988. A general model of forest ecosystem processes for regional applications. I. Hydrologic balance, canopy gas exchange and primary production processes. *Ecological Modelling*, 42, 125–154.

Running S.W., Hunt R.E., 1993. Generalization of a forest ecosystem process model for other biomes, BIOME-BGC, and an application for global-scale models. *In: Scaling Physiologic Processes: Leaf to Globe* (J.R. Ehleringer, C.B. Field, eds). Academic Press, San Diego, pp. 141–158.

Ryan G.M., 1991. Effects of climate change on plant respiration. *Ecological Applications*, 1 (2), 157–167.

Sands P.J., 1995. Modelling canopy production II. From single leaf photosynthetic parameters to daily canopy photosynthesis. *Australian Journal of Plant Physiology*, 22, 603–614.

Santaren D., Peylin P., Viovy N., Ciais P., 2006. Optimizing a process-based ecosystem model with eddy-covariance flux measurements: A pine forest in southern France. *Global Biogeochemical Cycles*, 21 (2), GB2013.

Schober R., 1975. *Ertragstafeln wichtiger Baumarten bei vershiedener Durchforstung*. J.D. Sauerländer's Verlag, Frankfurt, 154 p.

Schulze E.D., 1986. Carbon dioxide and water vapour exchange in response to drought in the atmosphere and in the soil. *Annual Review of Plant Physiology*, 37, 247–274.

Sellers P.J., Dickinson R.E., Randall D.A., Betts A.K., Hall F.G., Berry J.A., Collatz G.J., Denning A.S., Mooney H.A., Nobre C.A., Sato N., Field C.B., Henderson-Sellers A., 1997. Modeling the exchanges of energy, water, and carbon between continents and the atmosphere. *Science*, 275, 502–509.

Shuttleworth W.J., Wallace J.S., 1985. Evaporation from sparse crops – an energy combination theory. *Quarterly Journal of the Royal Meteorological Society*, 111, 839–855.

Sievänen R., Burk T.E., 1993. Adjusting a process-based growth model to different sites through parameter estimation. *Canadian Journal of Forest Research*, 23, 1837–1851.

Valentine H.T., Gregoire T.G., Burkhart H.E., Hollinger D.Y., 1997. A stand-level model of carbon allocation and growth, calibrated for loblolly pine. *Canadian Journal of Forest Research*, 27, 817–830.

Wang Y.P., Jarvis P.G., 1990. Influence of crown structural properties on PAR absorption, photosynthesis and transpiration in Sitka spruce: Application of a model (MAESTRO). *Tree Physiology*, 7, 297–316.

Wang Y.P., Leuning R., 1998. A two-leaf model for canopy conductance, photosynthesis and partitioning of available energy. I: Model description and comparison with a multi-layered model. *Agricultural and Forest Meteorology*, 91, 89–111.

White A., Cannell M.G.R., Friend A.D., 2000. CO_2 stabilization, climate change and the terrestrial carbon sink. *Global Change Biology*, 6, 817–833.

Williams M., Rastetter E.B., Fernandes D.N., Goulden M.L., Wofsy S.C., Shaver G.R., 1996. Modelling the soil-plant-atmosphere continuum in a *Quercus-Acer* stand at Harvard Forest: the regulation of stomatal conductance by light, nitrogen and soil/plant hydraulic properties. *Plant, Cell and Environment,* 19, 911–927.

Wilson K.B., Baldocchi D.D., Hanson P.J., 2000. Spatial and seasonal variability of photosynthetic parameters and their relationship to leaf nitrogen in a deciduous forest. *Tree Physiology*, 20, 565–578.

Wollenweber G.C., 1995. Influence of fine scale vegetation distribution on surface energy partition. *Agricultural and Forest Meteorology*, 77, 225–240.

Chapter 10

How will the production of French forests respond to climate change? An integrated analysis from site to country scale

PHILIPPE CIAIS, DENIS LOUSTAU, ALEXANDRE BOSC, JÉRÔME OGÉE,
ÉRIC DUFRÊNE, CHRISTOPHE FRANÇOIS, HENDRIK DAVI, NICOLAS VIOVY,
FRANÇOIS DELAGE, SHILONG PIAO

Introduction

Climate change effects on tree growth have been evidenced in various forests over Europe and France in the last decades (Becker *et al.*, 1994; Spiecker, 1999). Increased growth in response to warming was reported in the northern range of species distribution area (Dhote and Hervé, 2000), but neutral or decreased growth was conversely observed in the southern range (Piovesan *et al.*, 2008). Forest production and carbon balance are limited in the French metropolitan area by a variety of environmental abiotic factors such as soil nutrients, temperature, CO_2 atmospheric concentration, and water balance. The potential change impacts are, therefore, complex to predict (Nellemann and Thomsen, 2001). With 25% of its area covered by a wide variety of forests, the French metropolitan zone provides an interesting case study from this point of view as an area of convergence of different climate influences and biomes. Moreover, future changes in temperature, air vapour saturation deficit (VPD), or rainfall are not expected to be uniform across the whole country (see Chapter 6), making it even more challenging to predict the future of forest growth. The case of French forests is interesting in a broader context as a test-bed of impacts and adaptation over Europe because of its ecotonal nature, spanning from summer-dry Mediterranean species to cold temperate species in the mountains and in the east of the country.

Growth and yield empirical models are widely used for decision-making in forestry. However, because they are mainly based on inventory data, their predictive power is limited to reproducing situations observed in the past, which prevents them predicting effects of un-experimented conditions (Garcia-Gonzalo *et al.*, 2007). Models based on the physical and biological processes involved in primary production and tree growth offer the potentiality for simulating the influence of climate variables outside their present range because they incorporate the response of processes such as photosynthesis, respiration, phenology or growth allocation to increased CO_2 concentration, higher temperature and VPD, and soil water deficit (see Chapter 9 of this volume; Loustau *et al.*, 2001, 2005; Milne and Van Oijen, 2005). However, the complex interactions between plant and ecosystem physiology and the environment are not fully understood, which poses a challenge for mechanistic models to reproduce faithfully the effects of environmental changes on forest ecosystem production and carbon balance. In addition, mechanistic models provide a good understanding of the response of fluxes to changes in climate and environmental factors, but they usually lack any description of forest management. This shortcoming hinders the evaluation of mechanistic models against the observed biomass stocks, as measured extensively by forest inventories or ecological studies (Masek and Collatz, 2006; Zaehle *et al.*, 2007; Ciais *et al.*, 2008).

This chapter investigates the effects of a climate scenario on the gross primary production (GPP), growth and evapotranspiration (ETR) of the main forest types over the French metropolitan area at a range of spatial scales from the site to the entire country. The mechanistic models used here are assessed in detail in Chapter 9 where their performance is discussed. Three modelling approaches have been carried out. First, site-based predictions were produced by the site-specific models GRAECO and CASTANEA for a range of forest sites and forest management scenarios parameterized from datasets obtained from observed forests covering a range of climate, soil, species and management conditions. Second, the global model ORCHIDEE was parameterized with spatially uniform values of soil and species and integrated over the entire metropolitan area to project the potential climate change effects at regional and national scales. Here, tree growth and canopy development were modelled to reach values in equilibrium with future climates so that the ORCHIDEE results shown are steady-state, potential values of future forest production and ETR, rather than the transient state of forest vegetation perturbed by climate and management actions.

We have restricted our approach to an analysis of the direct impacts of changes in climate variables and CO_2 concentration upon primary production and ETR. Other factors implied in global change, such as nitrogen and ozone depositions (Ollinger *et al.*, 2002; Felzer *et al.*, 2004), and the increase in diffuse/direct light fractions (Gu *et al.*, 1999; Roderick *et al.*, 2001; Niyogi *et al.*, 2004) are not considered. The study described in this chapter is thus rather an analysis of the sensitivity to climate factors and CO_2 concentration of the main ecophysiological processes that determine tree growth and production than a realistic prediction of the future of forests in France. The main questions we address in this study are: (i) Will climate change decrease or increase the productivity of French forests in the next century? (ii) What are the regional differences in the effects of rising CO_2 and climate change on forest productivity and ETR? (iii) How will water-use efficiency (WUE), defined as the ratio of CO_2 gains to water vapour losses, change in the future?

After describing the regional climate scenarios and the three models (see "Materials and methods"), we investigate the forest response to climate at selected sites (see "Results"), and the potential response over the whole country.

Materials and methods

Regional climate scenarios

Version number three of the Météo-France atmospheric model ARPEGE has been used to simulate present climate and 21st century climate through a 140-year numerical experiment (Gibelin and Déqué, 2003). The greenhouse gas and aerosol concentrations were prescribed by the so-called IPCC-B2 hypothesis or IS92a (IPCC, 2001). This scenario provides an intermediate amount of climate change among the IPCC scenarios (see Chapter 5). The global ARPEGE climate model covers the whole globe with a variable spatial resolution reaching a maximal value of 50 km over metropolitan France ($N = 360$ grid-points). The ocean surface temperatures are provided by a model with a coarser resolution that is coupled to an oceanic water circulation scheme (Royer *et al.*, 2002). The radiative forcing scheme includes four greenhouse gases (CO_2, CH_4, N_2O and CFC) in addition to water vapour and ozone, and five aerosol classes, land, marine, urban, desert and sulphate, respectively (Morcrette, 1990). The CO_2 curve between 1990 and 2100, cloudiness, precipitation and vertical diffusion are implemented according to the statistical approach described by Ricard and Royer (1993). The hydrology-soil-vegetation surface scheme is modelled according to the Interactions Soil-Biosphere-Atmosphere (ISBA) approach (Douville *et al.*, 2000).

Models description and comparative analysis

The models used in this study, CASTANEA (Dufrêne *et al.*, 2005), GRAECO (Porté, 1999) and ORCHIDEE (Krinner *et al.*, 2005), are described in Chapter 9. CASTANEA and GRAECO are specific forest growth models developed for site-scale studies, while ORCHIDEE is a dynamic global vegetation model designed to be run at the global scale and thus not specific to any site or species (Krinner *et al.*, 2005; Sitch *et al.*, 2008). CASTANEA and GRAECO include a detailed ecophysiological knowledge obtained locally for representative temperate forest species and were evaluated against long-term CO_2 and water flux data (Berbigier *et al.*, 2001; Medlyn *et al.*, 2002; Dufrêne *et al.*, 2005; Davi *et al.*, 2005) as well as growth data issued from long-term plots (see Chapter 9 of this volume; Davi *et al.*, 2006).

All three models include a description of the carbon and water cycles in the soil-vegetation-atmosphere system and the related processes (air-to-leaf CO_2 diffusion, photosynthesis, ETR, respiration, carbon allocation, growth, phenology, mortality and mineralization), together with their sensitivity to climate variables and air CO_2 concentration. ORCHIDEE includes the complete SVAT model SECHIBA (Ducoudré *et al.*, 1993) to calculate the hydrological and surface energy budget. ORCHIDEE describes the turbulent surface fluxes of CO_2, water and energy (i.e. transpiration, photosynthesis, respiration) using a big-leaf approximation, the dynamics of water and carbon pools (i.e. soil moisture budget and allocation, growth, mortality, soil carbon decomposition) and processes of longer-term ecosystem dynamics (i.e. fire, sapling establishment, light competition). No explicit nitrogen cycle is modelled. The fluxes of CO_2, water and heat are calculated each hour, and carbon pools each day in response to varying meteorology.

Tree structure is described by a combination of five functionally different parts: foliage, stems, branches and coarse and fine roots. A carbohydrate storage compartment is also considered for broadleaved trees (i.e. in ORCHIDEE and CASTANEA), equivalent to reserves.

The GRAECO allocation scheme is derived from the 3-PG modelling approach (Landsberg and Waring, 1997) with the following changes: (i) it distinguishes two canopy layers; and (ii) allocates net primary production (NPP) among individual trees according to their contribution to stand leaf area index (LAI). Allocation between above- and below-ground biomass is driven by a stress index whereas the above-ground NPP is distributed among stemwood, bark, foliage and branch according to allometric characteristics of the species (Porté *et al.*, 2002).

The allocation scheme in CASTANEA is made according to a fixed priority scheme that can be implemented seasonally (Dufrêne *et al.*, 2005). The allocation scheme in ORCHIDEE calculates dynamically the fraction of assimilates to be allocated to the different plant parts taking into account environmental influences (i.e. light availability, temperature and soil water) (see Friedlingstein *et al.*, 1998). Phenology is also fully prognostic and based on growing degree-days, chilling or soil water content indexes specific to each plant functional type (PFT) and calibrated on remote sensing data (Botta *et al.*, 2000).

For the soil organic carbon cycle, the CENTURY model was implemented in CASTANEA and ORCHIDEE. In the GRAECO model heterotrophic respiration is dependent on soil temperature and moisture and constrained by the average litter input.

Table 10.1 summarizes the objects and scales modelled in each simulation experiment. GRAECO and CASTANEA were used for modelling climate change effects at the local scale in Experiments 1 and 2, whereas ORCHIDEE was used here for spatially explicit predictions covering the entire national territory in Experiment 3.

Table 10.1. Objects modelled and scales assessed in the three simulation experiments.

Exp.	Model	Stand and tree growth	Management	Site conditions	Stand conditions	Area covered
1	CASTANEA	None (fixed canopy)	Standard	High fertility	Five species, mature stands	Four locations (CARBO-EUROPE main sites)
2	CASTANEA GRAECO	Yes, from regeneration to final cut	High, medium and low management intensities	High and low nutrient levels High and low water holding capacity	Three species, whole rotation	Seven locations covering species area in France
3	ORCHIDEE	Only equilibrium values shown	None	High and low water holding capacity	Three plant functional types	Whole of French metropolitan area

Simulations at representative forest sites and across the whole metropolitan area

Site selection

Experiment 1: site-level simulations without management

Experiment 1 was run over selected sites representative of ecosystems of the French plains. The sites include (i) broadleaved deciduous forests, sessile oak and beech stands at Fontainebleau (central: temperate) and beech stand at Hesse (north-east: temperate/continental); (ii) one broadleaved evergreen stand, a holm oak coppice at Puéchabon (south-east; Mediterranean: summer dry); and (iii) two temperate needle-leaved forests, a maritime pine stand at Le Bray (south-western: temperate oceanic) and a Scots pine stand at Fontainebleau. Among these, five sites are part of a network of sites studied by several European projects since 1996 (e.g. the CARBOEUROPE integrated project). For Experiment 1, the soil conditions were those measured at the sites.

The annual energy, carbon and water exchanges of the six selected forest stands were simulated over 140 continuous years. For this experiment, the canopy keeps a similar structure throughout the period. Canopy characteristics were fixed at their present values according to observed characteristics at six stands grown on four sites monitored on a long-term basis, that is, the vegetation does not grow or age. This experiment is thus called "without management", because forest canopy structure was prescribed: Le Bray (one maritime pine stand); Fontainebleau (three stands of beech, oak and Scots pine); Puéchabon (one holm oak stand); and Hesse (one beech stand). All were mature stands with LAI close to its maximum value (Table 10.2). The simulation of Experiment 1 represented, therefore, the sensitivity of ecophysiological processes to the climate scenario as described through time series of GPP, NPP, net ecosystem exchange (NEE), respiration and ETR.

Experiment 2: site-level simulations including management

Experiment 2 was run in northern France over three locations, Fontainebleau, Bar-le-Duc and Sarrebourg, forming a west-east transect where most production forest composed of social broadleaved species grow. In southern France, four locations covering the climate range of the maritime pine production forest were chosen, Aire-sur-Adour, Belin-Beliet, Marmande and Toulouse.

This experiment consisted of a simulation of four climate time slices in the late 20th century and the 21st century. For each climate time slice, the CASTANEA and GRAECO models were run from tree regeneration to final clear-cut, and rotation averages of carbon and water fluxes were calculated. Climate forcing data for the models were obtained by using 30-year-long slices of the climate scenario covering the periods 1961–90 (T0), 2001–30 (T1), 2031–60 (T2) and 2081–95 (T3). Forest management scenarios were defined for each species in terms of the age of trees at the final cut and the thinning regime (Table 10.3), according to standard management practices applied in the respective geographic zones under consideration. Only even-aged monospecific tree rotations were modelled. Additional sensitivity tests were performed and the results are discussed as a function of two integrated ecosystem parameters, the foliar nitrogen concentration (N in gN·g dry matter) and the soil water holding capacity (WH) (in mm). For each species, the sensitivity simulations

Table 10.2. Site and stand characteristics of forest sites modelled in the site-scale Experiments 1 and 2 carried out by the GRAECO and CASTANEA models.

Site name		Fontainebleau (NW)			Hesse (NE)	Puécha-bon (SE)	Bray (SW)
Species		Sessile oak	Scots pine	Beech	Beech	Holm oak	Maritime pine
		Quercus petraea	*Pinus silvestris*	*Fagus silvestris*	*Fagus silvestris*	*Quercus ilex*	*Pinus pinaster*
Latitude (west)		2°40'			1°10'	3°6'	− 0° 46'
Longitude (north)		48°23'			48°23'	43°75'	44° 43'
Elevation (m)		120	120	120	300	270	60
Soil water holding capacity	(kg H$_2$O·m^{-2})	90	115	136	180	127	110
Mean annual temperature	(°C)	10.2	10.2	10.2	9.2	13.6	12.9
Stand age	(year)	136	100	135	30	58	29
Standing stock	(trees·ha^{-1})	1025		622	3840	8500	533
Upper canopy LAI	(m^2·m^{-2})	7.1	5.0	4.5	7	2.3	3
Understorey LAI	(m^2·m^{-2})	0	0	0	0	0	1.5
Mean height	(m)	32	24	33	15	13	19
Basal area	(m^2·ha^{-1})	31.3	33.9	32.8	22.5		33.8
Mean circumference	(m)	1.57		1.07			0.92

consisted in imposing 2–3 nitrogen levels determined according to the range observed in available published data, and 2–3 classes of soil WH fixed from soil inventory data related to the species and area point considered.

Table 10.3.   Management scenarios implemented in Experiment 2.

Species		*Quercus petraea*	*Fagus silvatica*	*Pinus pinaster*
Area		**(N)**	**(N)**	**(SW)**
Short rotation	Duration (years)	Not implemented		30
	Number of thinnings			2
Standard rotation	Duration (years)	135	99	45
	Number of thinnings	12	7	5
Long rotation	Duration (years)	160	150	90
	Number of thinnings	11	9	5

Experiment 3: spatially explicit simulations without management

In this experiment set-up, the ORCHIDEE global dynamic vegetation model (ORganizing Carbon and Hydrology In Dynamic Ecosystems) (Krinner *et al.*, 2005) was integrated over the French metropolitan area to simulate the space and time distribution of the equilibrium values of CO_2 and water fluxes, for present-day and future climate and atmospheric CO_2 concentration conditions. No management was considered as in Experiment 2 and the age of the forests was set to 200 years for calculating equilibrium fluxes. By construction, the biomass pools have reached steady state and the forest does not grow in Experiment 3.

In the ORCHIDEE model, the vegetation was described by eight distinct PFTs over Europe. Of interest in this study are PFT4 (needle-leaved forest) representing Scots pine and maritime pine, PFT5 (evergreen broadleaved) corresponding to holm oak or similar Mediterranean tree species, and PFT6 (deciduous broadleaved forest) corresponding to sessile oak or beech. These three PFTs shared the same governing equations in the model but had distinct physiological parameter values (Krinner *et al.*, 2005). The equations controlling the onset and phenological development were, however, different between the PFTs, following Botta *et al.* (2000). The onset and senescence of foliage was modelled to depend on a critical leaf age, plus water and temperature stresses.

In the Experiment 3 setup, PFT4–6 were assumed to cover the whole domain, irrespective of plausible bioclimatic zones of existence. By doing so, the potential impact of climate change on productivity and transpiration could be compared between grid-points and regions. Assuming 100% coverage of each PFT in each simulation modified, however, the competition for surface water, which occurred in ORCHIDEE when different vegetation types coexisted in the same grid-point (de Rosnay and Polcher, 1998). Water reservoirs were calculated in Experiment 3 for each PFT separately, implying that CO_2 and water fluxes simulated for each PFT might have been (slightly) different from the configuration where different types of vegetation were assumed to coexist in each grid-point.

Model spin-up

No adjustments were made to the ORCHIDEE parameterizations in order to calibrate the response of vegetation to present-day climate variability. However, the model had been evaluated against eddy-flux measurements and shown to partition well each month the net surface radiation into latent and sensible heat (Krinner *et al.*, 2005). The seasonal and diurnal cycles of NEE were well reproduced by ORCHIDEE at the Hesse beech forest (PFT6) but underestimated a little at the Bray maritime pine forest (PFT4). At the live oak forest of Puéchabon (PFT6), the seasonal cycles of NEE and LE were also realistic with the characteristic maximum NEE uptake between April and June, just before the summer dry period. Similarly, the modelled LAI and soil water content were comparable with observations (see Chapter 9). This gave us confidence in the performances of the ORCHIDEE model to simulate the dynamics of CO_2 fluxes for present-day conditions, but no particular clues to the projection of future changes in these fluxes.

We performed four equilibrium simulations of CO_2 and water fluxes corresponding to time slices of 15 years duration each: T0 = 1961 – 75, T1 = 2001 – 15, T2 = 2041 – 55 and T3 = 2081 – 95, respectively. For each time slice, ORCHIDEE was forced by inter-annual variable climate and fixed CO_2. A model spin-up was calculated for 200 years by repeating the climate of each time slice, starting from bare soil in order to grow forests over each grid-point and let their biomass reach steady-state equilibrium with climate. Such a long spin-up was necessary for safe analysis of the modelled equilibrium GPP and NPP for present and future conditions. This is because the mass of the biomass pool controls the autotrophic respiration flux, and must hence reach its equilibrium value. However, a 200 years spin-up is still too short for soil carbon to reach its equilibrium, which led us to exclude any analysis of equilibrium values of NEE changes in this study.

Soil depth and field capacity sensitivity tests

ORCHIDEE calculates how much water is available for plants using a simple two-bucket scheme with a rain-fed surface bucket capping a deeper bucket. The size of the deep soil bucket, corresponding to soil depth, was uniformly prescribed to the value of 2 m (Ducoudré *et al.*, 1993). The dynamics of soil moisture, θ, in each grid-point was calculated on a daily time step from the modelled balance of transpiration, soil evaporation, runoff and canopy interception fluxes. The soil field capacity, θ_f, was another essential model variable because it defines the threshold sensitivity of GPP and stomatal closure to soil water stress (Krinner *et al.*, 2005). The soil water stress was modelled as a scaling factor of GPP and ETR, which remained equal to unity when the relative root extractable water (REW) > 0.3 (no stress) and decreased linearly with REW reaching zero when REW = 0 at the wilting point (maximum stress, θ min), where REW was defined by REW = $(\theta - \theta_f)/(\theta_f - \theta_{min})$. The modelled GPP and transpiration were multiplied each day by the water stress factor calculated in that manner, therefore being forced to zero at the wilting point.

In global simulations with ORCHIDEE, the field capacity θ_f was usually prescribed using globally gridded information on soil texture (Krinner *et al.*, 2005). However, such global soil texture maps did not capture realistically the spatial heterogeneities of forest soils in France (Arrouays *et al.*, 2001), and instead we preferred to use a spatially uniform value of θ_f over the entire domain. Two sensitivity simulations were done, with θ_f = 0.1 cm³·cm⁻³, that is a water reserve of 20 cm over a total soil depth of 2 m, and with θ_f = 0.2 cm³·cm⁻³, respectively. Generally, the smaller the prescribed field capacity, the lower the equilibrium GPP, but non-linear and seasonal carry-over effects of soil water stress also modulate this expected response of GPP to varying θ_f. It must be also be noted that our lower field capacity of 0.1 $cm^3 \cdot cm^{-3}$ may still exceed the observed θ_f in forest regions where the soil depth was very shallow (e.g. Landes forest, southern Massif Central, Burgundy and most of the Alps) (Arrouays *et al.*, 2001). Therefore, our simulations were most likely to be biased towards too high GPP because of an overestimated θ_f.

Results

Changes in fluxes at representative sites without management (Experiment 1)

Among the flux variables considered, the inter-annual variations are lower for respiration and largest for NEE and GPP (see Figure 10.1, Plate 6). The difference between sites is also large for GPP and respiration and least for NPP and carbon-use efficiency (CUE), defined as the ratio of NPP to GPP. The time courses of GPP, R_a and NEE over the 1960–2100 period show contrasted patterns among species and locations. Beech and sessile oak GPP increases continuously, with a small inflexion from 2040 to 2080, whereas pines and holm oak GPP level off from as soon as 2000. The deciduous species benefit from the increasing length of their growing season by up to + 25 days by the end of the 21st century, which explains a more pronounced enhancement of their GPP compared with the evergreen species. The autotrophic respiration, R_a, increases linearly from 1960 to 2100 with little difference between species or location. Note that none of the GRAECO or CASTANEA models account for acclimation of R_a to higher temperatures (Atkin and Tjoelker, 2003; King *et al.*, 2006). The heterotrophic respiration R_h curve exhibits a high inter-annual variability with no long-term trend apart from the beech stand at Hesse (data not shown). The NEE changes are contrasted among the five sites (Figure 10.1). The Hesse beech forest NEE shows a positive anomaly (sink) maintained throughout the 1960–2100 period, a similar pattern being predicted for the holm oak NEE at Puéchabon, although less pronounced. The maritime pine stand NEE at Bray is a continuously increasing sink from 1960 to 2000, but shows no clear trend afterwards. At Fontainebleau, the NEE of the Scots pine and sessile oak stands decreases sharply from 2000 to 2060 as opposed to beech that does not change during this period.

Among the five sites, CUE is both higher and less sensitive to climate change for younger forest stands (beech at Hesse and holm oak at Puéchabon), where values of CUE reach slightly above 0.50, than for older ones (oak and Scots pine at Fontainebleau). However, the CUE curve shows a higher responsiveness to climate change at the end of 21st century. CUE values in older stands fluctuate between 0.25 and 0.40. This differential response according to age is explained mainly by the stronger increase of R_a in older stands in response to warming. The ratio of GPP over ETR defines the annual amount of carbon assimilated per kilogram of water lost by plants and soils, usually referred to as the water-use efficiency (WUE), is predicted to increase from 0.7 gC·kg H_2O^{-1} to 1.2 gC·kg H_2O^{-1} between 1960 and 2100 (data not shown). This increasing trend in WUE is almost linear over 1960–2100 with little inter-annual fluctuation. The WUE enhancement is higher for oaks followed by beech and pines, whichever site is considered. This was attributed to the stimulation of photosynthesis by increased CO_2 and to the reduction in stomatal conductance due to lower soil water and higher VPD, all being well described in the models used.

Opposite to GPP, the ETR flux was more stable throughout the next century, despite large changes in VPD and temperature. The amplitude of the change in ETR reaches 50–100 mm, at maximum over the whole period 1960–2100, which corresponds to 10–15% of the mean value of that flux. The ETR of forest is controlled by stomatal and canopy conductances, which tend to compensate for the climate variations, for example, the VPD effect on evaporation (see Chapter 2 of this volume; Granier *et al.*, 2000). The differences in ETR between sites are large, with the holm oak at Puéchabon and the Scots pine at Fontainebleau having 30% lower ETR values than the maritime pine, sessile oak or beech stands. Such inter-site differences are explained by the difference in the modelled canopy structure and local climate. Globally, the model results may be explained primarily by distinct geographical variations of climate change, especially of the precipitation regime and VPD, and secondarily by the distinct sensitivity of tree species to climate and CO_2. The changes in climate are contrasted between north and south. The reduction in summer and autumn precipitations and the enhancement in mean annual VPD are more pronounced in the south and in the centre of the French metropolitan area, with a doubling of VPD in south-western France (see Chapter 6). The forest NEE is favoured most for the beech stand at Hesse in north-eastern France where limitations by low temperature and CO_2 are relieved by climate change and the annual amount in precipitation remains unchanged. In contrast, the drastic diminution of the Scots pine NEE at Fontainebleau in the centre of France from 2010 to 2040, can be attributed to the increased respiratory carbon losses not compensated for by GPP enhancement, the weakest among species and sites simulated. These changes are primarily attributed to increased VPD and temperature.

The differential stomatal response of temperate tree species to changes in CO_2 and water deficit explains a part of the simulated higher sensitivity of GPP at the sessile or holm oaks and beech stands compared with the pine stands. The stomatal response of *Pinus* to CO_2 is neutral whereas *Quercus* species close their stomata at elevated CO_2 concentration thus enhancing their WUE (Medlyn *et al.*, 2001).

Changes in fluxes and growth at representative sites including management (Experiment 2)

The effects of simulating several cycles of rotations show the large interactions between forest management, site conditions and future impacts of climate change. From that point of view, the most striking result is the contrast between southern and northern France illustrated by Figures 10.2 and 10.3, which show the 1980–2080 changes for fertile (high nitrogen status) site conditions.

Northern France

In the northern part, the change in production (GPP, NEE and mean annual increment (MAI)) shows two distinct phases. The first from 1960 to 2000 where production is enhanced dramatically, followed by a second phase up to the end of the 21st century where, under the effect of climate change, the mean forest production and carbon

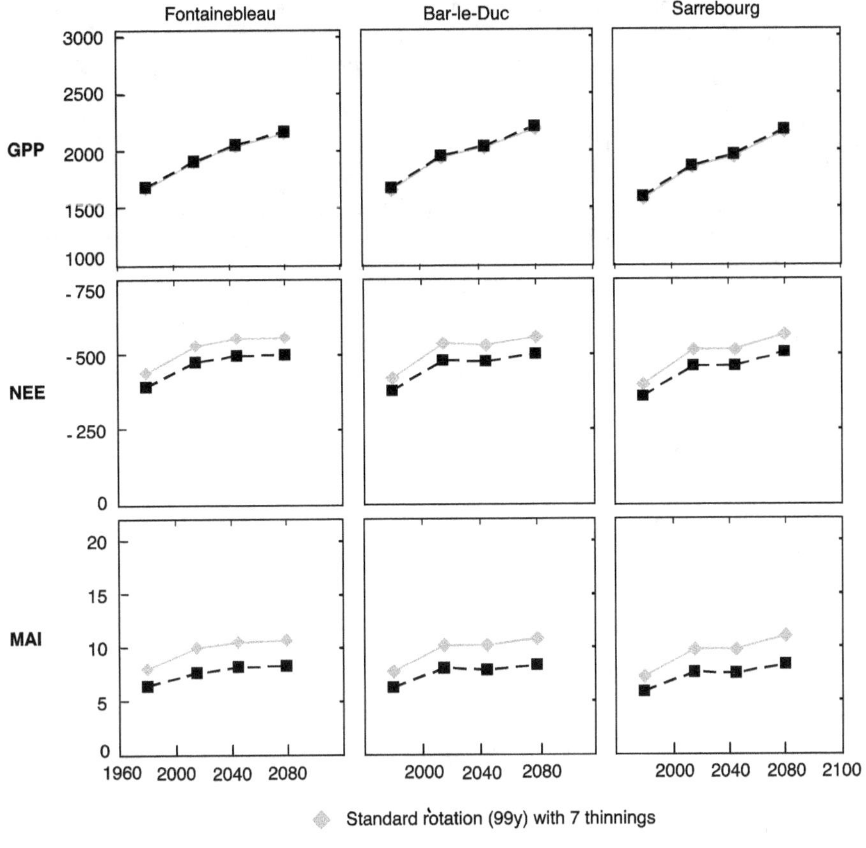

Figure 10.2. Evolution 1960–2080 of the gross primary production (GPP), net ecosystem exchange (NEE) (gC·m^{-2}·year^{-1}) and the net annual wood increment MAI (m^3 stem·ha^{-1}·year^{-1}) averaged over entire rotations for two management scenarios of beech forest at three different locations along a west-east transect across northern France. The predictions are those estimated for fertile site conditions (nitrogen = 2.2 gN·100g dm^{-1}, WH = 200 mm).

balance become neutral or slightly positive in accordance with results from Experiment 1. The net anomaly over the period 1960–2080 is positive in any case. The evolution of forest production shows some differences between eastern and western sites with minor interactions with the management strategy. This is well exemplified by Figure 10.2, which compares the mean GPP, NPP, NEE and mean biomass carbon stock along a west-east transect in northern France for fertile site conditions (high nitrogen and WH). The overall trend in Experiment 2 is a continuous increase in GPP whereas NEE and MAI tend to level off from 2000 onwards. The trend predicted for MAI is similar to NEE although between the two time slices 2045 and 2070, the change in MAI varied among sites from a slight increase in the east (Bar-le-Duc) to a neutral trend in the centre (Fontainebleau). At low WH (not shown), a slight decrease of MAI at the Fontainebleau location is predicted. Whereas the simulated management (rotation length) does not affect the mean GPP, the

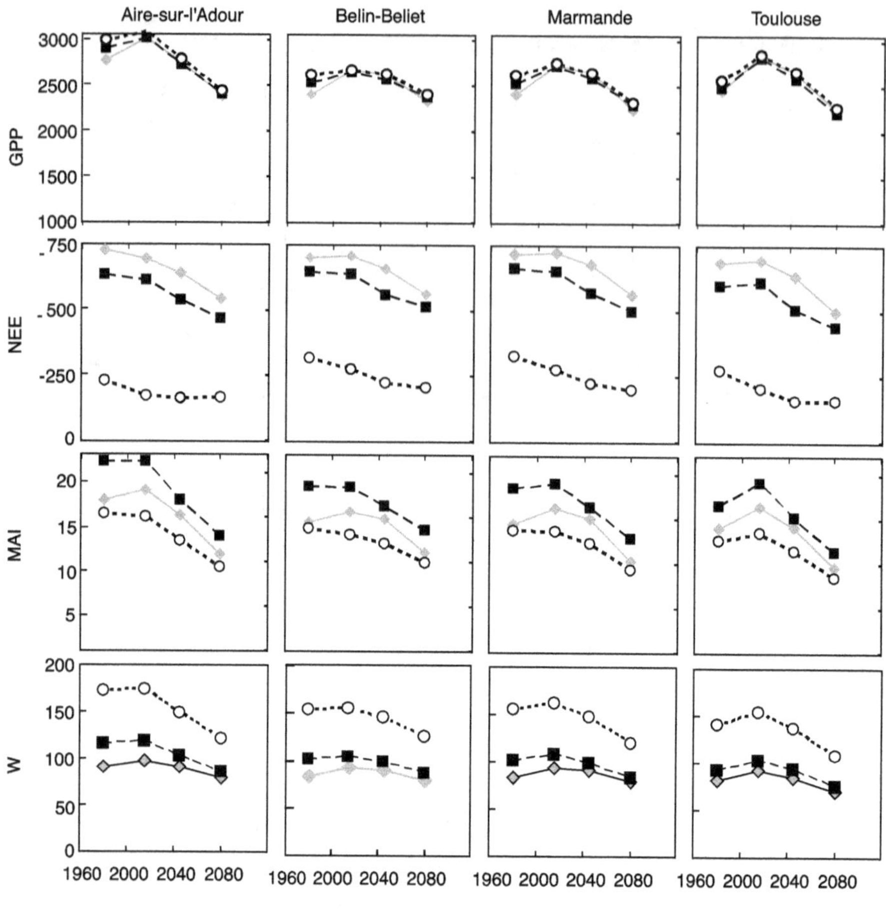

Figure 10.3. Evolution 1960–2080 of the gross primary production (GPP), net ecosystem exchange (NEE) (gC·m^{-2}·year^{-1}), net annual wood increment MAI (m^3 stemwood·ha^{-1}·year^{-1}) and biomass carbon stock, W (tC·ha^{-1}), averaged over entire rotations for three management scenarios of maritime pine forests at four different locations in south-west France. The predictions are estimated for fertile site conditions (nitrogen = 2.2 gN·100 g dm^{-1}, WH = 150 mm) by the GRAECO model.

mean NPP, NEE and MAI are higher for the standard rotation than for the fixed canopy of Experiment 1, caused by the sensitivity of autotrophic and heterotrophic respiration to temperature and to the lower standing biomass and litter input. There is no interaction between management options (i.e. rotation lengths) and climate effects, the time course of GPP, NEE or MAI being strictly parallel along the period 1960–2100.

Southern France

In the southern part, the situation is different from the north for both the predicted production trends and the interactions between climate, site and management (Figure 10.3). During the period 1960–2100, GPP, NPP and MAI increase until 2015–45 but decline subsequently and reach lower values in 2080 than in 1980. NEE declines or remains constant in a few cases from 2000 onwards, a consequence of the temperature-driven enhancement in autotrophic and heterotrophic respiration. Interestingly, MAI in the south of the French metropolitan area shows opposite trends among forest management scenarios between 1980 and 2015 with a systematic increase in short rotation, a stable evolution or increase in standard scenario and a decrease in all locations but Toulouse in the long rotation scenario. MAI decreases systematically after 2015 with a steeper slope in the standard scenario until 2045 and in the short rotation scenario until 2080. Overall, the net changes in GPP, NEE, NPP and MAI from 1980 to 2080 are systematically negative. This can be interpreted as an interaction between the management scenario and climate effects, the short and medium rotations scenario being more sensitive to climate. The MAI range between management options is therefore reduced by climate change.

Similarly, a net decrease over the period 1960–2080 is predicted for the biomass carbon stock, W, in all management options (Figure 10.3). Interestingly, the model predicts that the high carbon stock build-up in long rotations is the most vulnerable to climate effect among management scenarios and is drastically reduced at the end of the simulation period. This trend seems to be in contradiction with the lower sensitivity of the primary production to climate predicted for the long rotation. Actually, in this simulation, the average carbon stock in standing biomass for the intensive and standard management options is closely controlled by management that keeps constant the standing volume scenario whereas the amount of wood volume exported by thinning and harvest is reduced severely (not shown). For the long rotation, the carbon stock build-up averaged along the rotation declines dramatically as compared with the other management scenarios. This is because it depends more directly on primary production, that is, it is not modulated by thinning, which does not occur beyond an age of 40 years. The stock changes shown in Figure 10.3 must thus be regarded as a response to the combined effect of stand management and climate change rather than an ecological acclimation of biological processes.

An interesting interaction between site conditions, forest management and climate is illustrated for two representative locations, Bar-le-Duc (north-eastern France) and Marmande (south-western France) and species, beech and pine, in Figures 10.4 and 10.5, respectively. Whatever the location and species considered, the effects of management and climate are maximized under fertile conditions (i.e. high nitrogen and WH). NEE and MAI under short rotation at high nitrogen and WH are the most responsive variables to climate forcing as compared with GPP and NPP. Conversely, NEE and MAI under long rotations are less sensitive to climate, especially in the lower range of nitrogen and WH. At Bar-le-Duc, beech rotations show a higher sensitivity to soil WH, than to foliar nitrogen, which has a negligible effect. Conversely, nitrogen exerts a closer control on NEE and MAI than WH for maritime pine rotations at Marmande, which is consistent with the conclusions drawn by Loustau *et al.* (1999) in their review on the limiting factors of maritime pine growth in this area.

For both northern and southern France, the CASTANEA and GRAECO models predict that management does not differentiate much in GPP as compared with NEE, NPP and MAI (Figures 10.4 and 10.5). We conclude, therefore, that management does not create large differences in the mean LAI and radiation absorbed but rather reduces ecosystem

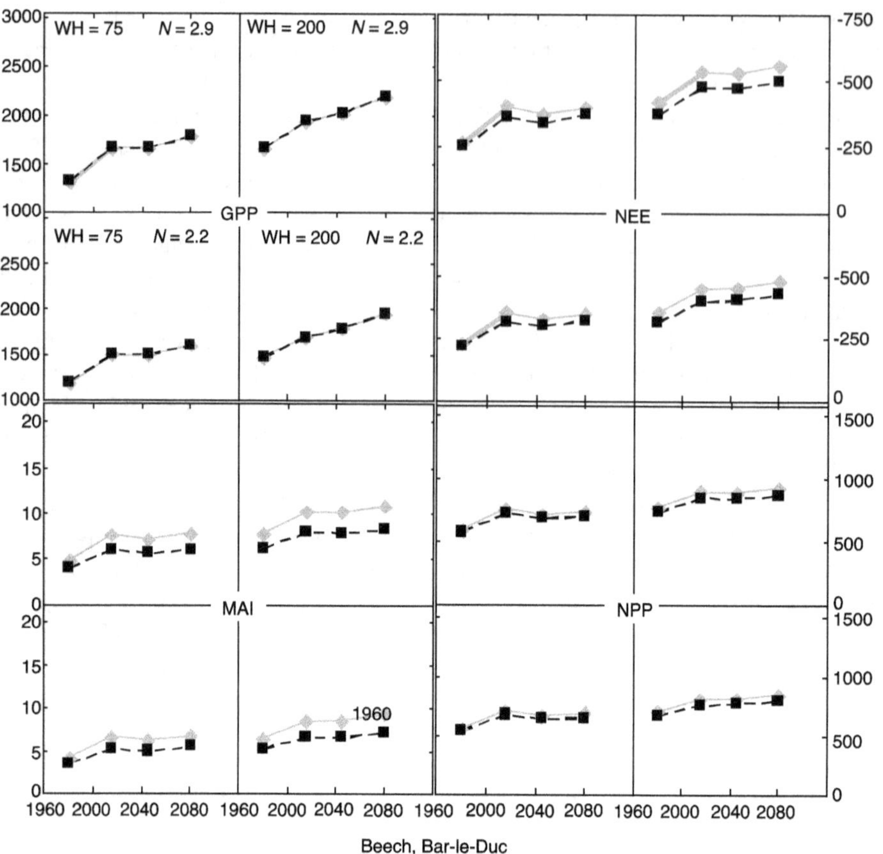

Figure 10.4. Values of the gross primary production (GPP), net ecosystem exchange (NEE), mean annual increment (MAI) and net primary production (NPP) averaged over complete rotations of beech forest at Bar-le-Duc for four consecutive time periods. For each variable, four plots corresponding to combinations of low and high soil water holding capacity (WH), 75 mm and 200 mm, respectively, and foliar nitrogen concentration, 2.2 gN·100 g dm^{-1} and 2.9 gN·100 g dm^{-1}, respectively, are shown. Labels are given only in the GPP plots for clarity.

respiration and increases allocation to stemwood. Indeed, because the standing stock is reduced by shorter rotations, the autotrophic respiration is attenuated accordingly.

Changes in fluxes over the metropolitan area (Experiment 3)

For analysis of the ORCHIDEE model response, France was sub-divided into five regions, north-west (NW), north-east (NE), south-west (SW), south-east (SE) and mountain (M) regions, respectively. The longitude and latitude boundaries separating between these regions are respectively, 3°E and 46°N. The mountain region was defined according to a mean annual temperature (MAT) criteria of MAT < 5°C. This encompasses roughly all grid-points in the Massif Central, Alps and Pyrénées mountains with an elevation higher than 1000 m.

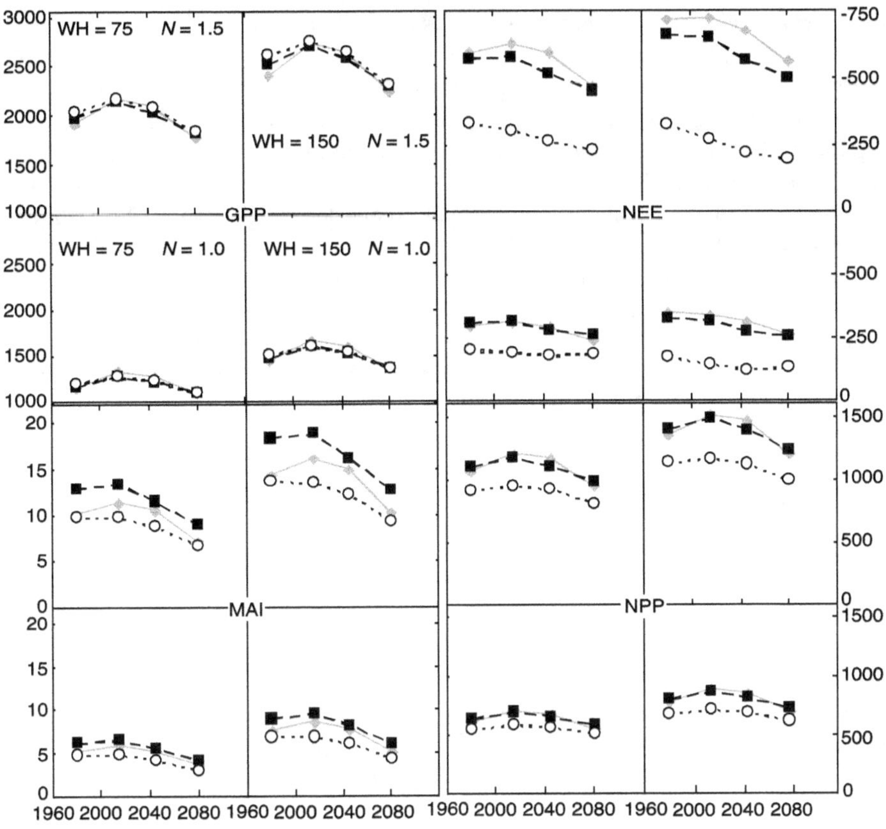

Figure 10.5. Values of the gross primary production (GPP), net ecosystem exchange (NEE), mean annual increment (MAI) and net primary production (NPP) averaged over complete rotations of maritime pine forest at Marmande and for four consecutive time periods. For each variable, four plots corresponding to combinations of low and high soil water holding capacity (WH), 75 mm and 150 mm, respectively, and foliar nitrogen concentration, 1.0 gN·100 g dm⁻¹ and 1.5 gN·100 g dm⁻¹, respectively, are shown. Labels are given only in the GPP plots for clarity.

GPP changes in response to climate and CO_2

Mean GPP changes in the metropolitan area

First, we look at equilibrium GPP changes, ΔGPP, between different time slices in the future and present day averaged across the model domain. Only potential, not actual, GPP changes are discussed, as we assumed full coverage of each region with the same type of vegetation (PFT). The results are shown in Figure 10.6 (Plate 7) for the difference between the 2080s (T3) and the 1960s (T0). The spatial and temporal patterns of ΔGPP look broadly similar between the three PFTs (see also Figure 10.7). The simulated mean GPP increases everywhere in response to higher CO_2 concentration, both through a reduction of stomatal conductance avoiding water losses and through increased carboxylation rates. Additionally, GPP is increased in spring and autumn by longer growing seasons. In our modelling set-up, higher CO_2 and longer growing seasons hence dominate over increased water stress during dryer summers. The projected future increase in GPP over France is significant, GPP being a factor 1.06 to 1.24 higher during T1 (2040s) and a factor 1.29 to 1.37 higher during T3 (2080s) than during T0 (1960s), the range spanning across the three different PFTs and five regions of France considered (see Figure 10.7). However, the magnitude of GPP changes in ORCHIDEE over each region is comparable with the changes found in the more realistic and better calibrated site-scale models, GRAECO and CASTANEA, as shown in Figure 10.1.

For the three forest PFTs modelled, GPP increases linearly with time between T0 and T3, suggesting no saturation effect. This apparent lack of saturation effect on GPP during the second half of the 21st century is obtained despite more frequent drier summers causing stress on photosynthesis. Hence, longer growing seasons in autumn and spring could compensate for increased summer stress. In total, 53% to 56% of the mean GPP increase between the late 20th century (T0) and the late 21st century (T3) is realized between T1 and T2.

Regional GPP changes

In our simulations, out of all the PFTs, the smallest regional GPP increase between T0 and T3 occurs in the mountain region (M) (Figure 10.7). In contrast, the largest GPP increase takes place in the north-east region (Figures 10.6 and 10.7). This reflects primarily the relieved temperature limitations on the length of the growing season (Berthelot *et al.*, 2002). As expected, a higher prescribed field capacity causes a higher ΔGPP value (Figure 10.8, Plate 8). Regional contrasts of ΔGPP are also greater in the experiment with a lower soil field capacity (see top panel of Figure 10.8), especially over the southern and western regions, which suffer more from water stress during 2081–95, under the low scenario $\theta_f = 0.1$ cm³·cm⁻³. Among these "hot-spot" areas where increased water stress is projected to limit the increase of GPP, the south-west of France between the Bay of Biscay and the Mediterranean sea, shows this clearly for all three PFTs (Figures 10.6 and 10.8). In that context, we note that forested regions which sustain high GPP today, that is, the western half of France and particularly the Landes forest, experience a lower ΔGPP than other regions, both between T0 and T2 and between T0 and T3 (Figure 10.8). Conversely, currently less productive forested regions in the north-east, that is, the Vosges forests, will realize a regionally higher ΔGPP.

Forests in the Landes and more generally in south-western France are projected to suffer from water stress in summer caused by a westwards shift towards the Atlantic of the Mediterranean climate regime. In this particular south-western ecotone region, it is seen from the climate model predictions (Figure 7.5 in Chapter 7) that an annual rainfall deficit of 130 mm·year⁻¹, that is, 10% less rainfall than at present, begins to develop during T2, and spreads further to the whole south-west of France during T3 near the end of the 21st century. Mountain regions (e.g. the Alps and Massif Central) which today have the lowest GPP (Figure 10.6), realize a systematically lower ΔGPP than anywhere else in the country. In other words, the geographical distribution of GPP, which is characterized today by a decrease of 20% along a continentality gradient between the south-west and the north-east of France (GPP values ranging from 2000 gC·m⁻²·year⁻¹ near the Atlantic down to 1600 gC·m^{-2}·year^{-1} in the east of France according to Figure 10.6), tends to become spatially more heterogeneous in the future.

Figure 10.7 shows the changes in GPP averaged over each of the five regions and for the three PFTs. It can be seen that the modelled GPP increase is linear for all PFTs. However, the rate of increase of GPP is slightly different between PFTs, so that by the end of the next century (T3), the GPP gradient over the metropolitan area between PFT6 (oak and beech) and PFT4 (maritime pine and Scots pine) becomes smaller than today (Figure 10.7).

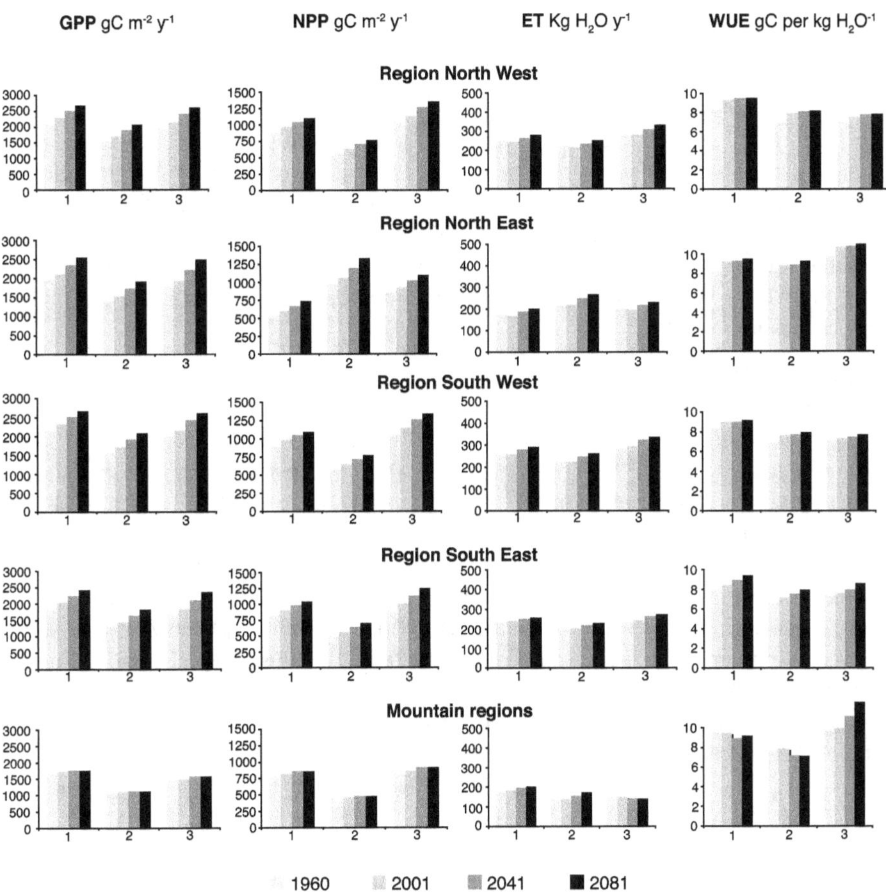

Figure 10.7. Mean values of the annual gross primary production (GPP), net primary production (NPP), evapotranspiration (ETR) and water-use efficiency (WUE) for the five French ecoclimatic regions and at four climate periods as simulated by the ORCHIDEE model. Each bar is the average value of the three forest functional types in each region. Soil water holding capacity is fixed at 0.1 cm^3·cm^{-3} everywhere.

NPP changes

Present-day spatial distribution of NPP versus GPP

Maps of the NPP present-day distribution and future changes, ΔNPP expressed as anomalies from the present, are shown in Figure 10.9, Plate 9. As for GPP, the simulated present-day NPP decreases with continentality along a diagonal from the south-west to the north-east of France. The simulated continentality NPP gradient follows broadly the MAT gradient, suggesting that MAT controls the annual value of NPP as well as GPP through the duration of the growing season. This modelled response of NPP, GPP to MAT is confirmed by GPP observations from 23 eddy-covariance data in Europe (Reichstein *et al.*, 2007) and by the more than 300 NPP observations of the large database of ecological measurements of Luyssaert *et al.* (2007). In ORCHIDEE, the autotrophic respiration R_a is expected to decreases inland with MAT, because R_a is parameterized as a linear function of MAT and sapwood biomass (for maintenance respiration) and as a function of GPP, which also decreases with MAT (for growth respiration). Thus, the continentality gradient of present-day mean annual NPP in ORCHIDEE is controlled by the concurrent sensitivities of two processes: the decrease of R_a with MAT and the decrease of GPP with MAT. We found that R_a decreases more than GPP for a given temperature gradient, which flattens the spatial distribution of present-day NPP compared with that of GPP (Figure 10.9). As a result, the present-day (T0) CUE, defined as the ratio of NPP to GPP, shows an increase from 40% to 45% from the south-west to the north-east of the metropolitan area (Figure 10.10, Plate 10). Over the Alps where the coldest climate conditions prevail, CUE is also the highest and reaches up to 47%.

Regional changes in CUE

The pattern of regional potential NPP changes (ΔNPP) between T3 and T0 is shown in Figure 10.9, for PFT4 (maritime pine and Scots pine). The modelled pattern of ΔNPP is parallel to that of ΔGPP (Figures 10.6 and 10.8). In a first-order approximation, the modelled autotrophic respiration, R_a, stays proportional to GPP under future climate conditions. This is an interesting result because it suggests that roughly the same fraction of GPP increase will remain available to increase NPP and the production of woody biomass. For instance, over the south-western forests, which become more vulnerable to water stress during the T3 time slice, the ΔNPP map (Figure 10.9) shows a pronounced regional minimum for PFT4 (also the other two PFTs), but this regional ΔNPP minimum remains roughly half of the ΔGPP minimum (Figure 10.6), suggesting that regional differences in warming (see Figure 6.3 in Chapter 6) might have a little influence on the ratio of R_a to GPP and on the CUE patterns.

Looking now at second-order effects, we estimated CUE for the five regions and three PFTs. We modelled that CUE tends to decrease everywhere by 1.5% to 0.7% as temperature increases (lower plots in Figure 10.10). The largest decrease in CUE is found over the Alps (–1.9%). Under a future warmer climate in France, CUE is hence predicted to decrease slightly more in the south-west than in the north-east, thereby amplifying the present-day spatial gradient of CUE (see above). The CUE spatial gradient is projected to increase from 38.5% to 44.3% (range 5.8%) between the Landes and Vosges forests during T3, compared with a present-day gradient from 40% to 45% (range 5%). Comparing regional changes in NPP summarized in Figure 10.7 together with GPP changes, we can see that NPP increases linearly with time, just like GPP. However, the between-PFT differences are different for NPP than for GPP. Over the north-east region with the largest warming signal, the NPP of PFT6 increases faster than for PFT4. On the other hand, over this region, PFT5 (green oak, which is not present today) is predicted to have the largest increase in NPP, suggesting that there is a potential for southern species to acclimate further north in the future.

Differential NPP changes between forest types

Figures 10.7 and 10.10 provide the present-day simulated NPP distribution and its changes in the scenario with $\theta_f = 0.1$ cm^3·cm^{-3} prescribed everywhere. The less productive PFT5 (live oak) shows a smaller ΔNPP than the others, except in the north-east region. On the other hand for PFT5, the negative impacts on ΔNPP of soil drying in southern France are less pronounced than for other PFTs. ΔNPP of PFT5 is even maximal along the Mediterranean coast where the climate scenario gives the most pronounced warming signal, unlike for the other vegetation types. Such a surprising model response (more warming in an already dry region causing greater NPP) is due to the fact that PFT5 has permanent LAI, allowing its NPP to immediately increase in winter and spring when the temperature is warmer but rainfall is still sufficient. A pine forest growing under very dry summer climate conditions in the Negev desert (Maseyk *et al.*, 2008) was indeed measured to grow mainly during the wet winter and spring, and to decrease foliage respiration during the summer through acclimation. Acclimation of respiration, although likely to be widespread for foliage (Atkin *et al.*, 2005) is not included in the ORCHIDEE fomulations, but we expect that this process would further enhance ΔNPP over summer-dry regions.

Evapotranspiration changes

Effects of different prescribed soil field capacities on ETR changes

Figure 10.11 (Plate 11) shows ETR changes for different future time slices. We can see that ETR increases everywhere over France in T2 (2040s) and T3 (2080s), with ΔETR increasing from 3 W·m^{-2} (+10%) near the Mediterranean to 9 W·m^{-2} (+20%) over central France (Figure 10.11). Our simulations further indicate small differences in ΔETR between the low and high soil field capacities scenarios. This can be seen for instance by comparing the lower and upper row of plots in Figure 10.11. This result suggests that, even in the sensitivity experiment with a lower field capacity, there is enough water in the soil to sustain evaporative demand under increased VPD (see Figure 6.7 in Chapter 6) and warmer temperatures. Only when the warming and summer drying become very pronounced at the end of the 21st century (T3), do the differences in ΔETR between the experiments with $\theta_f = 0.1$ cm^3·cm^{-3} and with $\theta_f = 0.2$ cm^3·cm^{-3} become significant. The region where ETR increases more than all the PFTs is the ecotone area between the Bay of Biscay and the Mediterranean, including the Landes forest along the Atlantic coast. In this region, the enhancement of ΔETR seen in Figure 10.12 (Plate 12) closely parallels that of VPD, not surprisingly since atmospheric dryness exerts a strong control on ETR in ORCHIDEE via stomatal conductance (Ball *et al.*, 1987) and evaporative demand. This marked response of ETR to VPD is further confirmed by the fact that in the 2000s (T1), the VPD decreases below its values during T0 (i.e. slightly wetter conditions) and the ETR lock steps this decrease.

Differential ETR changes between vegetation types

For comparing ETR changes between the different PFTs as illustrated in Figure 10.7, it is interesting to note that PFT6 (beech and oak), which makes most of its LAI in summer, has a higher ΔETR and hence suffers more from water stress than PFT5 (live oak) and PFT4 (maritime pine and Scots pine). This is clearly seen in Figures 10.7 and 10.12 in the south-east and south-west regions, where ΔETR of PFT6 increases by 8–10 W·m^{-2}, against 6–7 W·m^{-2} for PFT4 and a smaller 4–5 W·m^{-2} for PFT5. Both PFT4 and PFT5 have the advantage of retaining a permanent LAI, hence better able to regulate their water losses all year round. A distinct response of ETR to soil water stress for each PFT is also apparent in Figure 10.12 when comparing ΔETR between the sensitivity experiments $\theta_f = 0.1$ cm^3·cm^{-3} and $\theta_f = 0.2$ cm^3·cm^{-3}. The larger the difference in ΔETR when θ_f is changed, the more limited is ETR by soil moisture. In contrast to PFT6, there is little difference in ΔETR between $\theta_f = 0.1$ cm^3·cm^{-3} and $\theta_f = 0.2$ cm^3·cm^{-3} for PFT5. PFT5 corresponding to forests like live oak, has, however, no specific drought tolerance parameterization such as stomatal closure (Reichstein *et al.*, 2002, 2003). Thus, its better resilience to a prescribed smaller field capacity comes entirely from the fact that a permanent LAI allows a better utilization of soil moisture during spring and autumn, while GPP and ETR are brought down to near zero values in the hot and dry summer. Therefore, on average over a year as indicated in Figure 10.12, we expect that PFT5 will accumulate less water stress on ETR (and on GPP) compared with deciduous broadleaved forests (FPT6) and even temperate conifers (PFT4).

Discussion and model limitations

One should keep in mind a number of limitations inherent to the application of the ORCHIDEE model at the metropolitan scale. Firstly, the maps of Figures 10.6– 10.10 are likely to be too smooth compared with the real situation. Local heterogeneities in soil WH are not resolved, and are likely to be much higher than the range bracketed by our scenarios ($\theta_f = 0.1$ cm^3·cm^{-3} to 0.2 cm^3·cm^{-3}). Secondly, trees in the real world are adapted to particular climate and soil conditions, whereas in ORCHIDEE, the basic vegetation response to climate change is described by PFT parameters that are spatially and temporally uniform. Thirdly, no disturbances or growth limitation factors (e.g. pathogens, lagged effects of extreme years, climate induced mortality, etc.) are accounted for in this version of ORCHIDEE. Fourthly, the model equilibrium simulations describe old-growth forests which have reached their maximum LAI and biomass everywhere. Doing so ignores the greater sensitivity of young trees to summer drought (see "Results"). It is likely the reason why in Figure 10.7 GPP and NPP increase linearly in each region × PFT whereas in GRAECO (Figure 10.3) GPP and NPP both begin to decrease after 2000. The ORCHIDEE response to CO$_2$ and climate is similar to that of the site-specific models in the northern regions, but overestimates the increase of production in southern regions. Finally, in the regional climate scenarios, in particular over the Mediterranean regions, the occurrence of extreme events and convective precipitations may be underestimated, producing a too smooth response of the vegetation to climate change.

Discussion: synthesis of the different modelling approaches

Despite differences, the results issued from the different modelling approaches at a range of spatial scales allow us to draw a consistent picture of the main trends expected according to this particular climate scenario. They show that the response of forests to the climate scenario 1960–2100 varies: (i) geographically; (ii) from one PFT or species to another; and (iii) is not a linear function of time.

The geographical pattern of future changes contrasts the north-east and the south of France, and to a lesser extent, the east and west. Expansion of a warmer and seasonally drier climate northwards induces a corresponding shift of forest potentialities. These geographical differences are superimposed on the spatially uniform increase of CO_2 concentration that favours GPP and growth. The resulting predictions are greater depending on local site and management conditions, which open opportunities to forest managers. More precise and local adaptation strategies will need, however, impact assessments at a finer scale, and better information on soil and site fertility.

In north-eastern France, the future climate will relieve limiting factors of tree growth, elevated CO_2 and temperature inducing a positive response, whereas drought or extreme temperatures do not appear to offset significantly this response. The management options analysed here differ very little and management scenarios show a similar behaviour in response to climate forcing. The broadleaved forests receive a significant amount of nitrogen and are less sensitive to fertility. For this region, at similar values of soil WH, the ORCHIDEE and site-based predictions converge very well for broadleaved species GPP whereas ORCHIDEE predicts a less severe increase in respiration than CASTANEA and consequently a larger elevation of NPP.

In south-western France, site-scale predictions were parameterized with a much lower soil water reserve, 75–150 mm, than ORCHIDEE, 200–400 mm. Moreover, annual ETR values predicted by ORCHIDEE are lower than in GRAECO or CASTANEA. These differences explain why site-based predictions show a much larger effect of drought than ORCHIDEE. Where droughts are expected to increase in frequency and severity, water increasingly becomes a limiting factor for growth and production and the potential effects of other controlling variables, such as higher atmospheric CO_2 levels, do not appear.

Again, it should be kept in mind that the predictions shown here were produced according to only one business-as-usual median climate scenario (B2), which has been seriously overtaken by the recent trends. Moreover, this particular climate scenario fails to predict extreme events of which both the frequency and severity are expected to increase. Our results must, therefore, be regarded as an optimistic version of the future of French forests.

Outlook: adapting forest management to future climate change

We found that the contrast of each management scenario in their sensitivity to climate is caused essentially by the age class structure. The larger sensitivity to climate of the short rotation scenarios is dominated by the sensitivity of the juvenile phase that precedes canopy closure. At this stage, the standing biomass and, therefore, respiration are low, while productivity is high. Competition is absent so that individual trees may improve resource capture through the growth of their root and foliage areas. Juvenile trees respond straightforwardly to any change in CO_2, light or water availability. Conversely, due to competition, root and foliage expansion is increasingly constrained by limiting resources in older stands after canopy closure: hydraulic limitations (Delzon *et al.*, 2004; Magnani *et al.*, 2000) and nutrient sequestration in biomass and soil organic matter. Therefore, a change in CO_2 or water availability will not greatly affect tree growth and GPP in old stands subject to long rotation scenarios, which are additionally constrained by soil nutrients. Hence, with increasing age, stand productivity decreases while the standing biomass stock increases.

The differential interactions of climate change with forest management between the northern and southern zones are explained mainly by the contrast between management scenarios in Experiment 2, which is extreme for the south-western pine forests and weakly contrasted for the broadleaved forests in the north. The short rotation scenario applied to the south-western pine forest corresponds to intensively managed forests, optimized toward fibre production, while the long rotation scenario exceeds by 40 years the date of final cut currently applied in this area. Difference in age at the final cut between these two extreme scenarios is 55 years (or about 60% of the final age of the long rotation) and the strong response to climate of the south-western pine forest is due mainly to an age effect. Conversely, the difference in age between the two management scenarios at the final cut is only 15 years (10%) and 50 years (33%) for the northern oak and beech, respectively.

Although a sustainable forest management adaptation strategy should rely primarily on site-specific characteristics and data, we think that our results provide some valuable and important knowledge about the regional forcing trends constraining the future evolution of the forest physical environment. Where climate change effects are beneficial to forest production, our results suggest that forest management should target the reduction of limiting factors effects, for example, through increased rooting depth and fertilization. Conversely, where detrimental effects of the future climate are expected, the enhancement of resistance to drought, for example, through species substitution, LAI optimization by intensive thinning regimes and site preparation may be appropriate for attenuating the drought impacts on forest growth.

The differential changes in MAI between sites between 1980 and 2080 suggest that maps of potential forest production are expected to evolve in the near future, at least for species like maritime pine, Scots pine and beech. At the country scale, the scenario will provoke a global shift of optimal climate conditions towards the north-east of France. At the local scale, the contrast between site fertility classes may be amplified in north-eastern France, where climate will favour growth, and reduced in other regions (i.e. in the centre and southern zones).

From these conclusions, several steps for improving future impact assessment studies may be suggested as a basis for decision-making in forestry. First, and most important, forestry should use these results for drastic innovations. The main message is that the forest environment is not stable, but changes continuously in a non-linear way. Management should adapt forest stands dynamically with a response time of less than a decade. Creating specific adaptive management scenarios seems to be a relevant option since climate effects depend strongly on the management options. Forest with a long rotation could be managed to allow a continuous adaptation to climate: the phenotypic responsiveness and genetic adaptation potential may be improved, for example, allowing different age classes to grow together and mixing species by introducing southern vicariants. Conversely, shortening and intensifying short rotations based on even-aged stands may be an option for minimizing adverse climate change effects over a rotation, and renewing forest species or varieties at increased frequency. Second, synthetic predictions combining local characteristics of soil and forest stands on GIS datasets must be operated using higher resolution climate scenarios. Sufficient knowledge is now available for model implementation at this finer scale (Nuutinen *et al.*, 2006). Finally, additional climate and CO_2 future scenarios should be utilized. Recent data show that the median B2 scenario used in our study underestimates currently observed trends (Canadell *et al.*, 2007), which tentatively means that our results should be regarded as a minimum conservative prognosis of the future changes. The real ones may occur much more rapidly and be more severe. Finally, models used for future impact assessments should consider nitrogen effects, with a likely increased deposition in the scenarios and an increase in soil mineralization.

References

Arrouays D., Deslais W., Badeau V., 2001. The carbon content of topsoil and its geographical distribution in France. *Soil Use and Management*, 17 (1), 7–11.

Atkin O.K., Bruhn D., Hurry V.M., Tjoelker M.G., 2005. The hot and the cold: unravelling the variable response of plant respiration to temperature. *Functional Plant Biology*, 32 (2), 87–105.

Atkin O.K., Tjoelker M.G., 2003. Thermal acclimation and the dynamic response of plant respiration to temperature. *Trends in Plant Science*, 8, 343–351.

Ball J.T., Woodrow I.E., Berry J.A., 1987. A model predicting stomatal conductance and its contribution to the control of photosynthesis under different environmental conditions. *In: Progress in Photosynthesis Research,* Volume 4 (J. Biggins, ed.). M. Nijhoff Publishers, Dordrecht, pp. 221–224.

Becker M., Nieminen T.M., Gérémia F., 1994. Short-term variations and long-term changes in oak productivity in Northeastern France. The role of climate and atmospheric CO_2. *Annales des Sciences Forestières*, 51, 477–492.

Berbigier P., Bonnefond J.-M., Mellmann P., 2001. CO_2 and water vapour fluxes for 2 years above Euroflux forest site. *Agricultural and Forest Meteorology*, 108, 183–197.

Berthelot M., Friedlingstein P., Ciais, P., Monfray P., Dufresne J.-L., Le Treut H., Fairhead L., 2002. Global response of the terrestrial biosphere to CO_2 and climate change using a coupled climate-carbon cycle model. *Global Biogeochemical Cycles*, 16, 4.

Botta A., Viovy N., Ciais P., Friedlingstein P., 2000. A global prognostic scheme of leaf onset using satellite data. *Global Change Biology*, 6, 709–726.

Canadell J.G., Le Quere C., Raupach M.R., Field C.B., Buitenhuis E.T., Ciais P., Conway T.J., Gillett N.P., Houghton R.A., Marland G., 2007. Contributions to accelerating atmospheric CO_2 growth from economic activity, carbon intensity, and efficiency of natural sinks. *Proceedings of the National Academy of Sciences of the United States of America*, 104 (47), 18866–18870.

Ciais P., Schelhaas M.J., Zaehle1 S., Piao1 S.L., Cescatti A., Liski J., Luyssaert S., Le Maire G., Schulze E.D., Bouriaud O., Freibauer A., Valentini R., Nabuurs G.J., 2008. Carbon accumulation in European forests. *Nature Geoscience*, 1, 425-429.

Davi H., Dufrêne E., François C., Le Maire G., Loustau D., Bosc A., Rambal S., Granier A., Moors E., 2006. Sensitivity of water and carbon fluxes to climate changes from 1960 to 2100 in European forest ecosystems. *Agricultural and Forest Meteorology*, 141 (1), 35–56.

Davi H., Dufrêne E., A. Granier A., Le Dantec V., Barbaroux C., François C., Bréda N., 2005. Modelling carbon and water cycles in a beech forest. Part II: Validation of the main processes from organ to stand scale. *Ecological Modelling*, 185 (2–4), 387–405.

Delzon S., Sartore M., Burlett R., Dewar R., Loustau D., 2004. Hydraulic responses to height growth in maritime pine trees. *Plant, Cell and Environment*, 27 (9), 1077–1087.

De Rosnay P., Polcher J., 1998. Modelling root water uptake in a complex land surface scheme coupled to a GCM. *Hydrology and Earth System Sciences*, 2 (2–3), 239–255.

Dhôte J.-F., Hervé J.-C., 2000. Changements de productivité dans quatre forêts de chênes sessiles depuis 1930 : une approche au niveau du peuplement. *Annals of Forest Science*, 57 (7), 651–680.

Douville H., Planton S., Royer J., Stephenson D., Tyteca S., Kergoat L., Lafont S., Betts R., 2000. The importance of vegetation feedbacks in doubled-CO_2 time-slice experiments. *Journal of Geophysical Research*, 105, 14841–14861.

Ducoudré N.I., Laval K., Perrier A., 1993. SECHIBA, a new set of parameterizations of the hydrologic exchanges at the land-atmosphere interface within the LMD atmospheric general circulation model. *Journal of Climate*, 6, 248–273.

Dufrêne E., Davi H., François C., Le Maire G., Le Dantec V., Granier A., 2005. Modelling carbon and water cycles in a beech forest. Part I: Model description and uncertainty analysis on modelled NEE. *Ecological Modelling*, 185 (2–4), 407–436.

Felzer B., Kicklighter D., Mellilo J., Wang C., Zhuang Q., Prinn R., 2004. Effects of ozone on net primary production and carbon sequestration in the conterminous United States using a biogeochemistry model. *Tellus*, 56B, 230–248.

Friedlingstein P., Joel G., Field C.B., Fung I., 1998. Toward an allocation scheme for global terrestrial carbon models. *Global Change Biology*, 5, 755–770.

Garcia-Gonzalo J., Peltola H., Gerendiain A.Z., Kellomaki S., 2007. Impacts of forest landscape structure and management on timber production and carbon stocks in the boreal forest ecosystem under changing climate. *Forest Ecology and Management*, 241 (1–3), 243–257.

Gibelin A., Déqué M., 2003. Anthropogenic climate change over the Mediterranean region simulated by a global variable resolution model. *Climate Dynamics*, 20, 327–339.

Granier A., Loustau D., Bréda N., 2000. A generic model of forest canopy conductance dependent of climate, soil water availability and leaf area index. *Annals of Forest Science*, 57, 755–765.

Gu L.H., Fuentes J.D., Shugart H.H., Staebler R.M., Black T.A., 1999. Responses of net ecosystem exchanges of carbon dioxide to changes in cloudiness: Results from two North American deciduous forests. *Journal of Geophysical Research –Atmosphere*, 104, 31421–31434.

IPCC, 2001. *Climate Change 2001. The Scientific Basis. Contribution of Working Group I to the Third Assessment Report of the IPCC* (J. Houghton, Y. Ding, D.J. Griggs, M. Noguer, P.J. Van der Linder, X. Dai, K. Maskell, C.A. Johnson, eds). Cambridge University Press, Cambridge, 881 p.

King A.W., Gunderson C.A., Post W.M., Weston D.J., Wullschleger S.D., 2006. Plant respiration in a warmer world. *Science*, 312 (5773), 536.

Krinner G., Viovy N., de Noblet-Ducoudré N., Ogée J., Polcher J., Friedlingstein P., Ciais P., Sitch S., Prentice I.C., 2005. A dynamic global vegetation model for studies of the coupled atmosphere-biosphere system. *Global Biogeochemical Cycles*, 19 (1), article 1015.

Landsberg J.J., Waring R.H., 1997. A generalised model of forest productivity using simplified concepts of radiation-use efficiency, carbon balance and partitioning. *Forest Ecology and Management*, 95, 209–228.

Loustau D., Bert D., Trichet P., 1999. Fonctionnement primaire et productivité de la forêt landaise : implications pour une gestion durable. *Revue Forestière Française*, 51, 571–591.

Loustau D., Bosc A., Colin A., Ogée J., Davi H., François C., Dufrêne E., Déqué M., Cloppet E., Arrouays D., Le Bas C., Saby N., Pignard G., Hamza N., Granier A., Bréda N., Ciais P., Viovy N., Delage F., 2005. Modelling climate change effects on the potential production of French plains forests at the sub-regional level. *Tree Physiology*, 25 (7), 813–823.

Loustau D., Pluviaud F., Bosc A., Porté A., Berbigier P., Déqué M., Pérarnaud V., 2001. Impact of a regional $2 \times CO_2$ climate scenario on the water balance, carbon balance and primary production of maritime Pine in Southwestern France. *In: Models for the Sustainable Management of Plantation Forests* (J.M. Carnus, R. Dewar, D. Loustau, M. Tomé, eds), EFI Proceedings No. 41D. European Cultivated Forest Institute (EFI subdivision), Bordeaux, pp. 45–58.

Luyssaert S., Inglima I., Jung M., Richardson A.D., Reichsteins M., Papale D., Piao S.L., Schulzes E.D., Wingate L., Matteucci G., Aragao L., Aubinet M., Beers C., Bernhoffer C., Black K.G., Bonal D., Bonnefond J.M., Chambers J., Ciais P., Cook B., Davis K.J., Dolman A.J., Gielen B., Goulden M., Grace J., Granier A., Grelle A., Griffis T., Grunwald T., Guidolotti G., Hanson P.J., Harding R., Hollinger D.Y., Hutyra L.R., Kolar P., Kruijt B., Kutsch W., Lagergren F., Laurila T., Law B.E., Le Maire G., Lindroth A., Loustau D., Malhi Y., Mateus J., Migliavacca M., Misson L., Montagnani L., Moncrieff J., Moors E., Munger J.W., Nikinmaa E., Ollinger S.V., Pita G., Rebmann C., Roupsard O., Saigusa N., Sanz M.J., Seufert G., Sierra C., Smith M.L., Tang J., Valentini R., Vesala T., Janssens I.A., 2007. CO_2 balance of boreal, temperate, and tropical forests derived from a global database. *Global Change Biology*, 13 (12), 2509–2537.

Magnani F., Mencuccini M., Grace J., 2000. Age-related decline in stand productivity: the role of structural acclimation under hydraulic constraints. *Plant, Cell and Environment*, 23, 251–263.

Masek J.G., Collatz G.J., 2006. Estimating forest carbon fluxes in a disturbed southeastern landscape: Integration of remote sensing, forest inventory, and biogeochemical modelling. *Journal of Geophysical Research*, 111, G01006.

Maseyk K.S., Lin T., Rotenberg E., Grunzweig J.M., Schwartz A., Yakir D., 2008. Physiology-phenology interactions in a productive semi-arid pine forest. *New Phytologist*, 178 (3), 603–616.

Medlyn B.E., Barton C.V.M., Broadmeadow M.S.J., Ceulemans R., De Angelis P., Forstreuter M., Freeman M., Jackson S.B., Kellomaki S., Laitat E., Rey A., Roberntz P., Sigurdsson B.D., Strassemeyer J., Wang K., Curtis P.S., Jarvis P.G., 2001. Stomatal conductance of forest species after long-term exposure to elevated CO_2 concentration: a synthesis. *New Phytologist*, 149, 247–264.

Medlyn B.E., Loustau D., Delzon S., 2002. Temperature response of parameters of a biochemically-based model of photosynthesis. I. Seasonal changes in mature maritime pine (*Pinus pinaster* Ait.). *Plant, Cell and Environment*, 25, 1165–1175.

Milne R., Van Oijen M., 2005. A comparison of two modelling studies of environmental effects on forest carbon stocks across Europe. *Annals of Forest Science*, 62 (8), 911–923.

Morcrette J., 1990. Impact of changes to the radiation transfer parameterizations plus cloud optical properties in the ECMWF model. *Monthly Weather Review*, 118, 847–873.

Nellemann C., Thomsen M.G., 2001. Long-term changes in forest growth: Potential effects of nitrogen deposition and acidification. *Water Air and Soil Pollution*, 128, 197–205.

Niyogi D., Chang H.I., Sexana V.K., Holt T., Alapaty K., Booker F.L., Chen F., Davies K.J., Holben B., Matsui T., Meyers T., Oechel W.C., Pielke R.A., Wells R., Wilson K., Yongkang X., 2004. Direct observations of the effects of aerosol loading on net ecosystem CO_2 exchanges over different landscapes. *Geophysical Research Letters*, 31, L20506.

Nuutinen T., Matala J., Hirvela H., Harkonen K., Peltola H., Vaisanen H., Kellomaki S., 2006. Regionally optimized forest management under changing climate. *Climatic Change*, 79 (3–4), 315–333.

Ollinger S.V., Aber J.D., Reich P.B., Freuder R.J., 2002. Interactive effects of nitrogen deposition, tropospheric ozone, elevated CO_2 and land use history on the carbon dynamics of northern hardwood forests. *Global Change Biology*, 8 (6), 545–562.

Piovesan G., Biondi F., Di Filippo A., Alessandrini A., Maugeri M., 2008. Drought-driven growth reduction in old beech (*Fagus sylvatica* L.) forests of the central Apennines, Italy. *Global Change Biology*, 14 (6), 1265–1281.

Porté A., 1999. Modélisation des effets du bilan hydrique sur la production primaire et la croissance d'un couvert de pin maritime (*Pinus pinaster* Ait.) en lande humide. Thèse de doctorat, université de Paris XI, Orsay, 160 p.

Porté A., Trichet P., Bert D., Loustau D., 2002. Allometric relationships for branch and tree woody biomass of maritime Pine (*Pinus pinaster* Aït.). *Forest Ecology and Management*, 158, 71–83.

Reichstein M., Papale D., Valentini R., Aubinet M., Bernhofer C., Knohl A., Laurila T., Lindroth A., Moors E., Pilegaard K., Seufert G., 2007. Determinants of terrestrial ecosystem carbon balance inferred from European eddy covariance flux sites. *Geophysical Research Letters*, 34, L01402.

Reichstein M., Tenhunen J., Ourcival J.-M. Rambal S., Miglietta F., Peressotti A., Pecchiari M., Tirone G., Valentini R., 2003. Inverse modelling of seasonal drought effects on canopy CO_2/H_2O exchange in three Mediterranean Ecosystems. *Journal of Geophysical Research*, 108, D23, 4726.

Reichstein M., Tenhunen J.D., Roupsard O., Ourcival J.-M., Rambal S., Miglietta F., Peressotti A., Pecchiari M., Tirone G., Valentini R., 2002. Severe drought effects on ecosystem CO_2 and H_2O fluxes at three Mediterranean evergreen sites: revision of current hypotheses? *Global Change Biology*, 8 (10), 999–1017.

Ricard J., Royer J., 1993. A statistical cloud scheme for use in an AGCM. *Annals of Geophysics*, 11, 1095–1115.

Roderick M.L., Farquhar G.D., Berry S.L., Noble I.R., 2001. On the direct effect of clouds and atmospheric particles on the productivity and structure of vegetation. *Oecologia*, 129, 21–30.

Royer J., Cariolle D., Chauvin F., Déqué M., Douville H., Planton S., Rascol A., Ricard J., Salas y Melia D., Sevault F., Simon P., Somot S., Tyteca S., Terray L., Valcke S., 2002. Simulation of climate change during the 21st century including stratospheric ozone. *Géosciences*, 334, 147–154.

Sitch S., Huntingford C., Gedney N., Levy P.E., Lomas M., Piao S.L., Betts R., Ciais P., Cox P., Friedlingstein P., Jones C.D., Prentice I.C., Woodward F.I., 2008. Evaluation of the terrestrial carbon cycle, future plant geography and climate-carbon cycle feedbacks using five Dynamic Global Vegetation Models (DGVMs). *Global Change Biology*, 14 (9), 2015–2039.

Spiecker H., 1999. Overview of recent growth trends in European forests. *Water, Air and Soil Pollution*, 116 (1–2), 33–46.

Zaehle S., Bondeau A., Carter T., Cramer W., Erhard M., Prentice I.C., Reginster I., Rounsevell M.D.A., Sitch S., Smith B., Smith P.C., Sykes M., 2007. Projected changes in terrestrial carbon storage in Europe under climate and land-use change, 1990–2100. *Ecosystems*, 10 (3), 380–401.

Part III

Climate change and forest vulnerability

Climate change and the biogeography of French tree species: first results and perspectives

VINCENT BADEAU, JEAN-LUC DUPOUEY, CATHERINE CLUZEAU, JACQUES DRAPIER, CHRISTINE LE BAS

Introduction

Currently, the scientific community generally acknowledges the ongoing climate change (IPCC, 2001), and the results obtained for the Holocene period have shown that considerable geographical shifts in the distribution of species have occurred as a consequence of former climatic changes (Bradshaw *et al.*, 2000; Brewer *et al.*, 2002; Bradshaw and Hannon, 2004). In this context, many studies have been undertaken to quantify the species-environment relationships in an attempt to assess the impact of environmental changes on species distributions (Walther *et al.*, 2002; Parmesan and Yohe, 2003). The renewed interest in highlighting the main drivers of species distribution concerns all types of living communities (e.g. mammals, birds, reptiles, plants, parasites and pathogens, etc.) at the species level (Iverson and Prasad, 2002; Pearson *et al.*, 2002; Vetaas, 2002), the community level (Gignac *et al.*, 2000; Iverson and Prasad, 2001), or the biome level (Kirilenko *et al.*, 2000; Enquist, 2002; Malcolm *et al.*, 2002; Peñuelas and Boada, 2003).

A great deal of literature has also been published on the methodologies and mathematical data management tools (e.g. Manel *et al.*, 1999; Guisan *et al.*, 2000; Pearce and Ferrier, 2000; Lehmann *et al.*, 2002, 2003; Guisan, 2003; Thuiller, 2003; Rushton *et al.*, 2004). Guisan and Zimmermann (2000) published a synthesis of the methodological aspects of "predictive habitat distribution models". Two types of models – either static or dynamic – are generally used to study species-climate relationships. The dynamic models try to reproduce phenomena like competition, migration or mortality whereas the static models (also called biogeographical models or niche models) aim at reproducing the potential niche of species "at equilibrium", that is to say, the presence/absence of species for a given set of ecological parameters. The static models can integrate physiological or ecophysiological processes, but commonly, the niches are estimated by using environmental variables taken as surrogates for physiological variables (Thuiller *et al.*, 2003). So the static models are often statistical models.

The mechanistic models make it possible to understand the distribution of species through physiological aspects. However, to use these models, the biology of the species has to be known (e.g. phenology, resistance to frost and drought, carbon allocation, migration rates, etc.), but unfortunately, these data are still very incomplete for the majority of forest species.

The geographical scales vary greatly from one study to the next, from the local or regional scales (e.g. Jelaska *et al.*, 2003) to the global scale (e.g. Korner and Paulsen, 2004). Objectives may include biodiversity conservation or the development of management tools (Margules and Pressey, 2000; McKenzie *et al.*, 2003, Thuiller *et al.*, 2005). Generally speaking, predictive modelling of species distribution is limited by the available data, both for dependent and independent variables. Consequently, studies simulating shifts in species (or habitats) distribution have concerned small regions with a high geographical resolution (McKenzie *et al.*, 2003; Pinto and Gégout, 2005), or larger areas with a loss in precision (Lenihan, 1993; Huntley *et al.*, 1995; Sykes *et al.*, 1996; Pearson *et al.*, 2002).

In 2002, when we carried out this work, no result was available at our national scale: France was just a small part of European maps (Sykes and Prentice, 1995, 1996; Sykes *et al.*, 1996; Sykes, 2001; Bakkenes *et al.*, 2002; Pearson *et al.*, 2002; Thuiller, 2003). So the published maps were not sufficiently detailed for French foresters (in an attempt to anticipate tree reaction to global change) and not representative enough of French biogeographical diversity. If no study had ever been undertaken at the national scale in France, it was mainly due to the lack of data. Until recently, species distribution maps for France were relatively approximate. The distribution maps in *Atlas Florae Europaeae* are published at a 50 × 50 km spatial resolution (Jalas and Suominen, 1972–94); Dupont (1990) uses a 20 × 20 km grid, but does not consider all the French species. Such grids are too coarse for accurate full-territory extrapolations, especially in mountainous areas. For some regions, more precise maps are available: 10 × 10 km grids for Loire-Atlantique and Vendée (Dupont, 2001); 5 × 5 km for Burgundy (Bugnon *et al.*, 1998); 5 × 3.5 km for the Drôme (Garraud, 2003). However, as their names indicate, these publications concern only small regions. These remarks are also true for French climatic data and, as a result, previous works may not be very accurate at the national scale (e.g. 10' gridded meteorological datasets).

Thus, our study is an original contribution to the topic because we wanted to combine a "whole French territory" approach with the finest possible geographical resolution. To work at a high resolution was possible due to, on the one hand, the data from the French National Forest Inventory (NFI), and on the other hand, the gridded climatic data from the National Weather Service (Météo-France). The results presented here are the first we obtained with these datasets and the main objectives for the CARBOFOR project was as follows:
– to identify the species whose geographical distribution is primarily controlled by climatic constraints;
– to calibrate niche models to estimate current species occurrences;

– to draw the potential shift of the most sensitive species to climate by using a global warming scenario for the 21st century;
– to provide, if possible, a supplementary tool for forest managers to help them to select appropriate species for reforestation.

Materials and methods

Species data

The tree species occurrences were extracted from two NFI databases: the "tree database" and the "floristic database". Usually used at a regional scale, these data had never been aggregated on the national scale, so a first homogenization step was necessary.

The "tree database" includes tree measurements and is used by the NFI to assess the wood resources in French forests. Data are available for the whole country, and many of the 96 French administrative divisions (departments) have been surveyed several times (at 12-year intervals on average). However, these observations were not used generally because of the following constraints:
– the tree database gathers observations on 41 individual species and 22 species groups (e.g. willows, birches, ashes, fruit trees, maples, etc.). Some groups include a very large number of species (i.e. fruit trees, willows and other indigenous broadleaved with 31 different species). In these groups, many species, with different distributions and ecology, are unfortunately put together under the same heading (e.g. *Fraxinus angustifolia* and *Fraxinus excelsior*; *Tilia cordata* and *Tilia platyphyllos*; *Acer campestre* and *Acer monspessulanus*, etc.);
– as the NFI survey is designed to supply information for wood production estimates, only adult trees with a circumference of at least 22.5 cm diameter at breast height (DBH) are taken into account. This size threshold can sometimes lead to a serious underestimate of species occurrence at the regional scale (e.g. Mediterranean regions).

The "floristic database" supplies more precise information regarding the occurrence of species. Each one is clearly identified, so the tree database species groups can be clarified. The species are observed whatever their size, thus, on the one hand, the floristic database supplies observations on shrubs, which are subjected less to human impacts than commercial tree species. On the other hand, if a tree species is present in the database we cannot know if it is an adult tree or a seedling.

The first floristic inventories were performed in the middle of the 1980s in the north-eastern quarter of France and in the Mediterranean region, but they have been systematic for the whole territory only since 1992. When we began this work, the "standard floristic database" was available for 74 departments (80,087 survey points). Thus, some important gaps existed, notably for the north-west, Burgundy, the western Pyrénées, the Massif Central mountains and the Limousin region. To supplement the standard floristic observations, we used older floristic inventories carried out by the NFI teams in Montpellier and Nancy. These earlier data concern eight departments and 9061 survey points (Pyrénées-Orientales, Ariège, Aude, Tarn, Bouches-du-Rhône, Alpes-Maritimes and Corsica). For these points, the inventory methods do not correspond exactly to the standard protocol adopted in 1992 (e.g. inventory on 100 m² versus the current standard 700 m² surface).

For the other departments without floristic inventories, the original tree database field notes were revised in order to split species groups into individual species. This point concerned 14,614 sites and 12 departments (i.e. Somme, Aisne, Eure, Eure-et-Loire, Loiret, Côte-d'Or, Saône-et-Loire, Haute-Vienne, Corrèze, Cantal, Puy-de-Dôme and Creuse).

This makes a total of 104,259 sites surveyed from 1985 to 2001 (i.e. an accuracy of one sample site for every 135 ha of forest) (Figure 11.1, Plate 13), and data were analysed for the occurrence of 67 different species (Table 11.1).

Table 11.1. Occurrence of the 67 studied species in the 104,259 survey points.

Species	Number of presences	Species	Number of presences
Abies alba	15,907	*Laurus nobilis*	225
Acer monspessulanum	2662	*Mespilus germanica*	1204
Acer opalus	3172	*Olea europaea*	414
Acer platanoides	1807	*Ostrya carpinifolia*	285
Acer pseudoplatanus	11,072	*Phillyrea angustifolia*	2517
Alnus incana	266	*Phillyrea latifolia*	2155
Alnus viridis	330	*Picea abies*	15,290
Amelanchier ovalis	5448	*Pinus cembra*	242
Arbutus unedo	2589	*Pinus halepensis*	2820
Betula pendula	19,446	*Pinus pinaster*	10,162
Betula pubescens	1197	*Pinus pinea*	234
Buxus sempervirens	7144	*Pinus sylvestris*	19,674
Carpinus betulus	27,153	*Pinus uncinata*	1206
Castanea sativa	19,846	*Pistacia lentiscus*	937
Cedrus atlantica	749	*Pistacia terebinthus*	1122
Cedrus brevifolia	14	*Prunus brigantina*	42
Celtis australis	23	*Quercus ilex*	8351
Cercis siliquastrum	17	*Quercus petraea*	31,544
Cotinus coggygria	519	*Quercus pubescens*	15,849
Cupressus sempervirens	84	*Quercus pyrenaica*	2211
Erica arborea	2557	*Quercus robur*	37,762
Erica scoparia	4936	*Quercus suber*	1148
Euonymus latifolius	140	*Rhamnus alaternus*	2181
Fagus sylvatica	36,748	*Rhamnus alpinus*	544
Ficus carica	62	*Rhus coriaria*	45
Frangula alnus	8196	*Sambucus racemosa*	2945
Fraxinus angustifolia	365	*Sorbus aria*	13,578
Hippophae rhamnoides	69	*Sorbus aucuparia*	9363
Juniperus communis	11,946	*Sorbus domestica*	1457
Juniperus oxycedrus	3389	*Sorbus mougeotii*	382
Juniperus phoenicea	702	*Tilia cordata*	3424
Laburnum alpinum	59	*Ulmus glabra*	902
Laburnum anagyroides	1184	*Viburnum tinus*	1196
Larix decidua	2280		

Current climatic data

We recovered climatic maps from the National Weather Service's AURELHY model (Benichou and Le Breton, 1987). These maps give 30-year monthly norms for precipitation, minimum and maximum temperatures (period 1961–90), and the number of days with frost ($T_{min} < 0°C$, $T_{min} < - 5°C$ and $T_{min} < - 10°C$) (period 1971–2000). These parameters are available on a 1×1 km grid (i.e. 551,716 coordinates for the whole French territory).

We also recovered surface solar irradiance from the Space Observation Meteorological Processing and Storage Service (SATMOS Météo-France/National Centre for Scientific Research (CNRS)). The monthly means of global irradiance are obtained from the visible wavelength channel of Meteostat sensors using a method derived from Gautier *et al.* (1980). Solar irradiance was available from 1996 to 2002 on a 0.034° grid (~3 km grid). For each coordinate of the AURELHY grid, we retrieved a monthly solar irradiance value from the SATMOS grids using the nearest neighbour method of ArcMap. Other parameters, like wind speed or relative humidity, are important to take into account in the description of climates, however, such data are not available at the national scale.

From all these data, other indices were calculated in order to describe better the habitats:

– mean monthly temperature: $T_{moy} = (T_{max} + T_{min})/2$;

– mean annual temperature and extremes for minimum, maximum and average temperatures;

– temperature ranges: highest T_{max} – lowest T_{min}, highest T_{min} – lowest T_{max}, highest T_{mean} – lowest T_{mean};

– monthly potential evapotranspiration (PET) according to Turc's formula;

– monthly precipitation deficits (P – PET) and occurrence of water deficit (1 if (P – PET) < 0 and 0 if (P – PET) ≥ 0).

Cumulative values, two-monthly and quarterly means complete the dataset (see Table 11.2 for an exhaustive list of the climatic variables used in the project).

Future climatic data

Future climatic data were extracted from the ARPEGE climate model (Gibelin and Déqué, 2001; see Chapter 6 of this volume), forced by an effective greenhouse effect corresponding to the provisional Intergovernmental Panel on Climate Change (IPCC) B2 scenario (Special Report on Emission Scenarios – www.ipcc.ch).

Daily outputs were summarized into monthly 30-year norms: (i) the current period (1961–90); (ii) a near future (2020–49); and (iii) a distant future (2070–99).

AURELHY and SATMOS data were plotted against the ARPEGE current period. We noticed a systematic bias between the observed data and the modelled data and poor correlations, whatever time period was used (1961–90, 1971–2000 or 1996–2002). Thus, for each of the 240 ARPEGE coordinates, anomalies between future periods and the current period were calculated. Then, the future climatic conditions for the whole territory were estimated by adding the ARPEGE anomalies to the current AURELHY and SATMOS values.

Table 11.2. Summary of the climatic variables used in the study. The two-monthly and quarterly values correspond to the average (or the sum) of 2- or 3-month values consecutive with a 1-month shift throughout the year (e.g. Jan./Feb., Feb./Mar., Mar./Apr., etc. and Jan./Feb./Mar., Feb./Mar./Apr., Mar./Apr./May, etc.).

	Number of values					Total
	Monthly	**Extreme**	**Two-monthly**	**Quarterly**	**Annual**	
Minimum temperatures	12	2	12	12	1	39
Maximum temperatures	12	2	12	12	1	39
Mean temperatures	12	2	12	12	1	39
Temperature range					3	3
Irradiance	12		12	12	1	37
Precipitations	12		12	12	1	37
Potential evapotranspiration	12		12	12	1	37
Precipitation deficits	12		12	12	1	37
Drought index	12		12	12	1	37
Number of days of frost ($T < 0°C$)	12		12	12	1	37
Number of days of frost ($T < -5°C$)					1	1
Number of days of frost ($T < -10°C$)					1	1
Total	108	6	108	108	14	344

Soil characteristics

Since 1992 the NFI gives a systematic description of the environmental characteristics of each survey plot (e.g. altitude, slope, aspect, bedrock, soil and humus type, etc.). However, like the floristic inventories, such descriptions still lacked considerable portions of the French territory when we began this work.

With the objective of filling these gaps, we wanted to use the French soil database (1: 1,000,000-scale/Infosol Unit/French National Institute for Agricultural Research (INRA), Orléans (Jamagne *et al.*, 1995). Unfortunately, the comparison between the two datasets revealed that the NFI soil types (when available), do not usually correspond to the national soil map. This can be partially explained by the fact that the data of the French soil database come mainly from observations performed on agricultural soils.

So, using the 1: 1,000,000-scale soil database to rapidly estimate forest soil characteristics was impossible. However, to obtain a general indication of soil characteristics for the whole territory, which was easy to integrate into our models, we used pedotransfer rules to transform the national soil map into a three-level soil acidity map (i.e. calcareous, acidic and neutral soils) (Figure 11.2, Plate 13).

Data analysis

Analysis of species distribution maps

Species distribution maps were drawn and compared with those published in the *Atlas Florae Europaeae* (Jalas and Suominen, 1972–94) and the *Flore Forestière Française* (Rameau *et al.*, 1989; 1993). This step allowed us: (i) to verify the coherence of maps; and (ii) to check if the prevalence of species was sufficient or not for individual modelling. For example, the olive tree (*Olea europaea*) and the stone pine (*Pinus pinea*) are two Mediterranean species but their presence in forest areas is occasional. So the NFI data are unsuitable to calibrate a niche model for these two species.

However, even if the individual frequency is weak, species can supply information on regional climates, once grouped together. Thus, we grouped together the 67 ligneous species into 13 chorological groups according to a visual assessment of the individual distribution areas (Figure 11.3, Plate 13).

Defining homogeneous biogeographical zones

A principal components analysis (PCA) of the 344 climatic variables × the 104,259 NFI points was carried out. The PCA allowed us to perform some classical analyses of the correlations between the variables or the variance explained by each factorial axis, and to classify each NFI point along these axes.

We calculated the average position of each species along the principal components of the PCA in order to verify whether or not the 13 chorological groups defined earlier remained coherent with the climatic variability.

Finally, discriminant function analyses were used to determine which climatic variables are the best predictors of the chorological groups. A first series of analyses was performed on each chorological group against all the others to identify which populations were the most characteristic from a climatic point of view. Another series of analyses was then performed to determine which variables discriminate between all (or a subset) of the chorological groups. The discriminant function analyses were performed with the entire set of climatic variables with no preliminary selection (e.g. after a PCA as for Bakkenes *et al.*, 2002).

Species distribution modelling

Logistic regression is one of the classic methods used for analysing binary data, and particularly to characterize species/environment relationships (Austin *et al.*, 1984; Ter Braak and Looman, 1986; Coudun and Gégout, 2005). The presence/absence of five species was analysed with a "logit" link (logistic procedure, SAS 8.2, 1999). The 344 climatic variables and the soil acidity were taken as explanatory variables and were selected stepwise according to Chi^2 values. The spatial autocorrelation was not taken into account at this step.

Selecting the best models was the most sensitive part of the procedure. Owing to the sample size (104,259 survey points), the majority of the climatic variables were significantly correlated with the presence/absence of the species. Numerous explanatory variables could be added to the models with a significance level lower than 0.0001 and the use of a subset to calibrate the statistical models (from 100% to 25% of the original NFI dataset) had no influence on the results (best correlated variables, estimates and confusion matrix). Significant differences could appear if the calibration set was smaller than 25%, but the differences were then due to re-sampling bias within the geographical zones where the real probability of occurrence of a species decreases substantially.

As classical criteria (Akaike's information criterion, pseudo R^2) were not efficient for an automated selection of the models (Shtatland *et al.*, 2001, 2002), an expert-based selection procedure was used. We subjectively and strongly limited the number of explanatory variables in the models by avoiding collinearity between them. At each step, we checked if the addition of a new variable had improved or not the concordance between the modelled and observed occurrences: (i) by studying how effective the model was in reclassifying species (confusion matrix, receiver-operator characteristic curves (ROC) and area under the ROC curves (AUC)); and (ii) by a visual assessment of the produced maps.

The confusion matrix depends on the selected threshold probability at which the presence of a species is accepted (e.g. 0.5) (Manel *et al.*, 1999, 2001). Furthermore, the distribution curve of the probabilities varies from one species to the next, which makes the comparison between species difficult. The ROC curves were used to measure overall model performance (Zweig and Campbell, 1993; Fielding and Bell, 1997). They are obtained by plotting sensitivity (defined as the percentage of correctly predicted presences (true presences/total number of presences in the sample) versus 1 - specificity (true absences/total number of absences in the sample). The AUC indicates the performance of the model: usually AUC values from 0.5 to 0.7 indicate a low accuracy, 0.7 to 0.9 indicate an average accuracy and values > 0.9 indicate high accuracy.

Sensitivity and specificity were also plotted in a raw form against the whole range of probability values to take into account the intermittent character of the species. Indeed, the observed species distribution does not necessarily correspond to the potential species range. This is especially true for commercially managed tree species and for NFI data, which concern forested areas only. As a result, cases of "false absences" in our data were modelled as "presences" and integrated into the tests as "false presences", lowering the results of the model's predictive power (compare the example of beech in Figure 11.11). This problem was exacerbated by the fact that we modelled occurrences using inventory points rather than a continuous grid (e.g. 10×10 km raster), which would have buffered artificial spatial variations in species presence/absence (e.g. forest management) (Pearce and Ferrier, 2000).

Once we determined that a model was rather robust, it was used to calculate species occurrence probability based on the climatic variables simulated for the near and distant future.

Results

Principal component analysis of current climatic variables

The PCA reduced our 344 climatic variables $\times$ 104,259 points to six main factors, which accounted for more than 90% of the total variance (respectively, 55%, 16%, 11%, 4%, 3% and 2%). The first four factors were easily recognizable:
– the first axis was essentially correlated with temperatures (positive correlations with T_{min}, T_{max} and PET and negative correlations with the number of days of frost);

– the second axis was correlated with irradiance levels, particularly for the months from September to March;

– the third axis was essentially correlated with precipitation and precipitation deficit (positive correlation), but also with irradiance levels and the number of days of frost (negative correlations);

– the fourth axis was linked to temperature range and, to a lesser extent, to maximum springtime temperature.

Figure 11.4 (Plate 14) shows the geographical distribution of the NFI points according to their location on these four factors. The first factor contrasts the Mediterranean region and the south-western quarter of France on the one hand with the mountainous regions and the north-eastern part of the country on the other hand. The second factor structures the NFI points according to latitude; this corresponds to the increased irradiance intensity from north to south. The third factor essentially puts in evidence regions with high precipitations. Factor 4 shows a contrast between the Atlantic coastal regions and the interior of the country. The patterns observed do tend to show the robustness of climatic parameters in spatially discriminating the survey points and the major climatic zones in France.

Species classification and main climatic factors

Figure 11.5 presents the average position of the 67 species along the first two factors of the PCA presented above. Two sets of species are clearly opposed on the first factorial axis. On the one hand, we found the mountainous species (*Pinus cembra* and *P. uncinata*, *Alnus viridis, Larix decidua, Sorbus mougeotii, Prunus brigantina* and *Rhamnus alpinus*); on the other hand, we found the Mediterranean species (*Quercus ilex* and *Q. suber, Pinus halepensis, Olea europaea*, etc.). This contrast is coherent with the definition of the first factor based on a temperature gradient.

For the second factor, both sets of species mentioned above (Mediterranean and mountainous) contrast with *Carpinus betulus, Betula pubescens* and *B. pendula, Quercus petraea* and *Q. robur*, and *Tilia cordata*. These species are completely absent in the Mediterranean regions and southern Alps. Thus, the second factor separates species from regions with high levels of solar irradiance and species from the northernmost regions under study.

Using this result, we verified the coherence of the 13 chorological groups defined earlier from the individual distribution maps (Figure 11.3).

The first three chorological groups correspond to mountainous species.

• Group 1 is mostly represented by species from the inland Alps (*Pinus cembra, Alnus viridis, Laburnum alpinum* and *Prunus brigantiaca*) but also includes species that may be found as far afield as the Massif Central (*Sorbus mougeotii*) or the eastern Pyrénées (*Pinus uncinata, Rhamnus alpina*). This group corresponds to the "subalpine species".

• Group 2 includes the species mostly present in the mountain zones in the Alps, but at lower altitudes than Group 1 (*Acer opalus, Alnus incana, Euonymus latifolius* and *Hippophae rhamnoides*), the eastern Pyrénées, the south-western edge of the Massif Central and Jura mountains.

• Group 3 corresponds to the species that are commonly found in all the mountainous regions in France (the Alps, Pyrénées, Massif Central, Jura, Vosges and the Ardennes) and at foothill elevations in the north-eastern quarter of the country (*Abies alba, Picea*

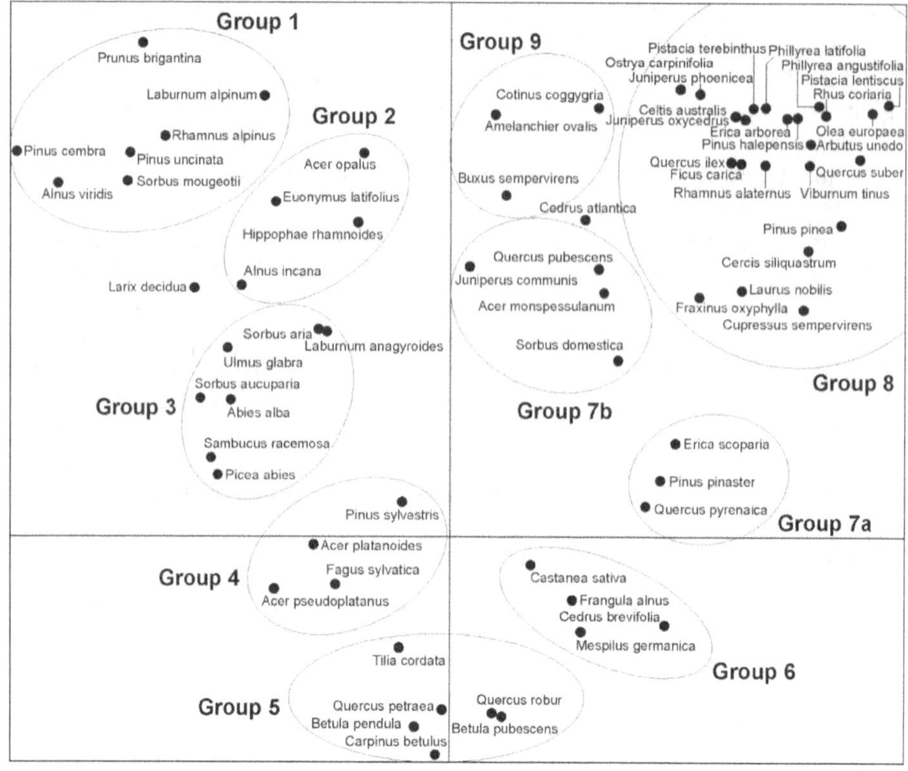

Figure 11.5. Average positions for the 67 studied species on the first factorial space calculated by the principal component analysis (PCA) of the climatic variables.

abies, Sorbus aria, S. aucuparia, Sambucus racemosa, Laburnum anagyroides and *Ulmus glabra*). Silver fir and spruce, in particular, have been introduced into the north-west.
• Larch (*Larix decidua*), although typical of Group 1, was kept apart since the introduction of the species at foothill elevations has shifted its area of distribution toward an intermediate position between Groups 1 and 3.
• Group 4 is an extension of Group 3. It is made up of species that are common in mountainous regions (up to mountain elevations), yet are also very often found in the plains of northern France (*Fagus sylvatica, Acer platanoides, A. pseudoplatanus* and *Pinus sylvestris*).
• The species in Group 5 are mostly found in foothill regions but may also climb to mountain elevations (except in the southern Alps). They are very common throughout a considerable part of the territory excluding the Mediterranean region (*Quercus robur, Q. petraea, Carpinus betulus, Betula verrucosa, B. pubescens* and *Tilia cordata*).
• Group 6 is an intermediate between Groups 5 and 7. It contains species from the foothill zone that are frequent in the south and west but scarcer in the north and north-east (*Castanea sativa, Mespilus germanica* and *Frangula alnus*).
• Group 7 includes species that mostly occur in western France. Sub-group 7a includes all the species that commonly grow in the western half of the country (mainly in the Landes de Gascogne region, the mountains of Brittany and the Sologne region) but that may extend as far as southern France (*Pinus pinaster, Erica scoparia* and *Quercus pyrenaica*).

• Sub-group 7b covers a similar distribution area to Sub-group 7a but the species are differentiated by a relatively lower frequency of occurrence in the west (and in the Landes de Gascogne in particular) and a relatively higher frequency of occurrence in the Mediterranean area. The two main species in this sub-group are *Quercus pubescens* and *Sorbus domestica*, which share the same distribution area. The two sub-species *Juniperus communis* and *J. nana* are also included in this sub-group.

• Group 8 corresponds to the Mediterranean species. If Figure 11.5 does not allow any sub-group to be distinguished, a visual assessment of the species distribution maps revealed four different sets.

– Sub-group 8a: species are strictly limited to the Mediterranean Basin (*Juniperus oxycedrus, Pinus halepensis, Phillyrea angustifolia, Pistacia lentiscus, Juniperus phoenicea, Olea europaea, Ostrya carpinifolia, Pinus pinea, Cupressus sempervirens, Rhus coriaria* and *Celtis australis*);

– Sub-group 8b: species that extend beyond the Mediterranean basin into the Aveyron, Lot, Tarn or Tarn-et-Garonne departments (*Erica arborea, Phillyrea latifolia, Pistacia terebinthus, Ficus carica* and *Cercis siliquastrum*);

– Sub-group 8c: Mediterranean species that are fairly frequent in Aquitaine and Poitou-Charente (*Rhamnus alaternus, Viburnum tinus, Fraxinus oxyphylla* and *Laurus nobilis*);

– Sub-group 8d: species that extend even farther west to the Loire region (Pays de la Loire), mostly along the coastline, but that do not reach the frequency levels of the species in Group 7 (*Quercus ilex* and *Q. suber, Acer monspessulanum* and *Arbutus unedo*).

• Finally, *Buxus sempervirens, Amelanchier rotondifolia* and *Cotinus coccygia*, species found mostly in the southern Alps in the supramediterranean vegetation belt make up Group 9, which is an intermediate group between Mediterranean and mountainous species groups.

Defining homogeneous biogeographical zones as a function of current climatic parameters

The first series of analyses showed that some groups could not be easily distinguished from others. Sub-groups 8a, 8b, 8c and 8d cannot be differentiated, which confirms the above result, and perhaps, that the visual assessment of these sub-groups was flawed. It was not possible to distinguish Sub-group 7b (which covers the same climatic zones as 7a and 8), or Group 9 (which overlaps groups 1 + 2 + 3 and 8). As Group 5 extends over a wide range of climatic conditions the NFI points were often re-classified into other groups (especially Groups 3, 4, 6 and 7a). Conversely, some groups were easy to separate from the others with only a limited number of climatic variables. The July PET was enough to separate 8a + 8b ($N = 7838$) from all the other groups. This single parameter allowed the correct re-classification of 95% of the points of these two groups with a probability threshold ≥ 0.5 (Figure 11.6(a), Plate 15). For groups 1 + 2 ($N = 7146$), three parameters were required to re-classify correctly 72% of the NFI points: the annual number of days with minimum temperatures below -5°C, the February solar irradiance and the November PET (Figure 11.7(a), Plate 15).

In a second step, by putting all the groups together, the best differentiation was obtained using chorological Groups 1, 2, 3, 4, 6, 7a and 8a. The use of five climatic variables allowed the correct reclassification of 40% of the NFI points. Table 11.3 summarizes the correct reclassification percentages for each chorological group and

Table 11.3. Discriminant analysis results for seven chorological groups. Number of points in each group, correct reclassification percentages and inter-group distances are shown.

Groups	N	Reclassified binary response						
		G1 (%)	G2 (%)	G3 (%)	G4 (%)	G6 (%)	G7a (%)	G8 (%)
G1	4122	**55**	21	12	8	2	0	2
G2	3536	12	**52**	12	7	2	0	14
G3	38308	16	13	**32**	22	10	2	5
G4	30183	12	14	20	**25**	21	3	6
G6	25748	0	4	6	11	**44**	27	7
G7a	12082	0	1	0	0	24	**56**	19
G8	11663	0	7	0	0	3	15	**74**

Groups	N	Inter-group distances						
		G1 (%)	G2 (%)	G3 (%)	G4 (%)	G6 (%)	G7a (%)	G8 (%)
G1	4122	–	3.5	2.8	3.7	8.6	12.8	13.8
G2	3536	3.5	–	1.7	1.8	5.3	8.6	7.2
G3	38308	2.8	1.7	–	0.2	3.2	7.3	9.9
G4	30183	3.7	1.8	0.2	–	2.0	5.5	8.5
G6	25748	8.6	5.3	3.2	2.0	–	1.0	5.3
G7a	12082	12.8	8.6	7.3	5.5	1.0	–	3.6
G8	11663	13.8	7.2	9.9	8.5	5.3	3.6	

shows generalized squared distances between groups. Groups 1 (subalpine), 7a (west) and 8 (Mediterranean) have the highest percentage of correctly reclassified points (55%, 56% and 74%, respectively); Groups 1, 2 and 3 (mountain species) are very close and Group 4 is the least differentiated. The discriminant function includes the PET for January, the PET for June + July, the PET for November, the solar irradiance from May to July and the mean temperature for February.

When we applied the function to the 551,716 points of the AURELHY grid, we obtained a map of the French broad biogeographical regions (Figure 11.8, Plate 16). This map was globally coherent with previous published results (Lamarck and De Candolle, 1805 *in* Ebach and Goujet, 2006; Goujon, 1932 *in* Grandjouan, 1982; Rameau *et al.*, 2000) and seven zones can be clearly distinguished: a Mediterranean zone (corresponding to Group 8a), three mountainous zones (Groups 1, 2 and 3), an Aquitaine zone (Group 7a), an Atlantic zone (Group 6) and a more continental north-eastern zone (Group 4). It is useless to attempt a more precise interpretation of this map, particularly with regard to the delimitation of boundaries for each zone, since calculations were based on 40 ligneous species only sorted into seven coarse chorological groups. Figure 11.9 (Plate 16) shows the probabilities associated with the biogeographical map. For current climatic conditions, the Mediterranean area, the mountainous zones and the Aquitaine region were all associated with strong probability values (0.80 on average for Group 8, 0.78 for Group 1 and 0.56 for Group 7a). Conversely, probabilities associated with Groups 6, 3 and 4 were quite low (0.43, 0.41 and 0.38, respectively). Many of the grid-points were attributed to one or another of these three zones not because of a strong probability of occurrence in that zone, but only because that probability was somewhat higher than for the other groups; they were classified by default. Thus, the predictability of membership of one of the seven chorological groups is much better in mountainous areas and the Mediterranean region than in other regions.

Lastly, the discriminant function was used for mapping the chorological groups for a forecasted climatic scenario (Figure 11.8(b) and (c)). According to scenario B2, the greatest increase should occur for Group 7a (17% of the territory under current climatic conditions versus 46% in 2100) and for Group 8 (9% of the territory under current climatic conditions versus 28% in 2100). All the other chorological groups can be expected to shrink, particularly in the mountainous regions with Groups 1, 2 and 3 decreasing from 16% of the territory to 6%.

Reaction of individual species to climate

The maps in Figures 11.10–11.14 (Plates 17 to 21) show the results obtained after analysing the presence/absence of five tree species with logistic regressions. Each figure presents: (i) the current species distribution as observed by the NFI; (ii) the modelled species occurrence as a function of current climatic parameters; (iii) the current climatic niche obtained after applying the model to the whole AURELHY grid; (iv) and (v) the two potential climatic niches modelled for predicted future climatic variables. The curves in the graphs show the distribution for sensitivity and specificity.

Evergreen oak (Quercus ilex)

As in the previous exercise (chorological Groups 8a and 8b), the PET of July proved sufficient to identify the survey points with evergreen oak along the Mediterranean region (positive indicator of oak occurrence). For this model the AUC = 0.967. By adding temperature range ($\max T_{mean} - \min T_{mean}$) and the number of days of frost with temperatures below $-10°C$ (negative effect of the two variables), we were able to decrease the species occurrence in the southern Alps and better estimate its presence along the Atlantic coast. However, the value of AUC was not much modified (0.973).

Using these three variables, the model was able to correctly reclassify more than 70% of the NFI survey points where this oak is present with a probability threshold ≥ 0.5. Nevertheless, to represent the occurrence of evergreen oak along the Atlantic coast, very low probability thresholds have to be taken into account. Conversely, because *Q. ilex* is present almost everywhere in the forests of the Mediterranean basin, the number of false presences is weak in this region, even if very low levels of probability are retained (compare the sensitivity and specificity curves).

The potential distribution by 2050 and 2100 modelled for evergreen oak corresponds to the results obtained for the chorological groups, and showed a strong extension from the Mediterranean area towards the Aquitaine region.

Beech (Fagus sylvatica)

Beech occurrence was very strongly (and negatively) correlated to the increase in cumulative precipitation deficit for the months of June + July (AUC = 0.77). With this variable, the model was able to estimate beech occurrence for the whole territory except for the Mediterranean region. For all the mountainous regions, as well as for the plains in the north-western quarter, the north, Normandy and Brittany, beech occurrence corresponds to a probability threshold of 0.3.

This result is coherent with those obtained in previous studies. According to Badeau (1995) and Granier *et al.* (1995), the total soil water deficit for June to August, calculated for the current and previous year, can explain more than 70% of beech annual radial growth. This result, obtained in Lorraine (north-eastern France), was confirmed by Lebourgeois (2005) at a national scale within the RENECOFOR network (level II plots). He noticed the strong influence of the early summer soil water deficit on beech radial growth and that the deficit in June alone explains a large part of the radial growth variability (mean value of 26.6%). Badeau and Bréda (1997) also showed a strong relationship between the intensity of summer water stresses and beech leaf loss the following year (besides the correlations with the fructifications).

By adding maximum temperatures for the month of October, we were able to balance occurrence predictions on the Atlantic coast and in the Garonne River valley and to increase probability rates for the north, Brittany and Normandy (AUC = 0.79). Although adding the variable improved the statistical robustness of the model, its effect was difficult to interpret from a functional/physiological point of view. The two-variable logistic model was able to predict more than 80% of the points where the species occurs at probability thresholds ≥ 0.3. In a comparable study including beech and other species in Switzerland, Bolliger *et al.* (2000) report a 70% prediction rate with $P > 0.21$. Visual verification of the maps revealed that a threshold value of 0.3 produced areas corresponding to the major beech current wood production regions. For other areas where probability threshold values were below 0.3 (blue zones on maps in Figures 11.11(c) and (d), beech is in fact much less frequent in the forested areas inventoried by the NFI even though it is always present (Jalas and Suominen, 1972–94; Rameau *et al.*, 1993; Dupont, 2001). An AUC equal to 0.79 indicates an average accuracy for the model from a statistical point of view: the value is lowered by the increasing number of false presences. Nevertheless, these false presences are mainly located in geographical areas where the beech is abundant. Thus, the AUC simply indicates that the statistical model cannot take into account the choices of the forest owner, even in the major beech producing regions and even if the climate is not a limiting factor.

Contrary to *Quercus ilex*, predictions for potential climatic niches showed that beech could decline considerably leaving the species confined to the north-eastern part of the country and mountainous areas. Without going into details, these maps are in accordance with those produced by Sykes and Prentice (1995, 1996) or Thuiller (pers. comm.).

Silver fir (Abies alba)

As for beech, silver fir occurrences were strongly correlated to precipitation deficit, but this time accumulated from April to June (AUC = 0.85). As was the case for beech, this result is in accordance with the inter-annual variations in radial growth for silver fir in the Jura and Vosges mountains (Bert, 1992; Becker, 1989). In these two studies, the correlation between the radial growth and water availability at the beginning of the growing season (May and June for years N and $N-1$) was clearly demonstrated.

The second variable selected in the model was the number of days of frost in January. It did not modify the value of the AUC (0.86 instead of 0.85), but on the one hand, fir disappeared from the predicted range in Brittany and Normandy, and on the other hand, the model was more accurate for all the southern half of France especially the Atlantic Pyrénées, the Vaucluse, the northern Alps (Alpes de Haute-Provence) and the coastal

Alps (Alpes-Maritimes). This led us to question the validity of including this variable in the model: (i) silver fir is actually present in the north-western plains, but it was artificially introduced; (ii) the frequency of frosts made this variable consistent with the species preference for mountain habitats.

Forecasted scenarios for 2050 and 2100 showed that, like the beech, silver fir could undergo a drastic reduction in its range, especially at lower altitudes.

Spruce (Picea abies)

Spruce shows similarities to silver fir in its geographical distribution but it is less frequent in the Pyrénées mountains and more common in the north-western plains. Spruce was artificially introduced into these two regions. The NFI data revealed that the current spruce occurrences extend far beyond the borders of its natural range (the Jura and the Alps and probably the Vosges mountains). Even so, a model including the cumulative rainfall for June and July (positive effect) and the mean temperatures from January to March (negative effect) predicted the occurrence of spruce throughout France (AUC = 0.87), even in areas where the species was mostly introduced (i.e. Morvan, Massif Central, Ardennes and the north-eastern plains).

The potential redistributions of spruce for 2050 and 2100 were very similar to the forecasts calculated for silver fir.

Maritime pine (Pinus pinaster)

As in the case of spruce, maritime pine has been widely introduced into France and its current distribution largely exceeds its natural range. However, whereas our climatic model was able to estimate accurately occurrence for spruce, this was not the case for maritime pine. The variable with the best correlation to maritime pine occurrences was January PET (positive effect). In a model with two variables, the second one was the mean temperature for January and February (positive effect). However, at this step: (i) this variable was closely correlated to January PET; and (ii) the positive effect of the previously entered PET variable turned into a negative effect. The third variable used in the model was frost frequency in March. Surprisingly, this variable appeared to have a positive effect on species occurrence for maritime pine. All three variables proved to be unsatisfactory predictors for well known ecological and physiological reasons (as evidenced by the maps produced). Nevertheless, from a statistical point of view, the AUC = 0.91 indicates the high accuracy of the model.

Using one, two or all three of the variables only modified the level of probability of occurrence and did not change the general shape of the distribution areas shown on the maps. We therefore chose to retain only one variable (PET for January) in the final model. Even though the geographical distribution produced by the model for maritime pine did not correspond to the actual observed distribution, it did broadly encompass the regions where maritime pine is present, from the Cotentin region to the surroundings of Nice. However, probability thresholds were very low and the strongest probabilities did not correspond to the current areas with the densest populations of maritime pine.

When future scenarios were modelled, the results were similar whether one, two or three variables were used: maritime pine was shown to spread north up the Rhone Valley and west to the Paris basin.

Discussion, conclusion and perspectives

One of the key points of this study was to group together different databases in order to create a tool for the analysis of the species-environment relationships: (i) at the national scale; and (ii) with a high level of detail.

As regards the environmental parameters, we can say that the AURELHY model is currently the most accurate in characterizing the climatic conditions of the territory. It provides only 30-year monthly norms; it covers only the French territory (and it is not free), but the high resolution grids are interpolated from the whole climatic database of the National Weather Service (Météo-France). This is not the case for others gridded datasets (e.g. Mitchell *et al.*, 2003; Hilmans *et al.*, 2005), making debatable the general quality of some parameters (e.g. precipitation). One of the most important climatic variables, from a vegetation point of view, is the incoming solar radiation, which drives temperature and PET and, as a result, soil water availability. For this parameter, ground data observations are not sufficient to draw accurate interpolated surfaces, but associated with satellite data one can obtain a realistic tool. So, the SATMOS data were useful for our study and their current spatial resolution of 3×3 km should evolve in years to come. Within the CARBOFOR project, we did not use a homogenous soil database. The NFI observations were not available for the whole territory and the French Soil Geographical Database cannot be used without caution for forest soils. Thus, in this work, we were not able to study the soil-dependent species (e.g. *Acer campestre*) (Coudun *et al.*, 2006).

As regards the occurrence of species, the NFI databases allowed us to work with 104,259 geo-referenced sites covering the entire French territory, evenly distributed throughout the country, and sampled with a standardized protocol. This accuracy had never been reached previously. Thuiller *et al.* (2005) worked at the European scale with *Atlas Florae Europaeae* maps, that is, with 2089 grid cells for Europe and only 291 for France. Coudun (2005) and Coudun *et al.* (2006) worked at the French national scale with 3286 sites from the EcoPlant database. Therefore, our analyses based on NFI data in conjunction with high-resolution climatic datasets represent a serious step forward in an attempt to objectively define and demarcate biogeographical zones and to determine the causes underlying species distribution in France.

On the other hand, if the NFI databases and the high-resolution climatic datasets offer merit in spatial precision, they do not allow us to avoid all possible bias. Even if several species present a climatic limit on French territory, we did not determine the whole climatic range for these species (both the southern and northern range). These restrictions in the environmental ranges can lead to an underestimation of the response curves (Thuiller *et al.*, 2004).

We chose a statistical species distribution model to analyse the data in our study. This classical approach assumes that species presence at any given point in space is determined by a finite number of climatic parameters and that any modification in these parameters immediately induces a change in vegetation (Kirilendo *et al.*, 2000). Such models do not include dynamic processes such as migration capability, competition with other species, dieback and decline or mortality, etc. However, we are forced to admit that, from a practical standpoint, we do not yet have precise enough knowledge of each species' autecology to interpret adequately current distributions at a national level (even by avoiding all the problems bound to climate change). So, although this approach has certain inherent limitations, among the numerous and widely used niche-based modelling techniques (Guisan and Zimmerman, 2000; Thuiller, 2003), it still remains "the best-known and the simplest method for predicting the equilibrium response of potential vegetation to climate change" (Peng, 2000). It can be quickly operated and enables the testing of a large number of working hypotheses (e.g. concerning species, species groups, climatic variables, etc.). We therefore felt this method would be the most adapted within the time scale of our project.

Even though a limited number of tree species, climatic parameters and statistical techniques were considered in this study, we were able to define several broad biogeographical zones; a not evident result because most of the studied species have been managed by foresters for centuries. We successfully modelled geographical distributions for several species. In some cases, the parameters retained in the models may also be related to tree physiology. For beech and silver fir, the climatic factors that determined their occurrence turned out to be the same variables shown to explain radial growth, at least at the regional level. This illustrates the importance of using bioclimatic indices and more explicitly physiological parameters to better explain species distribution. In other cases, the parameters have only a statistical significance: it is another limit to our work. It therefore seems essential to combine presence/absence analyses with more functional/ physiological approaches such as dendroecological analyses. Combined information taken from the NFI floristic inventories and tree measurements database should enable further progress in this direction by linking parameters such as stand characteristics, basal area and radial growth to species occurrence.

The results we presented here could be criticized on several points. Only a few statistical methods were used among all those that have been published. We did not compare the advantages and disadvantages of the statistical niche models with the mechanistic process-based models. We did not present how our projections agree or not with other published works, and so on. We agree that all these points are perfectly acceptable but all aspects of the problem could not be undertaken in the framework of the CARBOFOR project and work is still ongoing.

Even if we want to carry our approach further by applying deeper and more rigorous statistical analyses, the results already obtained exhibit the magnitude of potential shifts for climatic niches (although the B2 scenario is one of the most optimistic). Although we accept that the scenario used does indeed reflect the climate we can expect in 2100, these results do not constitute a forecast for our forests as many uncertainties still remain concerning the behaviour of forest species. Will they be capable of adapting to competition from new species? What balances will develop with new types of pathogens and mutualist organisms (see Chapter 12)? What will be the role of genetic variability? To what degree will species be able to colonize new climatic niches? The migration capability of forest species (not only the trees) will not be as fast as the niches shift, as shown by post-glacial recolonization dynamics (Brewer, 2002; Magri *et al.*, 2006; Giesecke, 2007; Muller *et al.*, 2007). For this century global warming will occur over a period shorter than the lifespan of a tree. Should we expect surges of massive dieback in our forests, or shall we see more diffuse and gradual mortality amongst the least adapted trees? Thus, in an attempt to help forest managers in their choices concerning the species most adapted to reforestation, we have to improve our knowledge of species autecology.

References

Austin M.P., Cunningham R.B., Fleming P.M., 1984. New approaches to direct gradient analysis using environmental scalars and statistical curve-fitting procedures. *Vegetatio*, 55, 11–27.

Badeau V., 1995. Étude dendroécologique du hêtre sur les plateaux calcaires de Lorraine. Influence de la gestion sylvicole. Thèse de doctorat, spécialité sciences et techniques biologiques, biologie forestière, université Nancy I, 224 p.

Badeau V., Bréda N., 1997. La récente crise de vitalité du hêtre en plaine semble liée aux déficits hydriques. *Les cahiers du DSF, La Santé des Forêts [France] en 1996*. Ministry of Agriculture and Fisheries (DERF), Paris, pp. 60–63.

Bakkenes M., Alkemade J.R.M., Ihle F., Leemans R., Latour J.-B., 2002. Assessing effects of forecasted climate change on the diversity and distribution of European higher plants for 2050. *Global Change Biology*, 8, 390–407.

Becker M., 1989. The role of climate on present and past vitality of silver fir forests in the Vosges mountains of northeastern France. *Canadian Journal of Forest Research*, 19, 1110–1117.

Benichou P., Le Breton O., 1987. Prise en compte de la topographie pour la cartographie des champs pluviométriques statistiques. *La Météorologie*, 19, 23–34.

Bert D., 1992. Influence du climat, des facteurs stationnels et de la pollution sur la croissance et l'état sanitaire du sapin pectiné dans le Jura. Thèse de doctorat, spécialité Sciences du bois, biologie végétale et forestière, Université Nancy I, 200 p.

Bolliger J., Kienast F., Bugmann H., 2000. Comparing models for tree distributions: concept, structures and behavior. *Ecological Modelling*, 134, 89–102.

Ter Braak C.J.F., Looman C.W.N., 1986. Weighted averaging, logistic regression and the Gaussian response model. *Vegetatio*, 65, 3–11.

Bradshaw R.H.W., Hannon G.E., 2004. The Holocene structure of North-west European temperate forest induced from palaeoecological data. *In: Forest Biodiversity* (O. Honnay, K. Verheyen, B. Bossuyt, M. Hermy, eds). The Global Network for Forest Science Cooperation/ CABI Publishing, Wallingford, pp. 11–25.

Bradshaw R.H.W., Holmqvist B.H., Cowling S.A., Sykes M.T., 2000. The effects of climate change on the distribution and management of *Picea abies* in southern Scandinavia. *Canadian Journal of Forest Research*, 30, 1992–1998.

Brewer S., Cheddadi R., de Beaulieu J.-L., Reille M., and data contributors, 2002. The spread of deciduous *Quercus* throughout Europe since the last glacial period. *Forest Ecology and Management*, 156, 27–48.

Bugnon F., Felzines J.C., Goux R., Loiseau J.E., Royer J.M., 1998. *Nouvelle flore de Bourgogne, tome III. Atlas de répartition, clés des groupements végétaux et suppléments aux tomes I et II. Bulletin Scientifique de Bourgogne*, 489 p.

Coudun C., 2005. Approche quantitative de la réponse écologique des espèces végétales forestières à l'échelle de la France. Thèse de doctorat, spécialité sciences forestières, École nationale du génie rural des eaux et des forêts (ENGREF), 128 p.

Coudun C., Gégout J.-C., 2005. Ecological behaviour of herbaceous forest species along a pH gradient: a comparison between oceanic and semicontinental regions in northern France. *Global Ecology and Biogeography*, 14, 263–270.

Coudun C., Gégout J.-C., Piédallu C., Rameau J.-C., 2006. Soil nutritional factors improve models of plant species distribution: an illustration with *Acer campestre* (L.) in France. *Journal of Biogeography*, 33, 1750–1763.

Dupont P., 1990. *Atlas Partiel de la flore de France*. Museum National d'Histoire Naturelle, Paris, 442 p.

Dupont P., 2001. *Atlas floristique de la Loire-Atlantique et de la Vendée. Tomes 1 et 2*. Société des Sciences Naturelles de l'Ouest de la France, Nantes.

Ebach M.C., Goujet D.F., 2006. The first biogeographical map. *Journal of Biogeography*, 33, 761–769.

Enquist C.A.F., 2002. Predicted regional impacts of climate change on the geographical distribution and diversity of tropical forests in Costa Rica. *Journal of Biogeography*, 29, 519–534.

Fielding A.H., Bell J.F., 1997. A review of methods for the assessment of prediction errors in conservation presence/absence models. *Environmental Conservation*, 24, 38–49.

Garraud L., 2003. *Flore de la Drôme. Atlas écologique et floristique*. Conservatoire botanique national alpin, Gap-Charance, 925 p.

Gautier C., Diak G., Masse S., 1980. A simple physical model to estimate incident solar radiation at the surface from GOES satellite data. *Journal* of *Applied Meteorology*, 19, 1005–1012.

Gibelin A.L., Déqué M., 2001. *Un scénario à l'échelle de l'Europe pour le climat de la fin du XXIᵉ siècle. Note de centre*, N° 79. Groupe de météorologie de grande échelle et climat, Météo-France, Toulouse.

Giesecke T., Hickler T., Kunkel T., Sykes M.T., Bradshaw H.W., 2007. Towards an understanding of the Holocene distribution of *Fagus sylvatica* L. *Journal of Biogeography*, 34, 118–131.

Gignac L.D., Halsey L.A., Vitt D.F., 2000. A bioclimatic model for the distribution of *Sphagnum*-dominated peatlands in North America under present climatic conditions. *Journal of Biogeography*, 27, 1139–1152.

Grandjouan G., 1982. Une méthode de comparaison statistique entre les répartitions des plantes et des climats. Thèse de doctorat, université Louis Pasteur, Strasbourg, 316 p.

Granier A., Badeau V., Bréda N., 1995. Modélisation du bilan hydrique des peuplements forestiers. *Revue Forestière Française*, No. sp, 59–68.

Guisan A., 2003. Simuler la répartition géographique des espèces et de la végétation (ou, "si De Candolle avait eu un ordinateur…"). *Saussurea*, 33, 79–99.

Guisan A., Edwards T.C., Hastie J.T., 2000. Generalized linear and generalized additive models in studies of species distributions: setting the scene. *Ecological Modelling*, 157, 89–100.

Guisan A., Zimmermann N.E., 2000. Predictive habitat distribution models in ecology. *Ecological Modelling*, 135, 147–186.

Hilmans R.J., Cameron S.E., Parra J.L., Jones P.G., Jarvis A., 2005. Very high resolution interpolated climate surfaces for global land areas. *International Journal of Climatology*, 25, 1965–1978.

Huntley B., Berry P.M., Cramer W., McDonald A.P., 1995. Modelling present and potential future ranges of some European higher plants using climate response surfaces. *Journal of Biogeography*, 22, 967–1001.

IPCC, 2001. *Climate Change 2001: The Scientific Basis. Third Assessment Report* IPCC/WMO/UNEP. Intergovernmental Panel on Climate Change, Geneva, 785 p. [www.ipcc.ch]

Iverson L.R., Prasad A., 2001. Potential changes in tree species richness and forest community types following climate change. *Ecosystems*, 4, 186–199.

Iverson L.R., Prasad A., 2002. Potential redistribution of tree species habitat under five climate change scenarios in the eastern US. *Forest Ecology and Management*, 155, 205–222.

Jalas J., Suominen J. (eds), 1972–94. *Atlas Flora Europaeae: Distribution of Vascular Plants in Europe*, Volumes 1–10. The Committee for Mapping the Flora of Europe/Societas Biologica Fennica Vanamo, Helsinki.

Jamagne M., Hardy R., King D., Bornand M., 1995. La base de données géographique des sols de France. *Étude et Gestion des Sols*, 2 (3), 153–172.

Jelaska S.D., Antonic O., Nikolic T., Hrsak V., Plazibat M., Krizan J., 2003. Estimating plant species occurrence in MTB/64 quadrants as a function of DEM-based variables. A case study for Medvednica Nature Park, Croatia. *Ecological Modelling*, 170, 333–343.

Kirilenko A.P., Belotelov N.V., Bogatyrev B.G., 2000. Global model of vegetation migration: incorporation of climatic variability. *Ecological Modelling*, 132, 125–133.

Körner C., Paulsen J., 2004. A world-wide study of high altitude treeline temperature. *Journal of Biogeography*, 31, 713–732.

Lebourgeois F., Bréda N., Ulrich E., Granier A., 2005. Climate-tree-growth relationships of European beech (*Fagus sylvatica* L.) in the French Permanent Plot Network (RENECOFOR). *Trees*, 19, 385–401.

Lehmann A., Overton J., Austin M.P., 2002. Regression models for spatial prediction: their role for biodiversity and conservation. *Biodiversity and Conservation*, 11, 2085–2092.

Lehmann A., Overton J., Leathwick J.R., 2003. GRASP: generalized regression analysis and spatial prediction. Erratum. *Ecological Modelling*, 160, 165–183.

Lenihan J.M., 1993. Ecological response surfaces for North American boreal tree species and their use in forest classification. *Journal of Vegetation Science*, 4, 667–680.

Magri D., Vendramin G.G., Comps B., Dupanloup I., Geburek T., Gömöry D., Latalova M., Litt T., Paule L., Roure J.-M., Tantau I., van der Knaap W.O., Petit R., de Beaulieu J.-L., 2006. A new scenario for the Quaternary history of European beech populations: palaeobotanical evidence and genetic consequences. *New Phytologist*, 171, 199–221.

Malcolm J.R., Markham A., Neilson R.P., Garaci M., 2002. Estimated migration rates under scenarios of global climate change. *Journal of Biogeography*, 29, 835–849.

Manel S., Dias J.M., Buckon S.T., Ormerod S.J., 1999. Alternative methods for predicting species distribution: an illustration with Himalayan river birds. *Journal of Applied Ecology*, 36, 734–747.

Manel S., Williams H.C., Ormerod S.J., 2001. Evaluating presence-absence models in ecology: the need to account for prevalence. *Journal of Applied Ecology*, 38, 921–931.

Margules C.R., Pressey R.L., 2000. Systematic conservation planning. *Nature*, 405, 243–253.

McKenzie D., Peterson D.W., Peterson D.L., Thornton P.E., 2003. Climatic and biophysical controls on conifer species distributions in mountain forest of Washington State, USA. *Journal of Biogeography*, 30, 1093–1108.

Mitchell T.D., Carter T.R., Jones P.D., Hulme M., New M., 2003. A comprehensive set of climate scenarios for Europe and the globe. *Tyndall Centre Working Paper*, No. 55.

Muller S.D., Nakagawa T., de Beaulieu J.-L., Court-Picon M., Carcaillet C., Miramont C., Roiron P., Boutterin C., Ali A.A., Bruneton H., 2007. Post-glacial migration of silver fir (*Abies alba* Mill.) in the south-western Alps. *Journal of Biogeography*, 34, 876–899.

Parmesan C., Yohe G., 2003. A globally coherent fingerprint of climate change impacts across natural systems. *Nature*, 421, 37–42.

Pearce J., Ferrier S., 2000. An evaluation of alternative algorithms for fitting species distribution models using logistic regression. *Ecological Modelling*, 128, 127–147.

Pearson R.G., Dawson T.P., Berry P.M., Harrison P.A., 2002. SPECIES: A Spatial Evaluation of Climate Impact on the Envelope of Species. *Ecological Modelling*, 154, 289–300.

Peng C., 2000. From static biogeographical model to dynamic global vegetation model: a global perspective on modelling vegetation dynamics. *Ecological Modelling*, 135, 33–54.

Peñuelas J., Boada M., 2003. A global change induced biome shift in the Montseny mountains (NE Spain). *Global Change Biology*, 9, 131–140.

Pinto P., Gégout J.-C., 2005. Assessing the nutritional and climatic response of temperate tree species in the Vosges Mountains. *Annals of Forest Science*, 62, 761–770.

Rameau J.-C., Gauberville C., Drapier N., 2000. *Gestion forestière et diversité biologique. Identification et gestion intégrée des habitats et espèces d'intérêt communautaire. France – domaine continental.* ENGREF/ONF/IDF, 119 p.

Rameau J.-C., Mansion D., Dumé G., 1989. *Flore forestière française. Tome 1, Plaines et collines.* IDF/DERF/ENGREF, 1785 p.

Rameau J.-C., Mansion D., Dumé G., 1993. *Flore forestière française. Tome 2, Montagnes.* IDF/DERF/ENGREF, 2421 p.

Rushton S.P., Ormerod S.J., Kerby G., 2004. New paradigms for modelling species distribution? *Journal of Applied Ecology*, 41, 193–200.

Shtatland E.S., Cain E., Barton M.B., 2001. The perils of stepwise logistic regression and how to escape them using information criteria and the output delivery system. *In: SAS Users Group International (SUGI) Proceedings, Seattle, 30 March–2 April 2001*, pp. 1–6.

Shtatland E.S., Kleinman K., Cain E., 2002. One more time about r^2 measures of fit in logistic regression. *In: NorthEast SAS Users Group (NESUG), 15th Annual Conference, 29 September–2 October 2002, Buffalo, New York*, pp. 1–6.

Sykes M.T., 2001. Modelling the potential distribution and community dynamics of lodgepole pine (*Pinus contorta* Dougl. ex. Loud.) in Scandinavia. *Forest Ecology and Management*, 141, 69–84.

Sykes M.T., Prentice I.C., 1995. Boreal forest futures: modelling the controls on tree species range limits and transient responses to climate change. *Water, Air and Soil Pollution*, 82, 415–428.

Sykes M.T., Prentice I.C., 1996. Climate change, tree species distributions and forest dynamics: a case study in the mixed conifer/northern hardwoods zone of Northern Europe. *Climate Change*, 34, 161–177.

Sykes M.T., Prentice I.C., Cramer W., 1996. A bioclimatic model for the potential distributions of North European tree species under present and future climates. *Journal of Biogeography*, 23, 203–233.

Thuiller W., 2003. BIOMOD: Optimising predictions of species distributions and projecting potential future shifts under global change. *Global Change Biology*, 9, 1353–1362.

Thuiller W., Araujo M.B., Lavorel S., 2003. Generalized models vs. classification tree analysis: predicting spatial distributions of plant species at different scales. *Journal of Vegetation Science*, 14 (5), 669–680.

Thuiller W., Brotons L., Araújo M., Lavorel S., 2004. Effects of restricting environmental range of data to project current and future species distributions. *Ecography*, 27, 165–172.

Thuiller W., Lavorel S., Araujo M.B., Sykes M.T., Prentice I.C., 2005. Climate change threats to plant diversity in Europe. *Proceedings of the National Academy of Science*, 102 (23), 8245–8250.

Vetaas O.R., 2002. Realized and potential climate niches: a comparison of four *Rhododendron* tree species. *Journal of Biogeography*, 29, 545–554.

Walther G.R., Post E., Convey P., Menzel A., Parmesan C., Beebee T., Fromentin J.M., Hoegh-GuldbergI O., Bairlein F., 2002. Ecological responses to recent climate change. *Nature*, 416, 389–395.

Zweig M.H., Campbell G., 1993. Receiver-operating characteristic (ROC) plots: a fundamental evaluation tool in clinical medicine. *Clinical Chemistry*, 39, 561–577.

Chapter 12
Simulating the effects of climate change
on geographical distribution
and impact of forest pathogenic fungi

MARIE-LAURE DESPREZ-LOUSTAU, VALÉRIE BELROSE, MAGALI BERGOT, GILLES CAPRON, EMMANUEL CLOPPET, CLAUDE HUSSON, DOMINIQUE PIOU, GREGORY REYNAUD, CÉCILE ROBIN, BENOÎT MARÇAIS

Introduction

Along with the presence and density of hosts (resource availability), climatic factors are key drivers in the dynamics of pathogen populations and their impact (Coakley, 1988). A large body of literature is available that deals with the relationships between meteorological conditions and disease development, explaining the between-year or geographical variation in disease severity (e.g. Houston and Valentine, 1988; Jahn *et al.*, 1996). The effects of climate on disease include not only direct effects on the pathogens, but also indirect effects through host physiology and parasite-associated organisms such as vectors or hyperparasites (Harrington *et al.*, 2001; Harvell *et al.*, 2002).

Ongoing climate change with its magnitude and speed is likely to induce severe perturbations, which may favour an amplification of parasite impact. Indeed, these organisms have numerous traits related to an r-strategy:
– high reproductive potential (e.g. short generation times, asexual reproduction and large populations) allowing greater adaptation potential than their host plants (Goudriaan and Zadocks, 1995; Coakley *et al.*, 1999; Davis and Shaw, 2001; Saxe *et al.*, 2001; Etterson and Shaw, 2001); the rapid development of races of parasites able to break down resistance in resistance-bred cultivars is a clear illustration of this high adaptation potential (Mc Donald and Linde, 2002);
– high dispersal potential, through different spore types, human-mediated transport (e.g. infected seeds and plants);

– high phenotypic plasticity, enabling them to develop within large ranges of environmental variation.

Climatic change may first affect current pathosystems. In particular, warming may increase fungal parasite development in temperate regions by relaxing constraints on the life cycles of these poïkilothermic organisms, that is, winter survival, growth, reproduction and dispersal (Harrington *et al.*, 2001). Most fungi have temperature optima in the range 20–25°C and minimum winter temperatures can be a limiting factor in survival (Harvell *et al.*, 2002).

A more dramatic effect of climate change may be the emergence of new pathosystems. Indeed, range shifts are an expected response of many species to climate change. Extensions or shifts in geographical distributions have been demonstrated in response to climatic changes during the Quaternary period (Davis and Shaw, 2001). Recent changes associated with changing climate in the last decades have already been observed for some species (Parmesan *et al.*, 1999; Parmesan and Yohe, 2003). Range shifts for a given species result both from changes in its climatic envelope (i.e. climatically suitable regions according to species requirements) and its own capacities for migration and establishment, among other factors. It seems, therefore, highly probable that climate change will have a differential impact on different species and will lead to rearrangements in communities (Walther *et al.*, 2002), in particular in the case of host-parasite interactions. Due to the high dispersal potential of many fungi, the colonization of new regions becoming climatically favourable may put them into contact with naive populations (genotypes or species), that is, with no co-evolution or co-adaptation history, with the same potentially dramatic consequences as those observed with introduced parasites (Harvell *et al.*, 2002). The pine processionary caterpillar, which has shown a recent latitudinal and altitudinal shift together with a host shift (first reported on Douglas fir), provides an example of such changes (Goussard *et al.*, 1999).

The aim of our study was: (i) to describe the current spatial distribution in France of some forest diseases caused by pathogenic fungi; (ii) to derive the main climatic drivers of pathogen development and the diseases they cause from an epidemiological analysis of these data complemented by experimental literature data; (iii) to simulate the effects of climatic change on the potential geographical range or local impact of parasite species by using several epidemiological or population-dynamic models with the "ARPEGE climate scenario (see Chapter 6).

Materials and methods

Disease data

The parasite species for study were selected from among those most frequently reported in France on oaks, pines, chestnut and poplar or because of their recent evolution or thermophilic habit. The main source of geographical distribution data was the database of the French Forest Health Service (Département de Santé des Forêts, DSF), created in 1989, which today contains more than 60,000 records on diseases, insect or abiotic damage. Most records are "spontaneous", that is, made by forest service observers on encountering disease damage, but a few are from specific surveys. Such a systematic survey was carried out for chestnut canker caused by *Cryphonectria parasitica* in

1996–97 on 1100 plots, corresponding to a sampling rate of one plot per 1000 ha in the affected area (south) and one plot per 500 ha in the zone considered to be disease-free at that time (the northern part of the country) (de Villebonne, 1997, 1998). The percentage of infected trees (PIT), estimated on 100 trees) and site characteristics (i.e. altitude and topography) were recorded. In the case of *Phytophthora cinnamomi*, data from the DSF were complemented by observations from the French National Institute for Agricultural Research (INRA) and the Plant Health Service (SPV) (Vettraino *et al.*, 2005). Finally, the distribution of the different *Melampsora* spp. causing poplar rust was established from a specific database of the SPV including systematic observations made in 311 nurseries monitored for a quarantine survey of *Melampsora medusae* from 1993 to 2003. Only detections made on non-discriminant poplar clones, that is, susceptible to all species, were taken into account. Data from a study aimed at monitoring first infection dates in larch (alternate host of *M. larici-populina*) since 1998 in Nancy were also analysed (Pinon, unpublished results).

Standardization of disease data

Data from the DSF database had to be standardized in order to reduce variation linked to host prevalence and observers' reporting activity. The general principle, as used in medical epidemiology, was to relate the case counts for a given disease to a reference constituted by the total number of records of all other diseases on the same host(s), taking into account age classes if an effect of age on that disease has been reported. When spatial distributions were studied, the aggregation unit for case counts and reference data was either administrative (departments) or ecological (National Forest Inventory (NFI) units). When temporal variations were studied, wider geographical areas were used as units and the annual reference was calculated as the sum of the mean annual numbers of records per observer over all the observers in each area.

Distribution data of poplar rusts from the SPV database also required standardization as only records of occurrence of each of three species were available. Records of *M. larici-populina*, which is widely distributed in the whole study area, were used as a reference for the two other species, *M. allii-populina* and *M. medusae,* assuming that the presence of *M. larici-populina* is a good indicator of the occurrence of poplar rusts in a nursery (poor control). In the absence of any reference data, no analysis could be done for *M. larici-populina*.

Specific models of pathogen response to climate

Statistical models

Oak powdery mildew, Erysiphe alphitoïdes

Reports from the DSF database were analysed using a generalized linear model (SAS genmod procedure). The model is as follows:

$$Y = R \cdot \exp(\eta) \tag{1}$$

where Y is the annual number of case counts for powdery mildew in the zone used for aggregation; Y is assumed to follow a Poisson distribution; R is the number of reports used as reference; η is the linear predictor, including explaining variables.

A limited number of climatic variables chosen for their epidemiological significance were tested. The first hypothesis is that the initiation of epidemics at budburst may be a key factor in the dynamics and final severity of the disease. Winter (January–February–March) temperatures and precipitation, which may explain the phenology of the host and parasite in spring, were therefore included, as well as variables relating to the time of budburst (i.e. mean temperature, number of frost days, precipitation in April–May). Climatic variables during the vegetative season were also considered (i.e. mean temperature and sum of precipitations in May–June–July). Climatic variables were taken from the meteorological station representative of the five geographical zones of aggregation within the Météo-France network. All variables were tested separately and in pairs and the best models were then studied more thoroughly.

Melampsora allii-populina and *M. medusae* on poplars

The proportion of clones affected in each nursery was analysed by a logistic regression with a "logit" link (SAS logistic procedure). The climatic variables tested were the trimestrial means over the 1961–90 period: minimum, maximum and mean daily temperatures, number of frost days, sum of precipitations, water index = sum of precipitations minus evapotranspiration and the annual number of days with a minimum temperature below –5°C and –10°C. Climatic data for each site were obtained from the Météo-France network using the AURELHY interpolation method. All models using one, two or three variables were tested. Those associated with the highest deviance decrease were studied in more detail (i.e. residuals and influent points). The best model with the least variables was selected.

Chestnut canker, *Cryphonectria parasitica*

A logistic regression was performed, as described for poplar rusts, using PIT per plot as the dependent variable and climatic variables as explaining variables. The data used to fit the model were a random sub-sample (432 points) of the 586 "positive" points (i.e. affected by disease), the remaining points being used for validation.

Mechanistic model: Phytophthora cinnamomi *on oaks*

Considering the tropical origin of this pathogen (Zentmyer, 1980), its sensitivity to frost and the differences in distribution of the diseases it causes on oaks and chestnuts (compare hereafter), Delatour (1986) hypothesized that lethal frost effects on the pathogen could be a major factor limiting disease range in oaks. Indeed, winter survival is likely to be more critical when the pathogen overwinters in the aerial system (in *Quercus*) than in roots (in *Castanea*), as freezing temperatures seldom occur in soils below 10–30 cm deep in temperate regions. A model of *P. cinnamomi* survival in oak trunks as a function of air temperature and tree characteristics was then developed (Marçais *et al.*, 1996). This model was used to map the climatic risk of ink disease in oaks by mapping the probability of survival of the parasite as a function of climatic data for the 1968–98 period, and by relating the survival probability to an actual disease index using the DSF survey data (Desprez-Loustau *et al.*, 2002; Marçais *et al.*, 2004). The mapped index ($F_{0.5}$) is the frequency of annual survival rate (ASR) of *P. cinnamomi* lower than or equal to 0.5. This threshold value is the lowest survival value allowing a perennial development of disease (cankers).

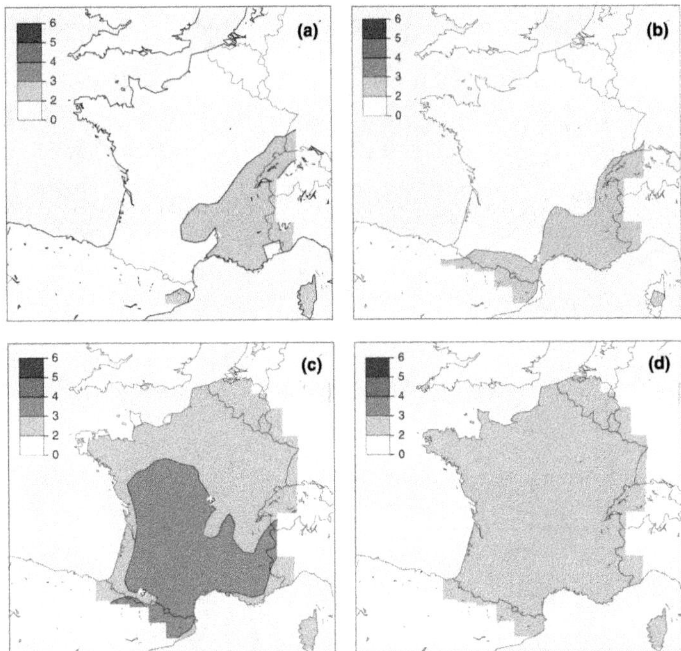

Figure 6-2. Temperature differences between 2070–99 and 1960–89 averages: (a) winter; (b) spring; (c) summer; (d) autumn. Contour interval 0.5 K. B2 simulation.

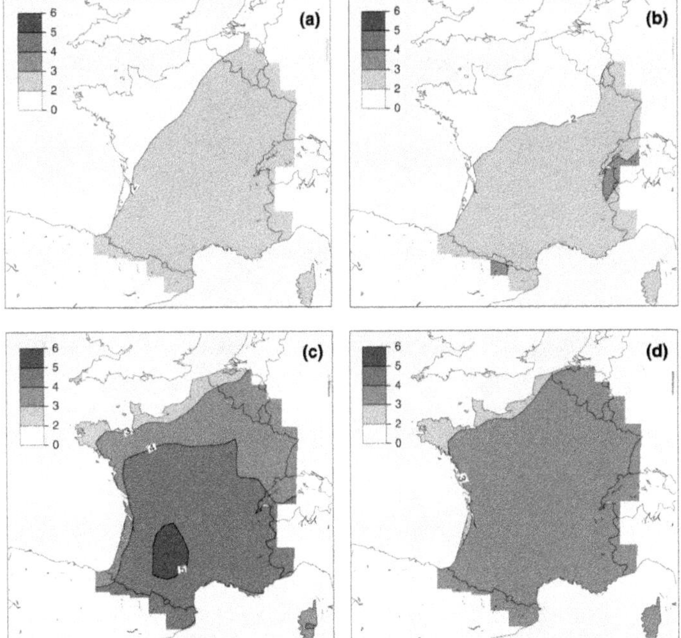

Figure 6-3. As for Figure 6.2 for the A2 simulation.

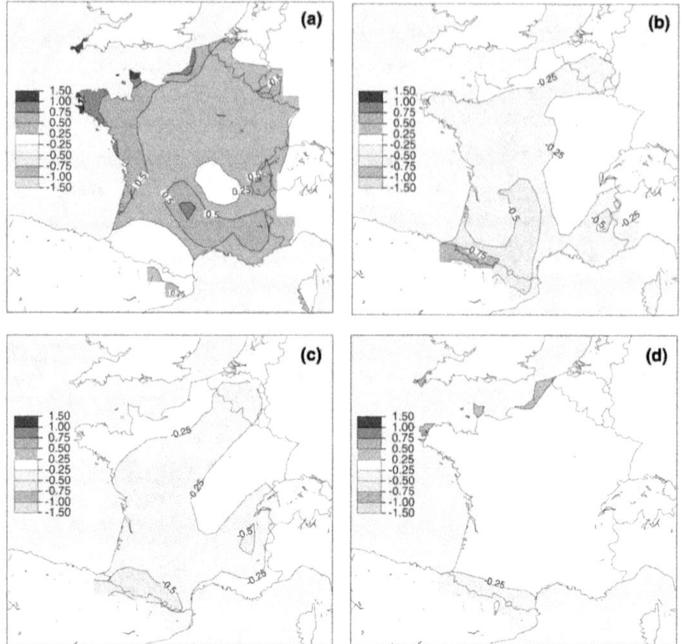

Figure 6-4. As for Figure 6.2 for precipitation. Contours ±0.25, 0.50, 0.75 and 1 mm · day^{-1}.

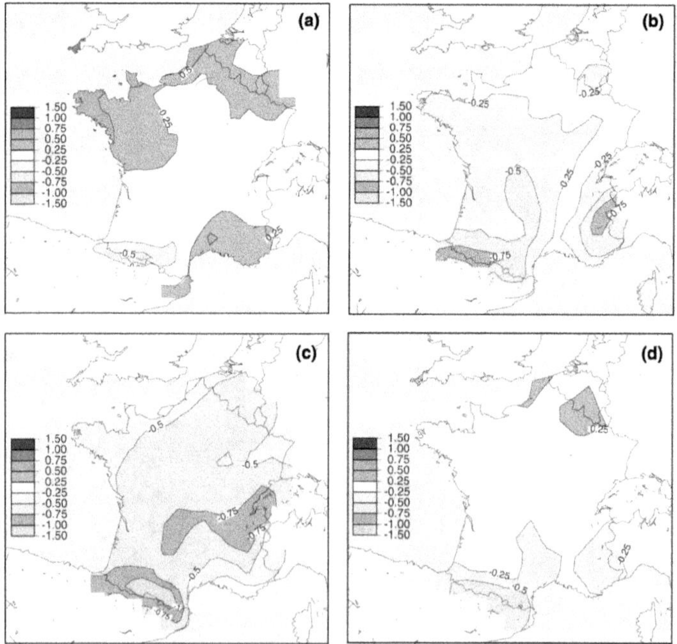

Figure 6-5. As for Figure 6.4 for the A2 simulation.

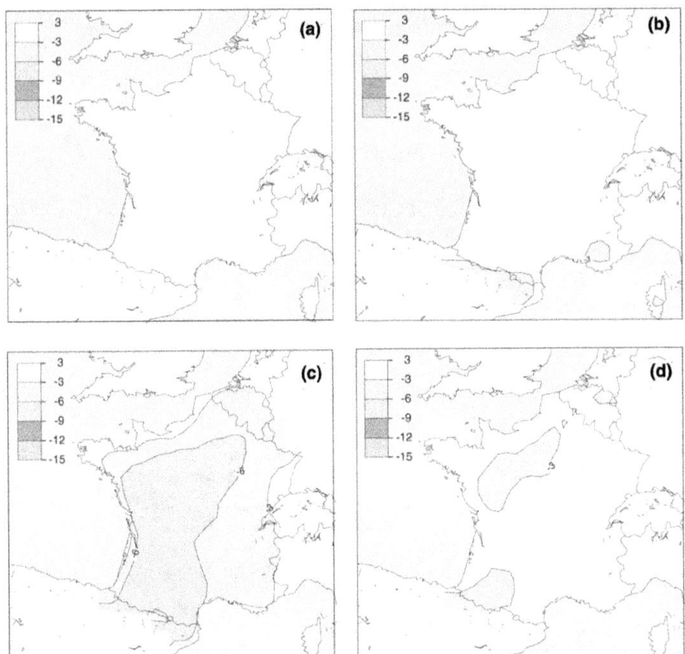

Figure 6-6. As for Figure 6.2 for surface (2 m) air relative moisture. Contours -3, -6, -9 and -12%.

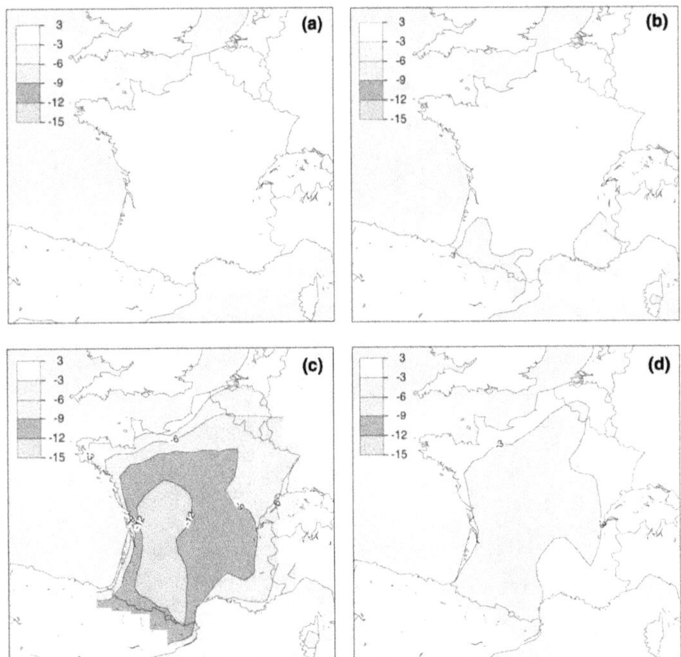

Figure 6-7. As for Figure 6.6 for the A2 simulation.

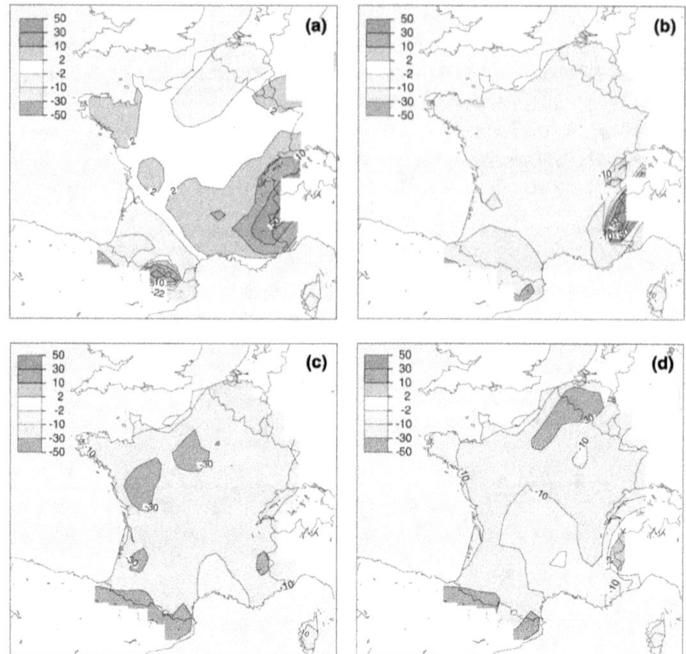

Figure 6-8. As for Figure 6.2 for liquid water soil content. Contours ± 2, 10, 30 and 50 kg · m^{-2}.

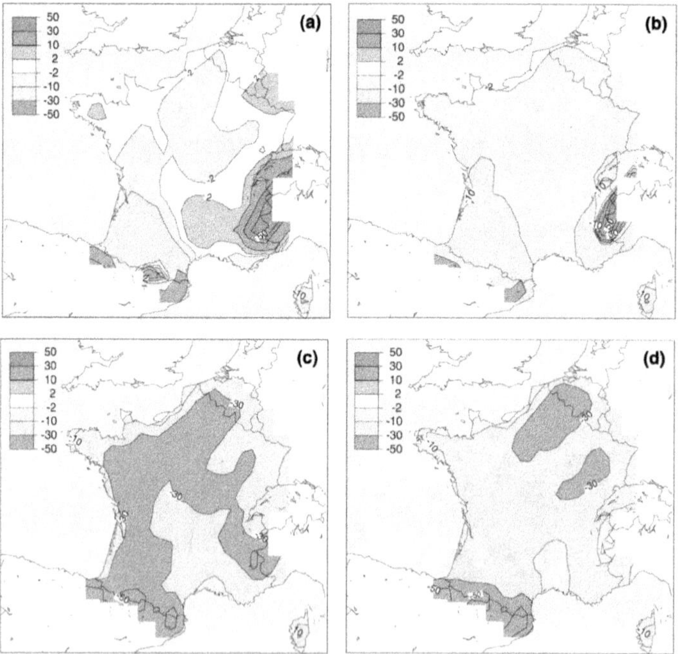

Figure 6-9. As for Figure 6.8 for the A2 simulation.

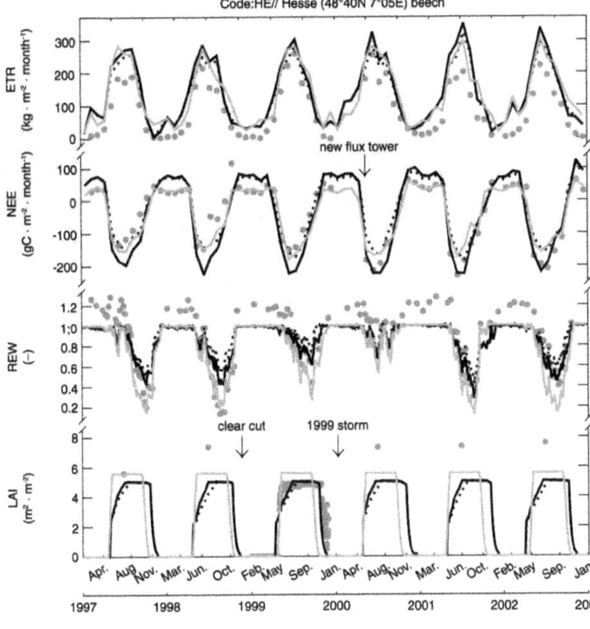

Figure 9-2. Monthly evapotranspiration (ETR), net ecosystem exchange (NEE), relative extractable water (REW) and leaf area index (LAI) at Hesse from 1997 to 2003. Simulations of CASTANEA and ORCHIDEE (both with standard and optimized versions) are compared with measured values obtained from eddy-covariance data.

Figure 9-3. Monthly evapotranspiration (ETR), net ecosystem exchange (NEE), relative extractable water (REW) and leaf area index (LAI) at le Bray from 1996 to 2003. Simulations of GRAECO and ORCHIDEE (both with standard and optimized versions) are compared with measured values obtained from eddy-covariance data.

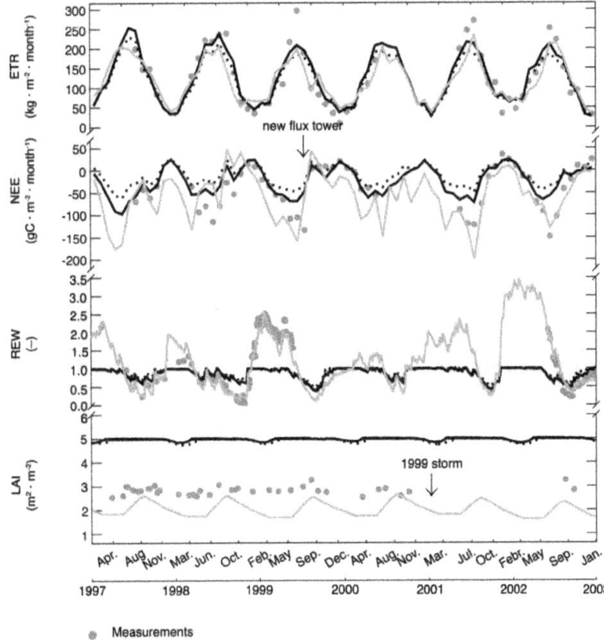

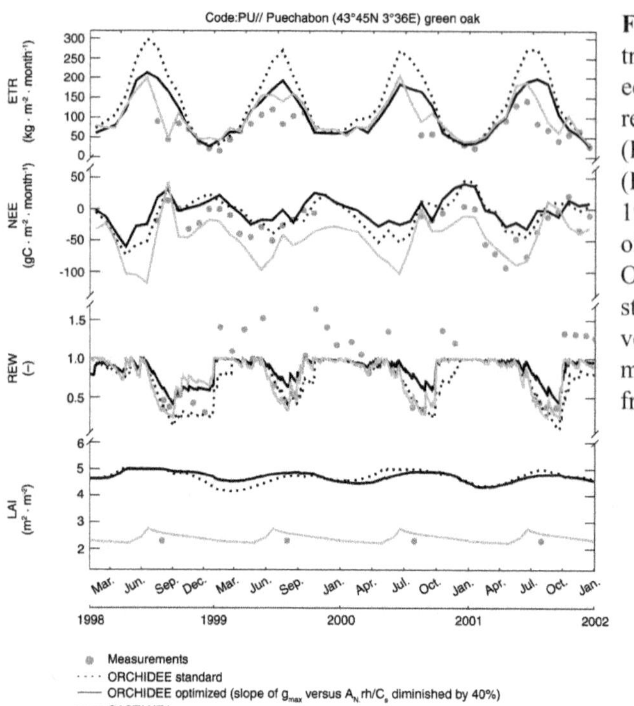

Figure 9-4. Monthly evapotranspiration (ETR), net ecosystem exchange (NEE), relative extractable water (REW) and leaf area index (LAI) at Puéchabon from 1998 to 2002. Simulations of CASTANEA and ORCHIDEE (both with standard and optimized versions) are compared with measured values obtained from eddy-covariance data.

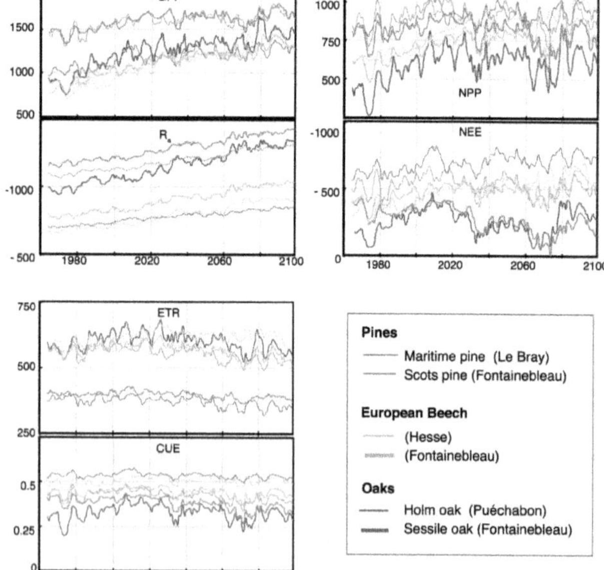

Figure 10-1. Carbon and water fluxes simulated by the CASTANEA model for six forest canopies as forced by the regionalized climate scenario 1960–2100. Gross primary production (GPP), R_a, net primary production (NPP) and NEE are in $gC \cdot m^{-2}.year^{-1}$, evapotranspiration (ETR) is in $kg\ H_2O \cdot m^{-2}.year^{-1}$ or $mm \cdot year^{-1}$, carbon-use efficiency (CUE) is unitless. All curves are sliding averages (N = 5 years).

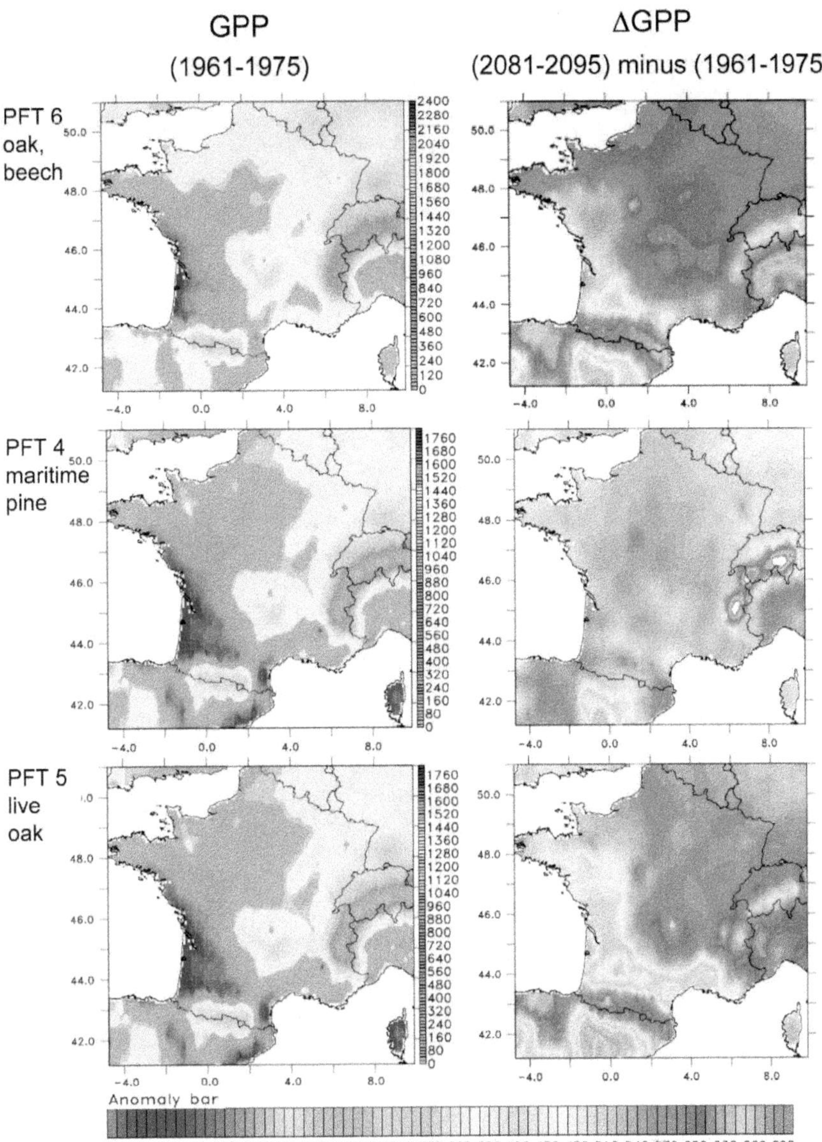

Figure 10-6. Left: mean value of the annual gross primary production (GPP) (gC · m^{-2} · year^{-1}) over the period 1961–75 for three forest functional types over France. Right: GPP anomaly between 2081 and 2095 and the period 1960–75. Soil water holding capacity is fixed at 0.1 cm^3 · cm^{-3} everywhere.

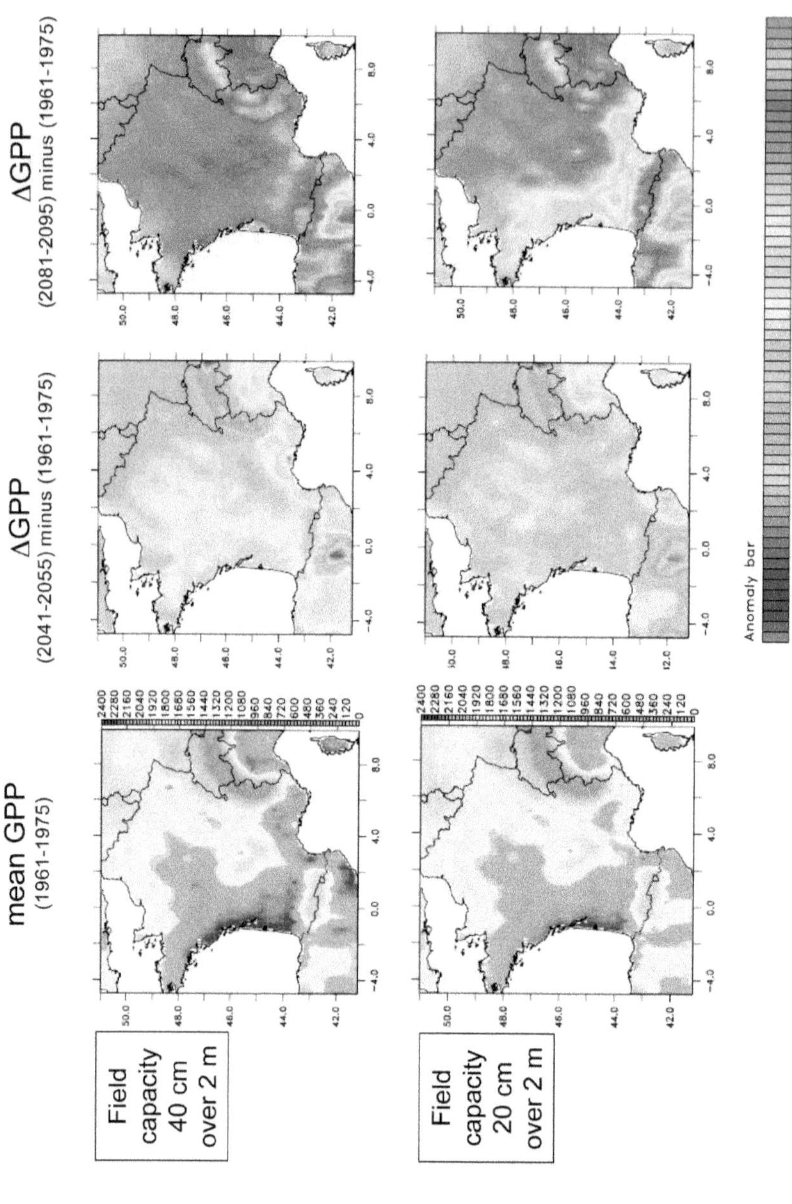

Figure 10-8. Mean value over 1961–75 of the annual gross primary production (GPP) (gC · m^{-2}·year^{-1}), for the live oak functional type over France and its anomaly in 2041–55 and 2081–95 for two levels of soil water holding capacity: 0.1 (upper) and 0.2 cm^3 · cm^{-3} (lower).

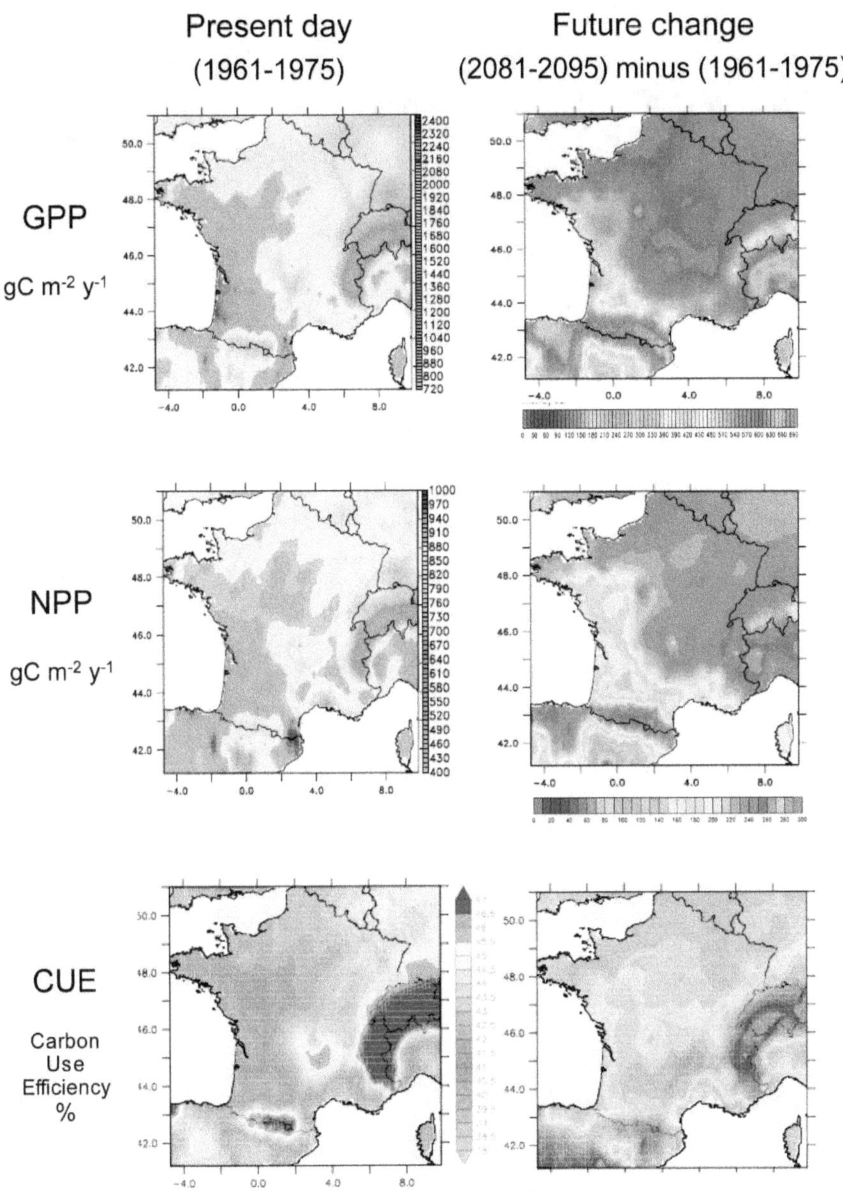

Figure 10-9. Mean value 1961–75 of gross primary production (GPP), net primary production (NPP) (gC · m⁻² · year¹) and carbon use efficiency (CUE), defined as the ratio NPP : GPP, for the oak-beech forest functional type over France and its anomaly in 2081–95. Soil water holding capacity is fixed at 0.1 cm³ · cm⁻³.

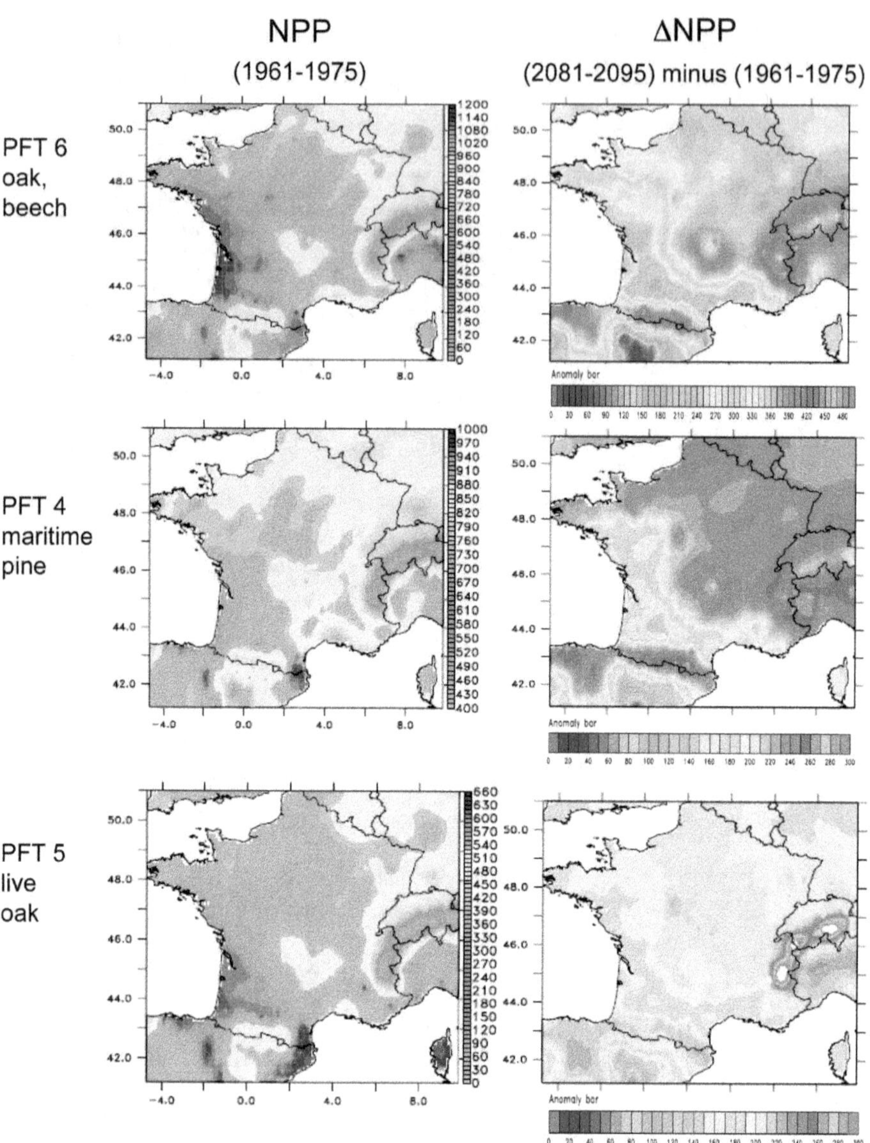

Figure 10-10. Mean value 1961–75 of net primary production (NPP) (gC · m^{-2} · year^{-1}) for three forest functional types over France and its anomaly in 2081–95. Soil water holding capacity is fixed at 0.1 cm^3 · cm^{-3}.

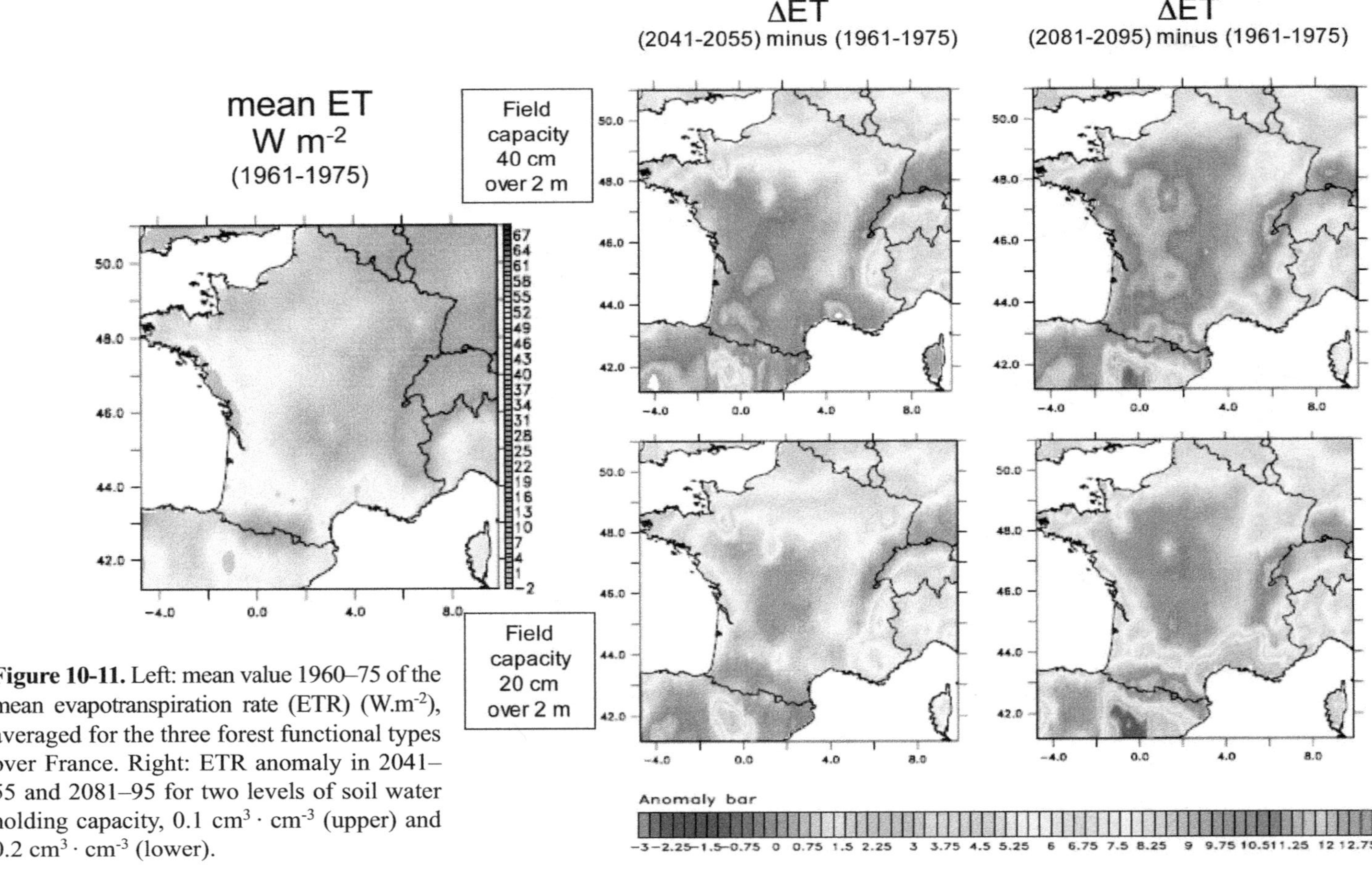

Figure 10-11. Left: mean value 1960–75 of the mean evapotranspiration rate (ETR) (W.m⁻²), averaged for the three forest functional types over France. Right: ETR anomaly in 2041–55 and 2081–95 for two levels of soil water holding capacity, 0.1 cm³ · cm⁻³ (upper) and 0.2 cm³ · cm⁻³ (lower).

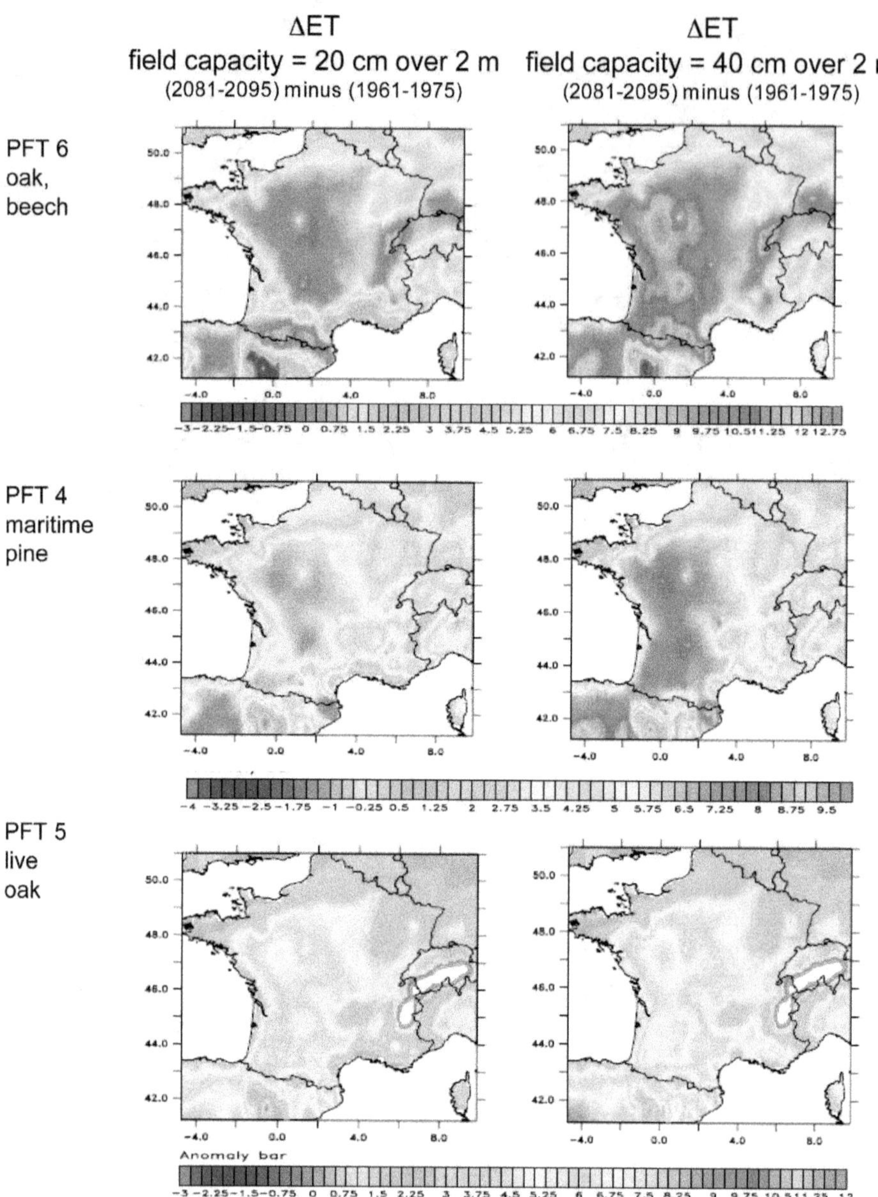

Figure 10-12. Anomalies in 2081–95 of the mean evapotranspiration rate (ETR) (W · m^{-2}) for three forest functional types over France and two levels of soil water holding capacity, 0.1 cm^3 · cm^{-3} (left) and 0.2 cm^3 · cm^{-3} (right).

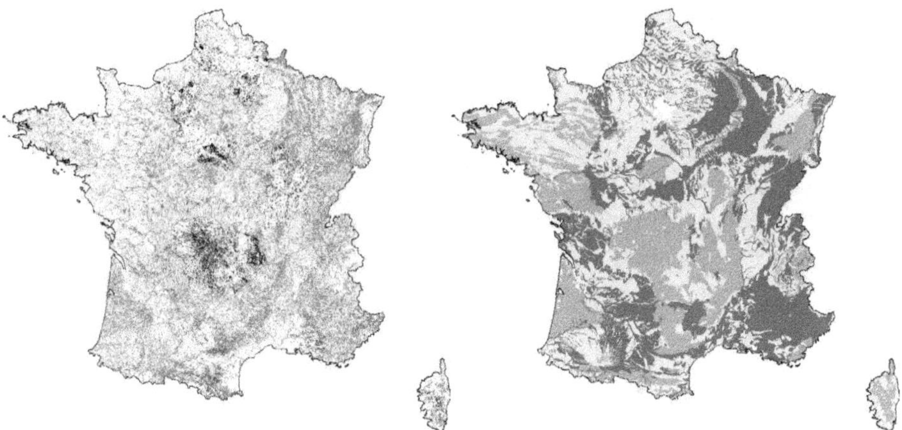

Figure 11-1. Location of the 104, 259 analysed forest sites from the French Forest National Inventory; red: floristic database; black: tree database; blue: older floristic surveys (see the text).

Figure 11-2. Soil acidity map calculated from the national geological soil database (1:1,000,000 scale): red = pH under 5; yellow = pH between 5 and 7; blue = pH over 7.

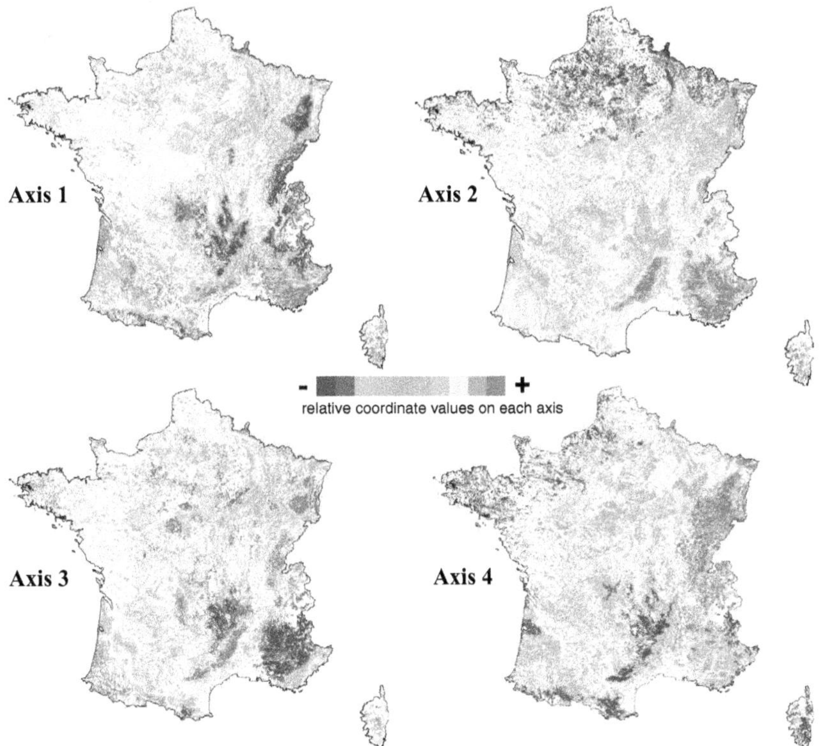

Figure 11-3. NFI points according to their positions on the first four factors extracted by the principal component analysis (PCA) of the climatic variables.

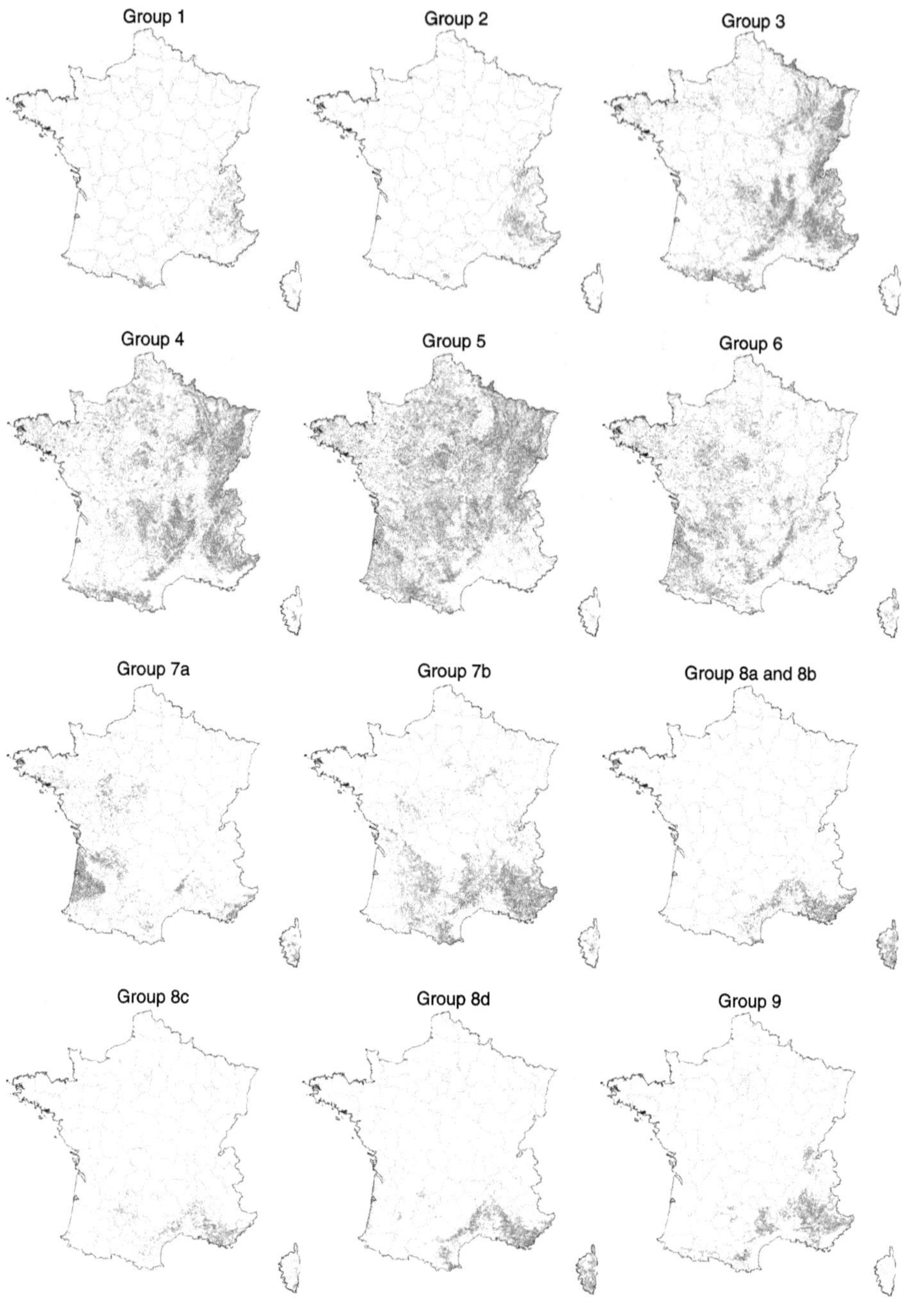

Figure 11-4. Distribution maps for the 67 studied species divided into 13 chorological groups, numbered 1 to 9.

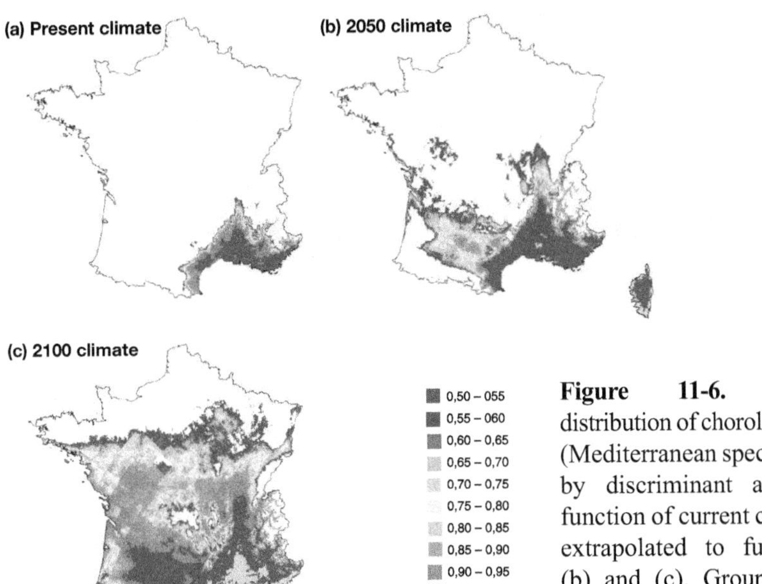

Figure 11-6. Geographical distribution of chorological Group 8 (Mediterranean species) estimated by discriminant analysis as a function of current climate (a) and extrapolated to future climates (b) and (c). Group membership probabilities are shown for values above 0.5. Grey color is used for areas where climate is beyond the present range.

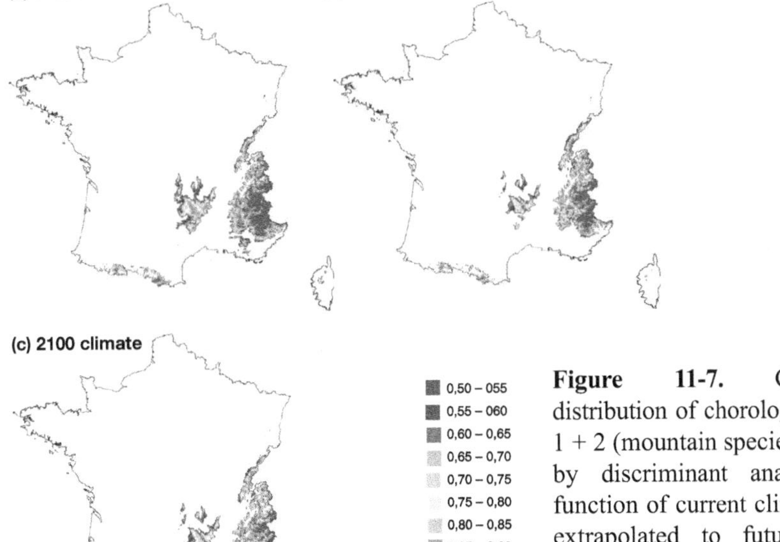

Figure 11-7. Geographical distribution of chorological Groups 1 + 2 (mountain species) estimated by discriminant analysis as a function of current climate (a) and extrapolated to future climates (b) and (c). Group membership probabilities are shown for values above 0.5.

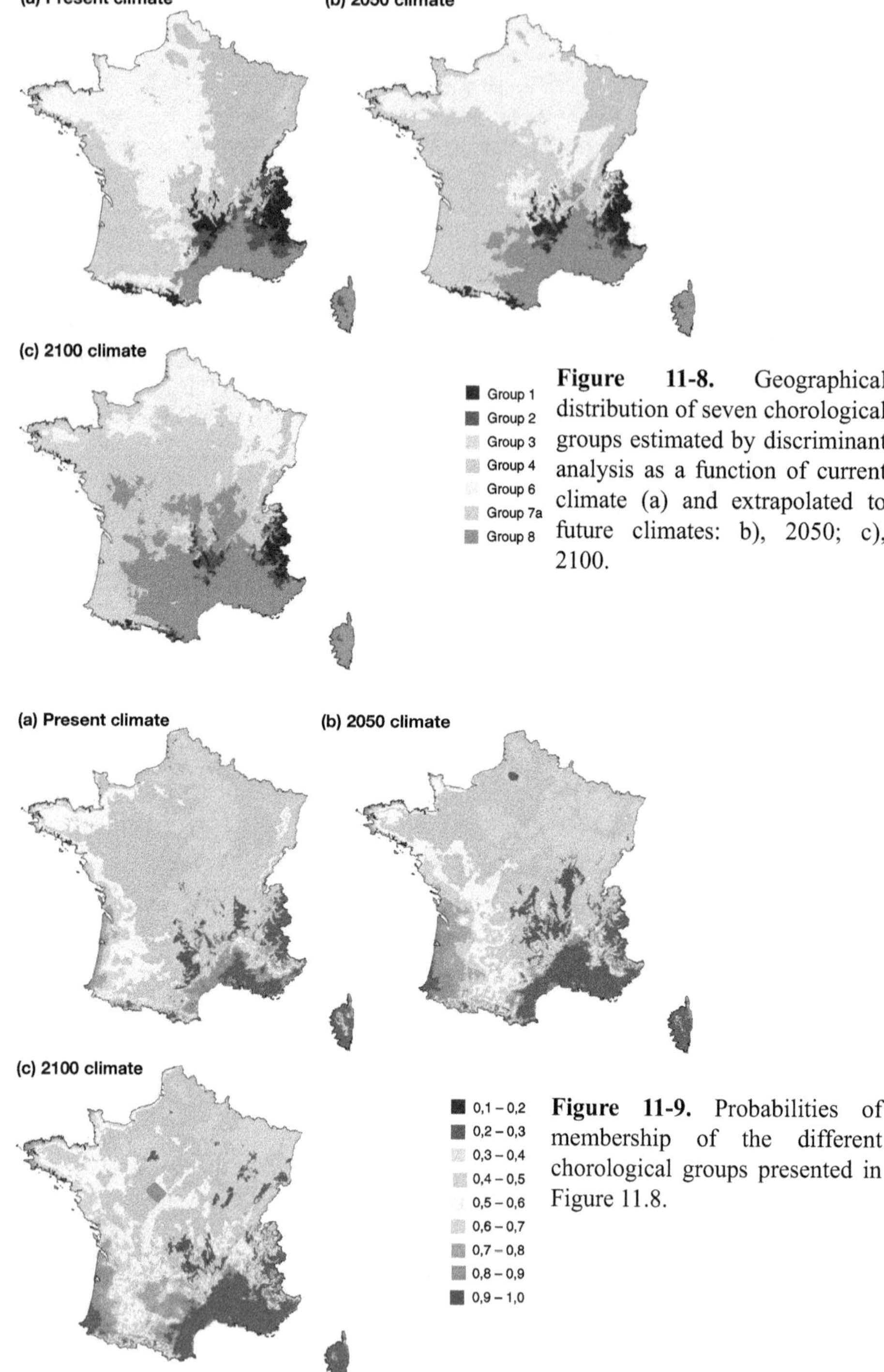

Figure 11-8. Geographical distribution of seven chorological groups estimated by discriminant analysis as a function of current climate (a) and extrapolated to future climates: b), 2050; c), 2100.

Figure 11-9. Probabilities of membership of the different chorological groups presented in Figure 11.8.

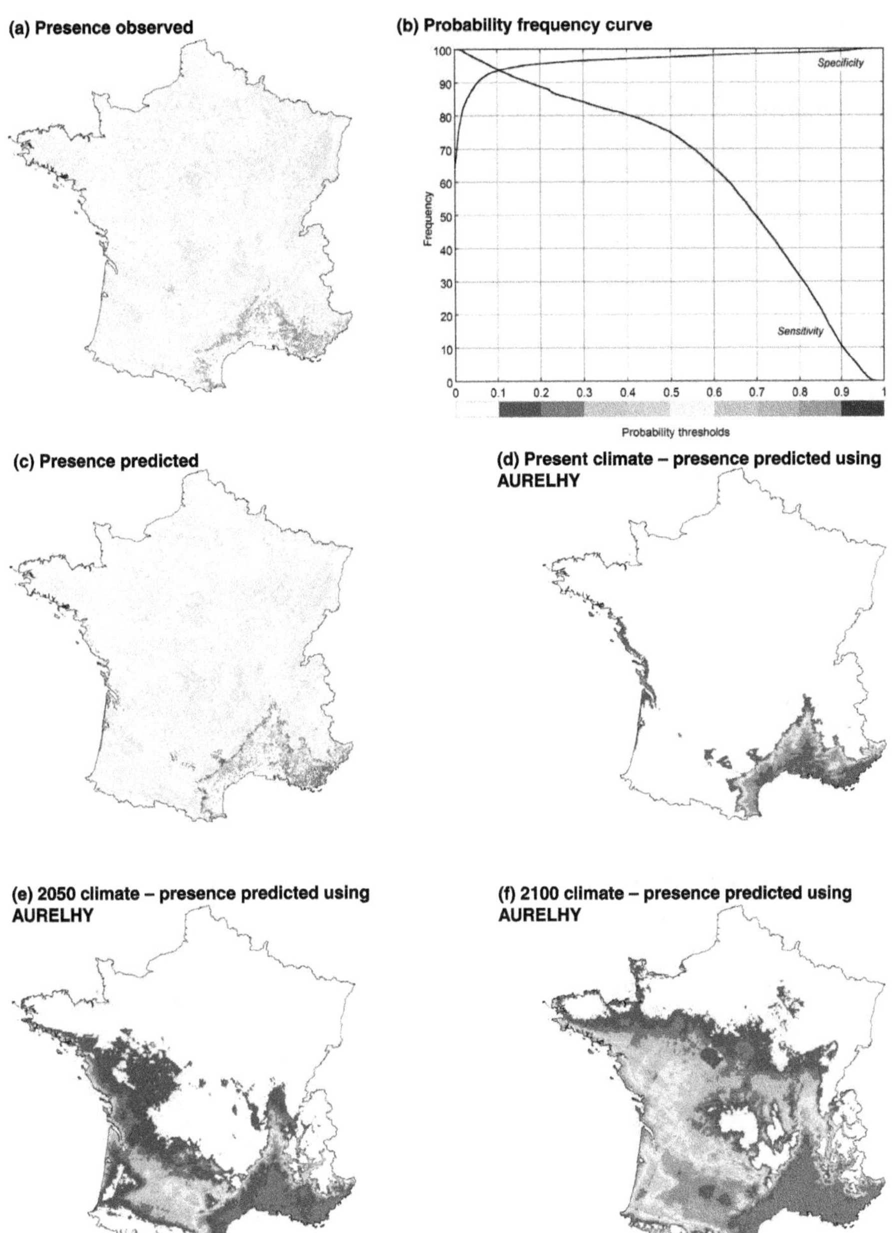

Figure 11-10. Modelled species distribution under present and predicted future climatic conditions for evergreen oak (*Quercus ilex*).

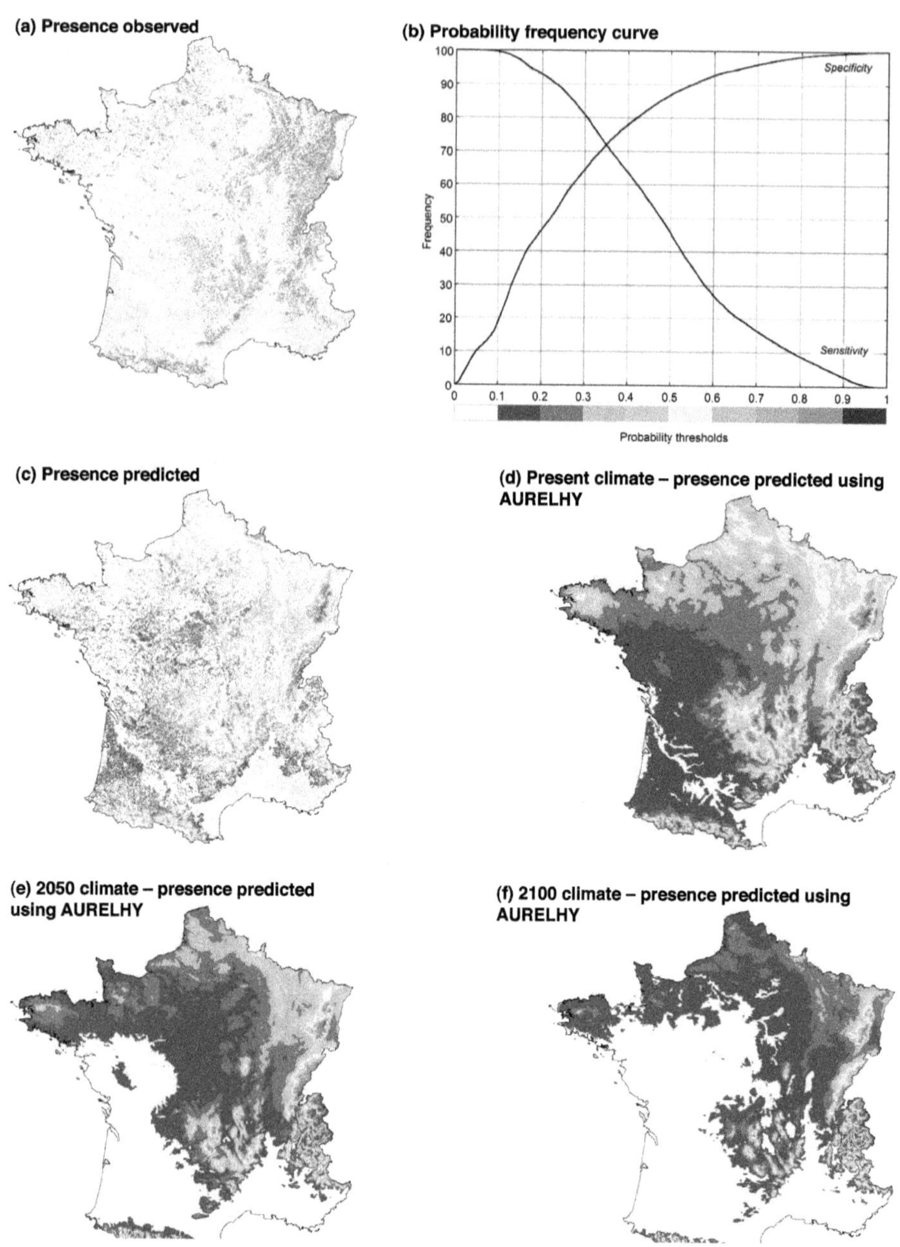

Figure 11-11. Modelled species distribution under present and predicted future climatic conditions for beech (*Fagus sylvatica*).

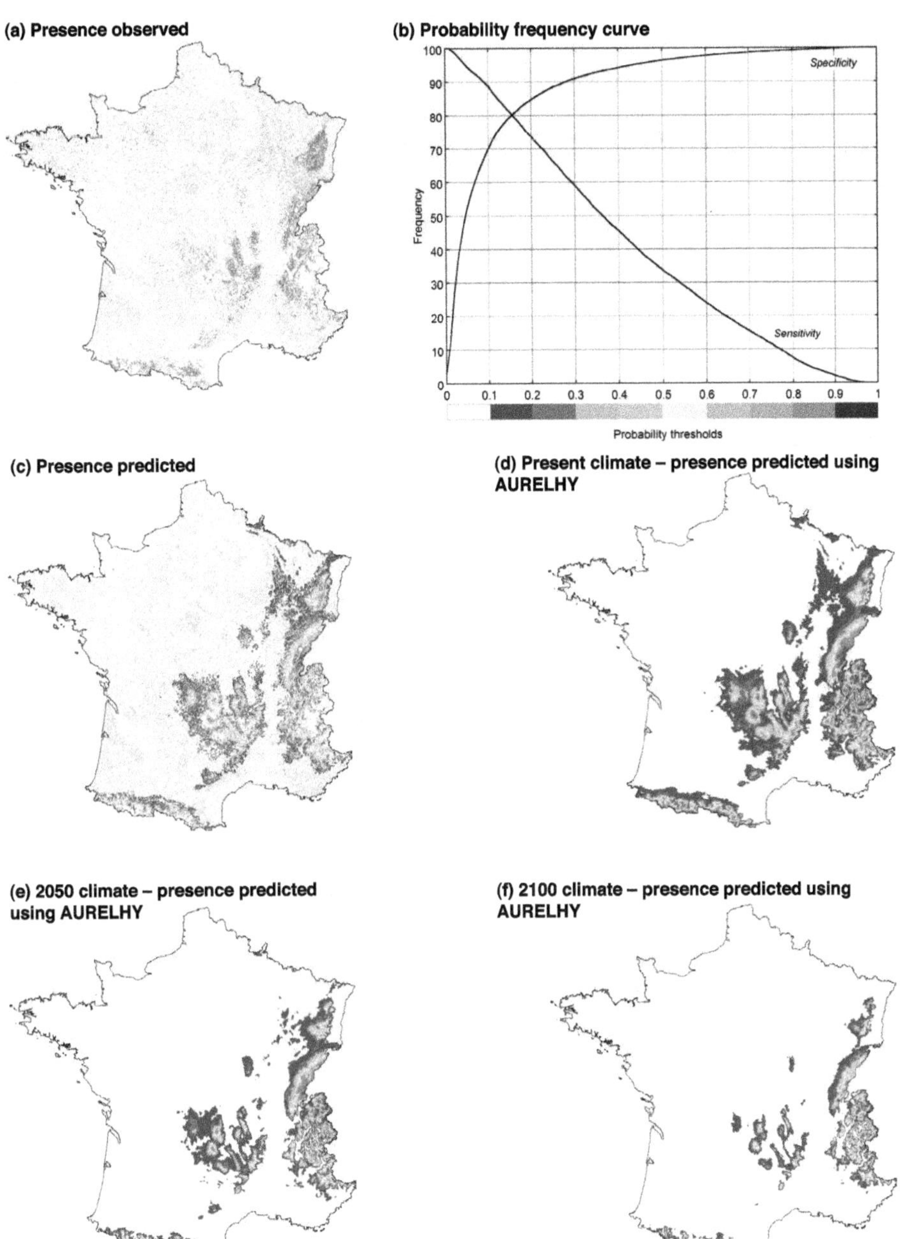

Figure 11-12. Modelled species distribution under present and predicted future climatic conditions for silver fir (*Abies alba*).

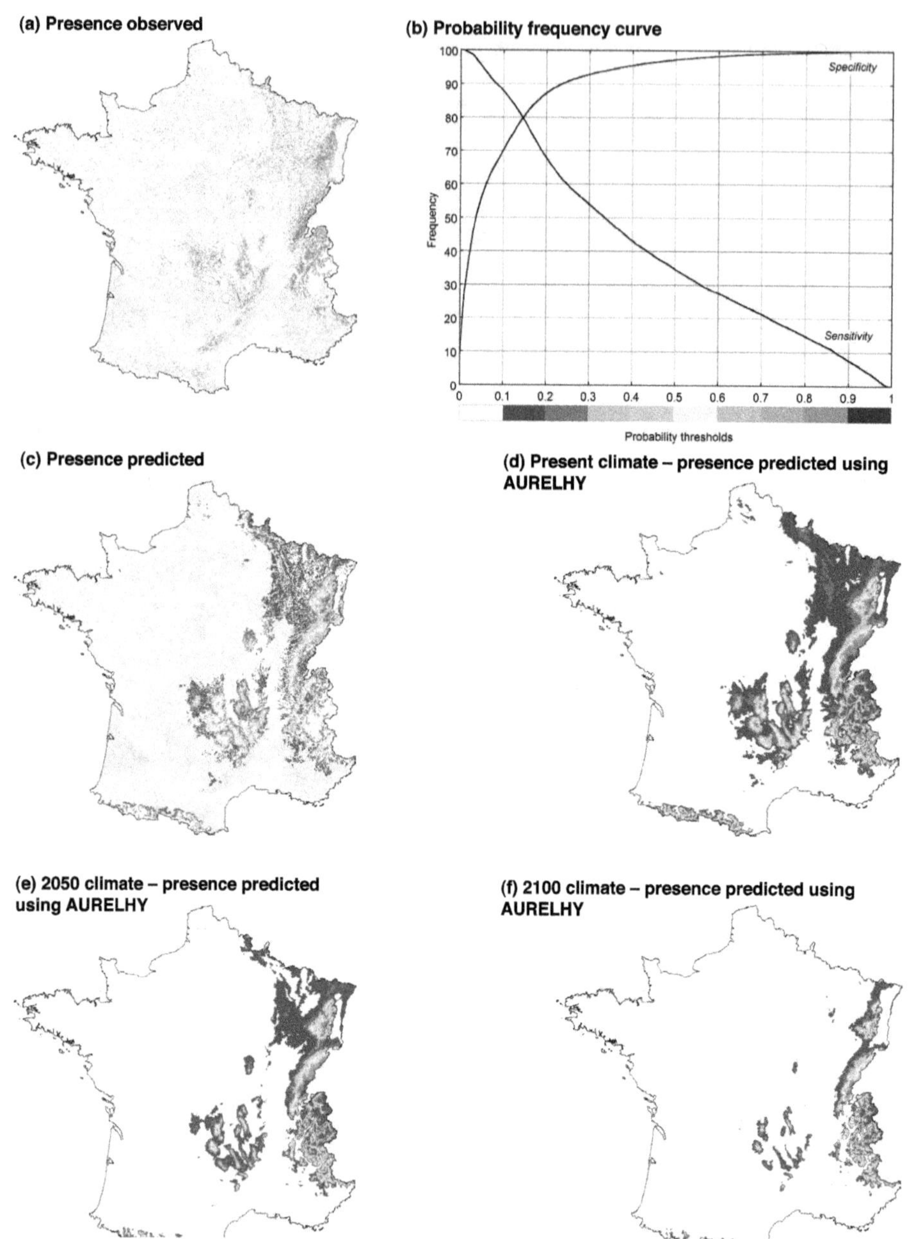

Figure 11-13. Modelled species distribution under present and predicted future climatic conditions for spruce (*Picea abies*).

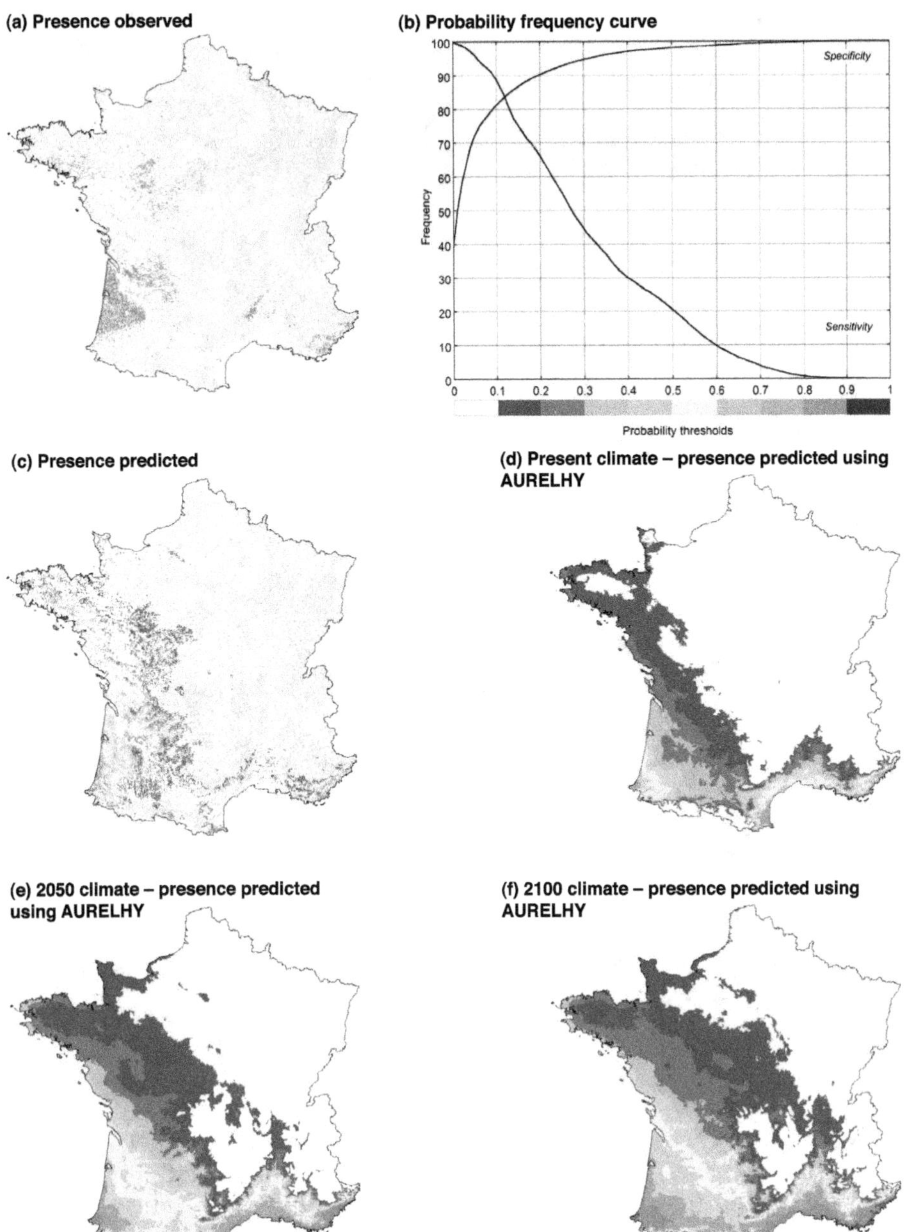

Figure 11-14. Modelled species distribution under present and predicted future climatic conditions for maritime pine (*Pinus pinaster*).

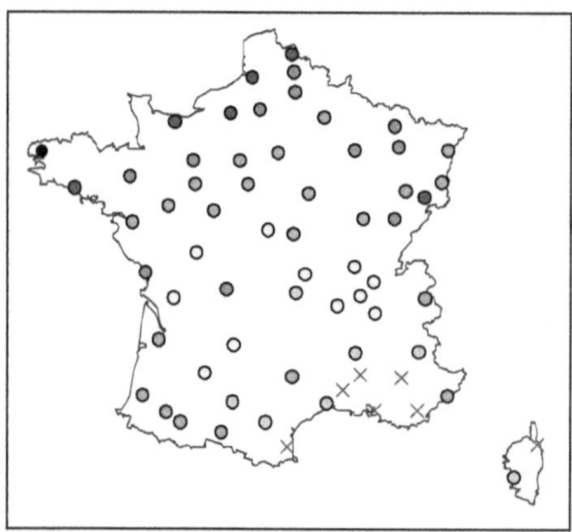
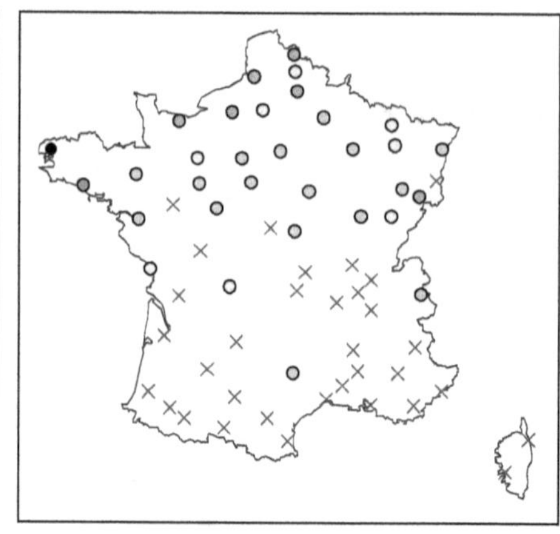

Figure 12-3. Ecoclimatic indices (EI) simulated with Climex for *Melampsora pinitorqua*, under current climate (1961–90) (left) and for the 2070–99 period using the ARPEGE climate scenario (right).

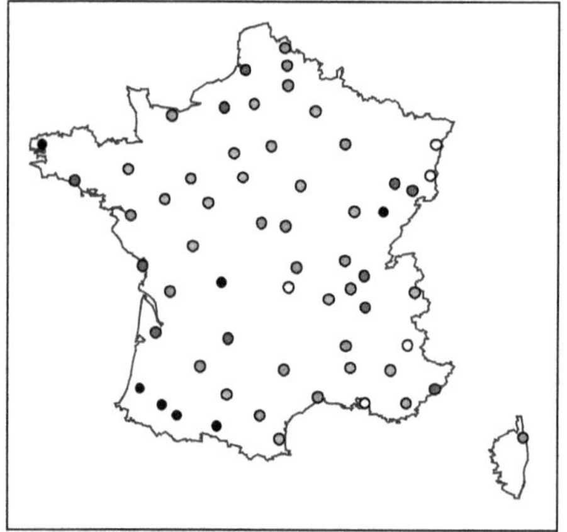
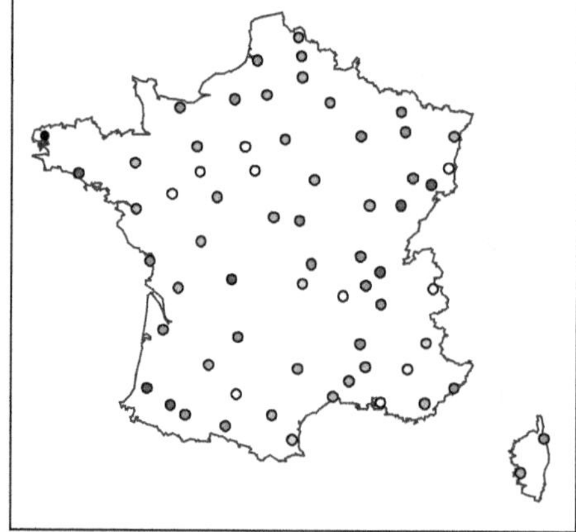

Figure 12-5. Ecoclimatic indices (EI) simulated with Climex for *Dothistroma pini* under current climate (1961–90) (left) and for the 2070–99 period using the ARPEGE climate scenario (right).

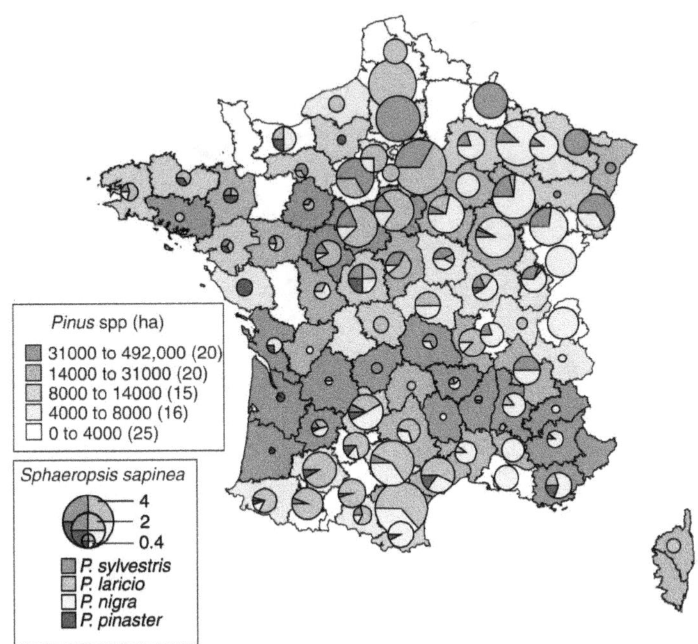

Figure 12-6. Distribution of *Sphaeropsis sapinea* reports (ratio of standardized disease rates (SDR) to the mean species rate).

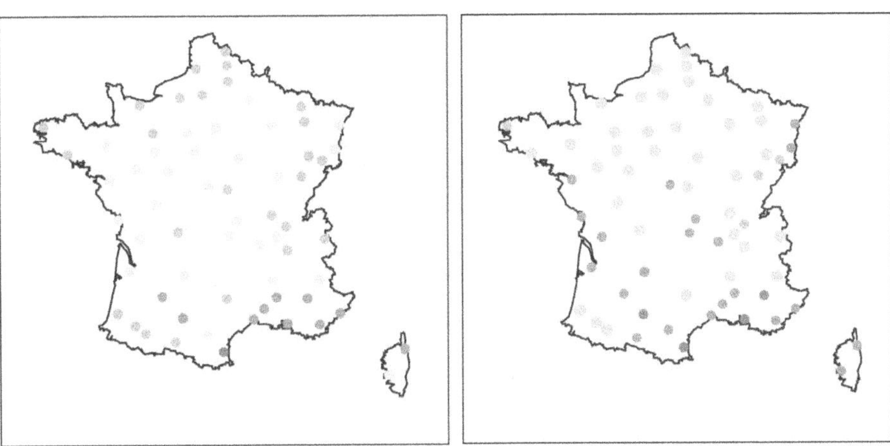

Figure 12-9. Ecoclimatic indices (EI) simulated with Climex for *Sphaeropsis sapinea* under current climate (1961–90) (left) and for the 2070–99 period using the ARPEGE climate scenario (right).

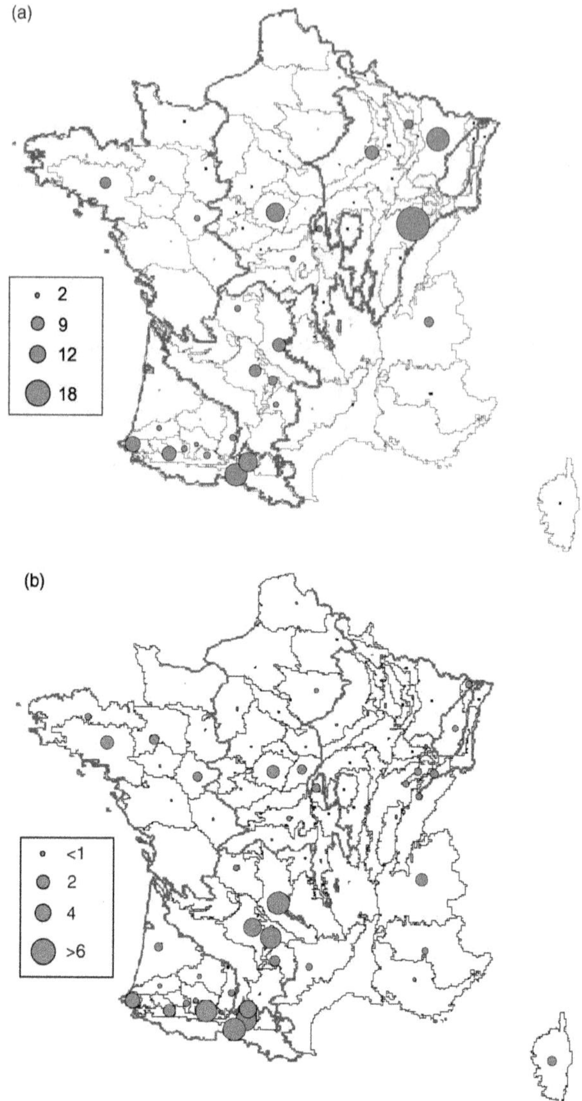

Figure 12-10. Distribution of *Erysiphe alphitoïdes* on oaks (from the DSF database, *Quercus petraea*, *Q. pubescens* and *Q. robur*, 1989–2002). (a) Total number of oak mildew reports (not associated with defoliation or frost) on mature trees; (b) standardized disease rates (SDR) (1 represents the average rate for France). The areas delimited with bold lines are those used for aggregation in the study of inter-annual variation of disease.

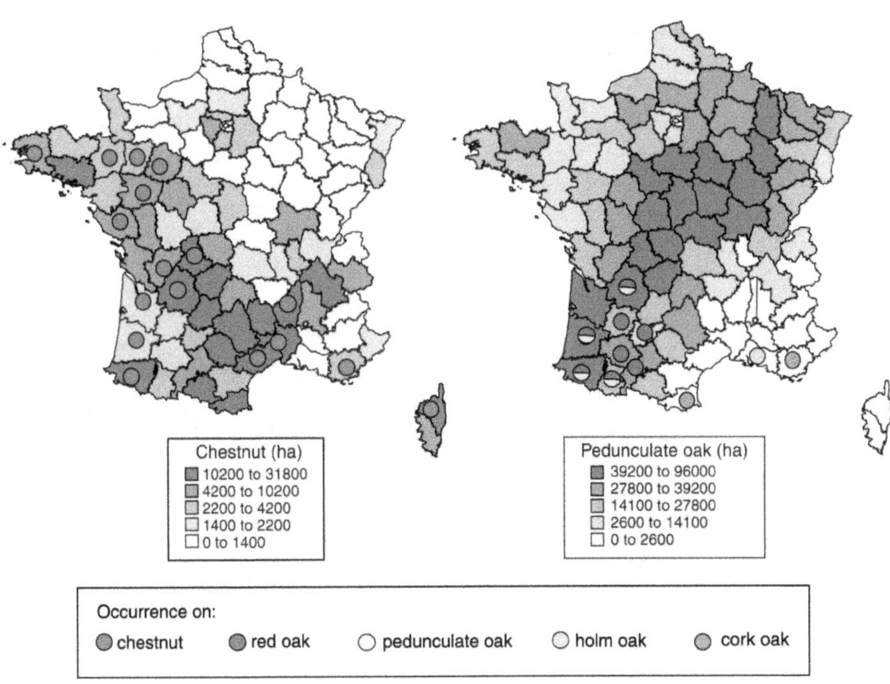

Figure 12-13. Distribution of ink disease *Phytophthora cinnamomi* reports on chestnut and oaks (disease data: DSF, INRA and SPV; forest data: NFI).

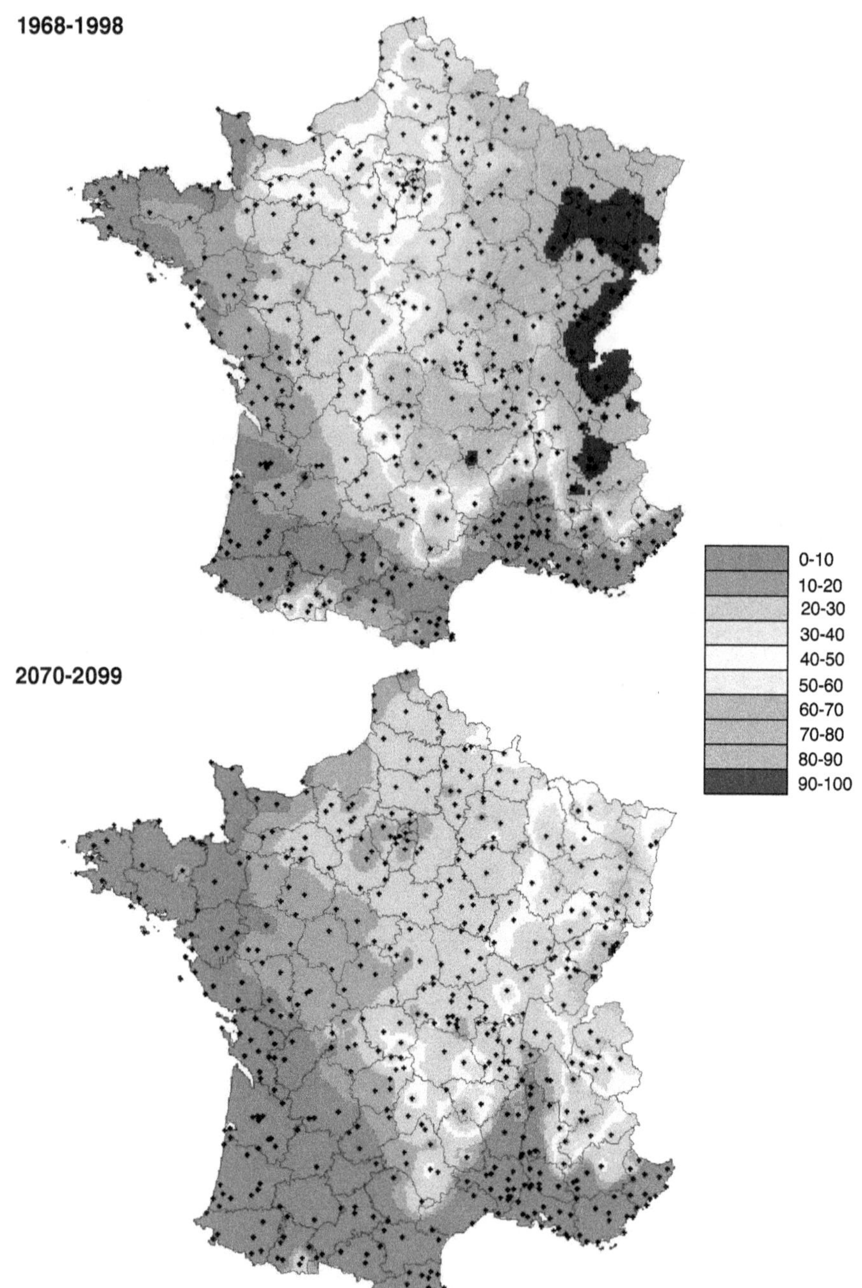

Figure 12-14. Risk mapping of ink disease on pedunculate oak: mapping of $F_{0.5}$ index = frequency of years when the predicted annual survival rate of *P. cinnamomi* (ASR) is lower or equal to 0.5 ($F_{0.5}$ <10%: high risk; 10% <$F_{0.5}$ <40%: moderate risk; $F_{0.5}$ >40%: low risk) (Bergot *et al.*, 2004).

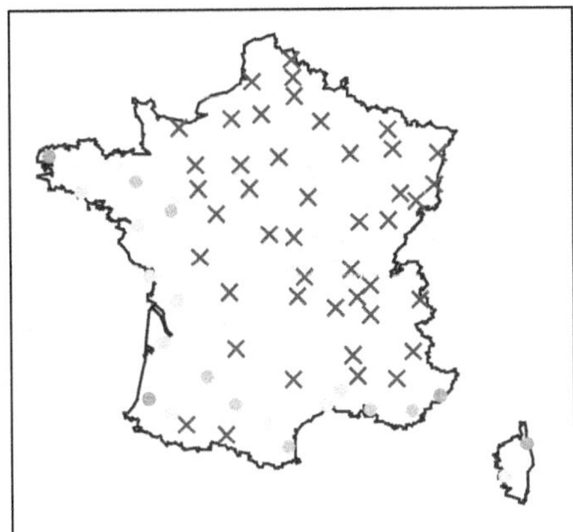
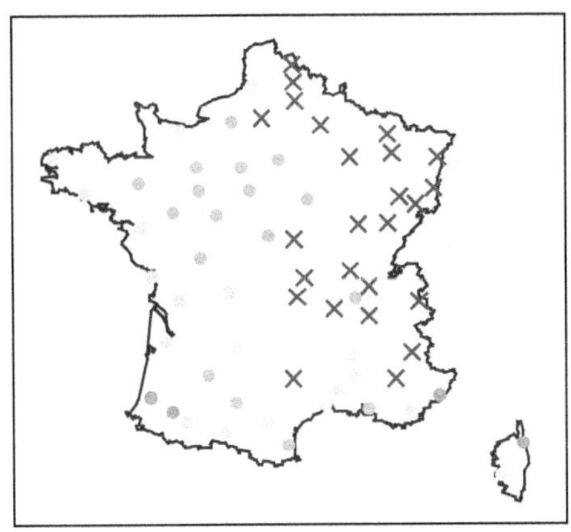

Figure 12-15. Ecoclimatic indices (EI) simulated with Climex for *Phytophthora cinnamomi* as a stem canker agent under current climate (1961–90) (left) and for the 2070–99 period using the ARPEGE climate scenario (right).

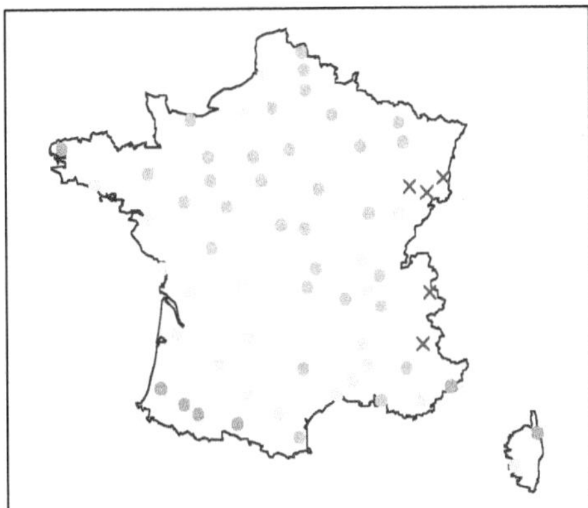

Figure 12-16. Ecoclimatic indices (EI) simulated with Climex for *Phytophthora cinnamomi* as a root necrosis agent under current climate (1961–90) (left) and for the 2070–99 period using the ARPEGE climate scenario (right).

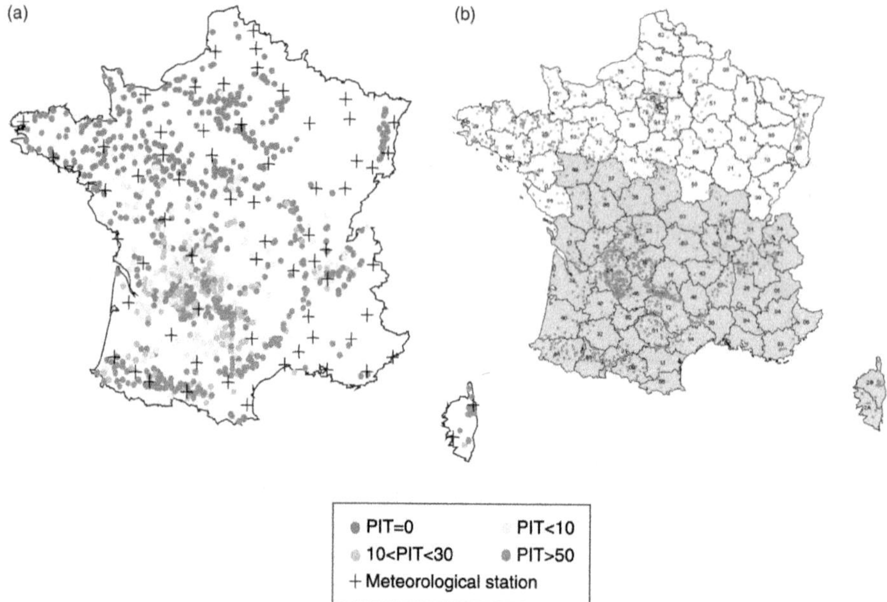

Figure 12-17. Distribution of *Cryphonectria parasitica* according to the systematic surveys completed in (a) 1996–97 and (b) 2002. PIT = percentage of infected trees per plot.

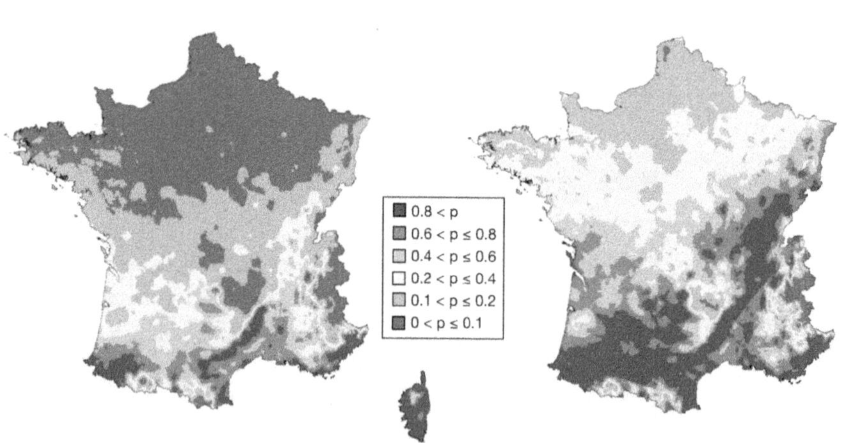

Figure 12-18. Estimated percentage of *Cryphonectria parasitica* (p) infected trees with the logistic model under current climate (1961–90) (left) and for the 2070–99 period using the ARPEGE climate scenario (right).

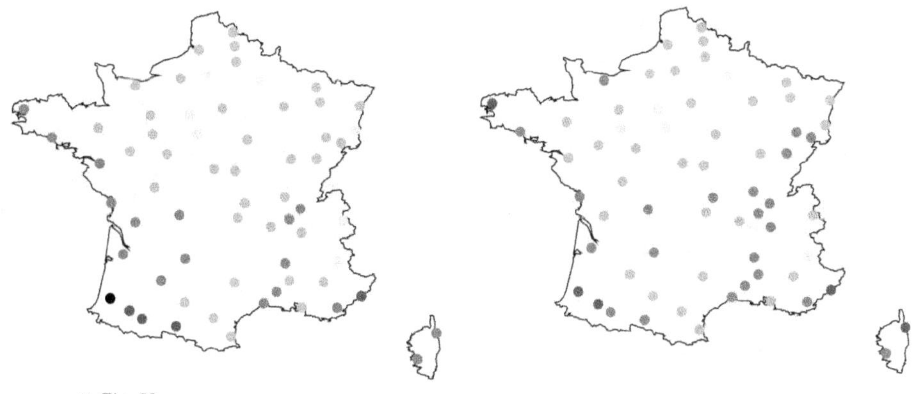

Figure 12-19. Ecoclimatic indices (EI) simulated with Climex for *Cryphonectria parasitica* under current climate (1961–90) (left) and for the 2070–99 period using the ARPEGE climate scenario (right).

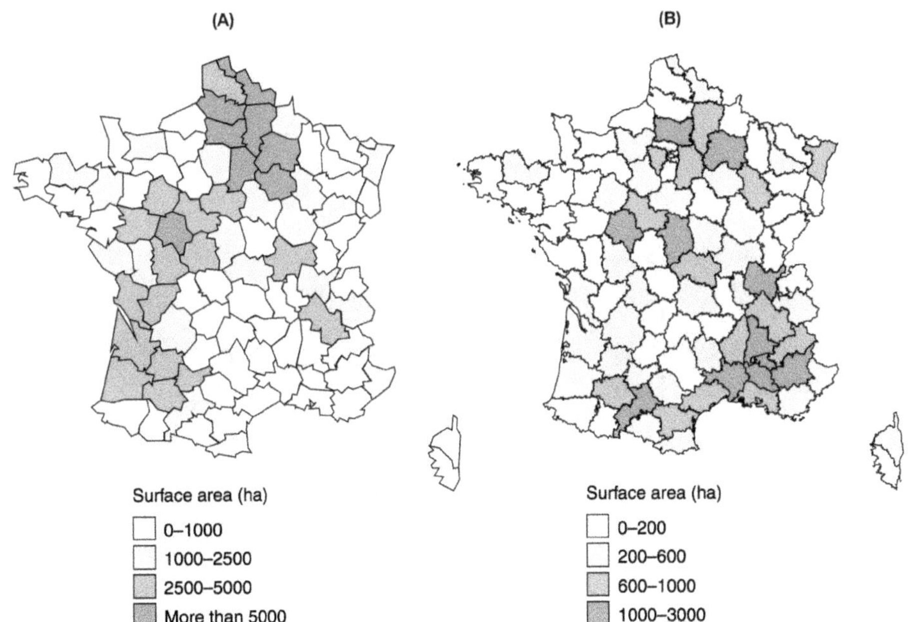

Figure 12-20. Distribution of (A) cultivated and (B) wild poplars stands (NFI data).

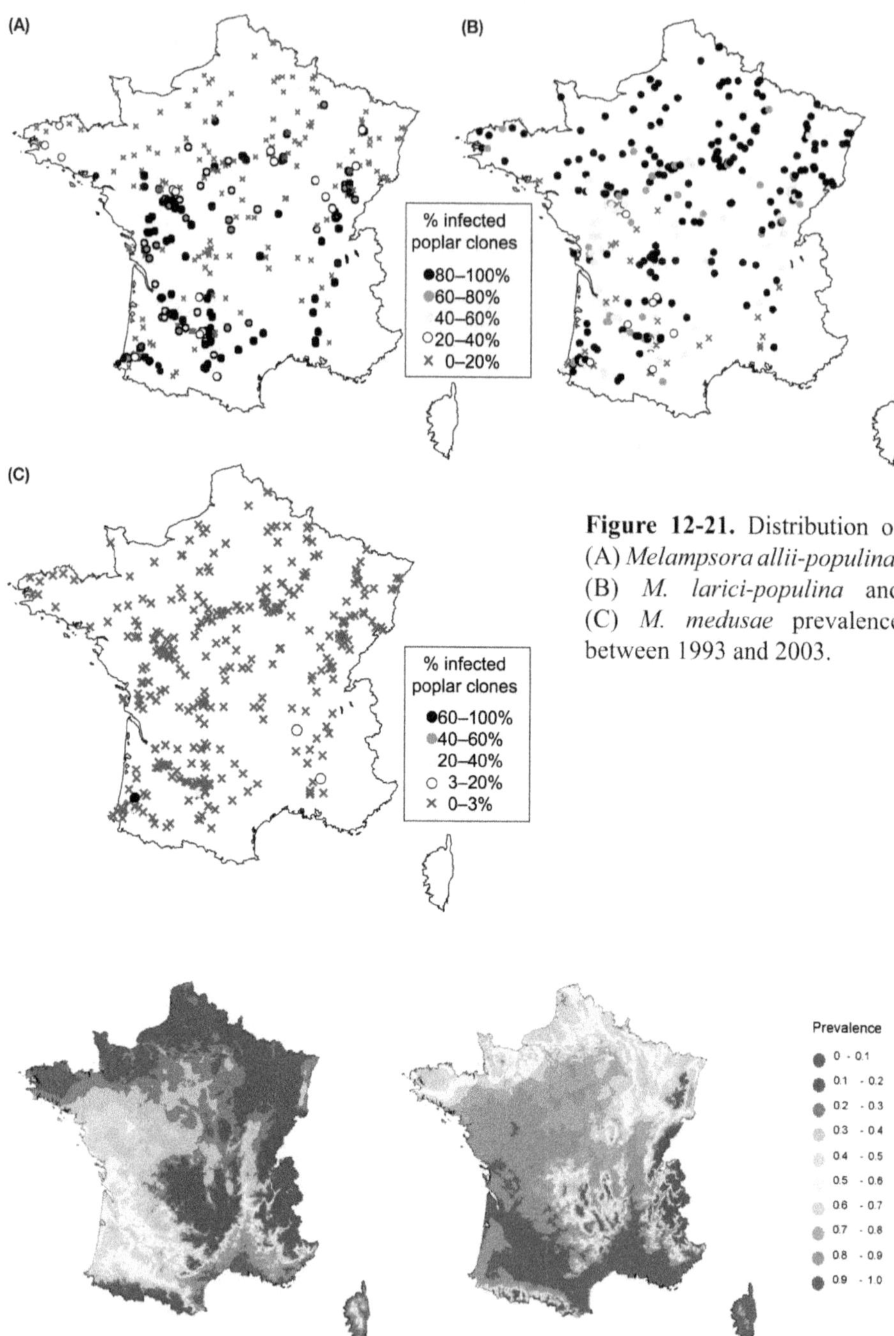

Figure 12-21. Distribution of (A) *Melampsora allii-populina*, (B) *M. larici-populina* and (C) *M. medusae* prevalence between 1993 and 2003.

Figure 12-22. Spatial simulated distribution of *Melampsora allii-populina* derived from a logistic analysis with current climatic data (1961–90) (left) and under the climatic scenario for the 2070–99 period (ARPEGE) (right).

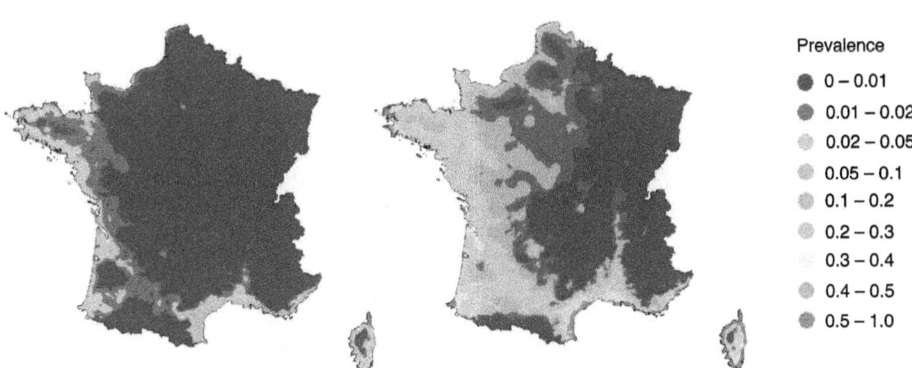

Figure 12-23. Spatial simulated distribution of *Melampsora medusae* derived from a logistic analysis with current climatic data (1961–90) (left) and under the climatic scenario for the 2070–99 period (ARPEGE) (right).

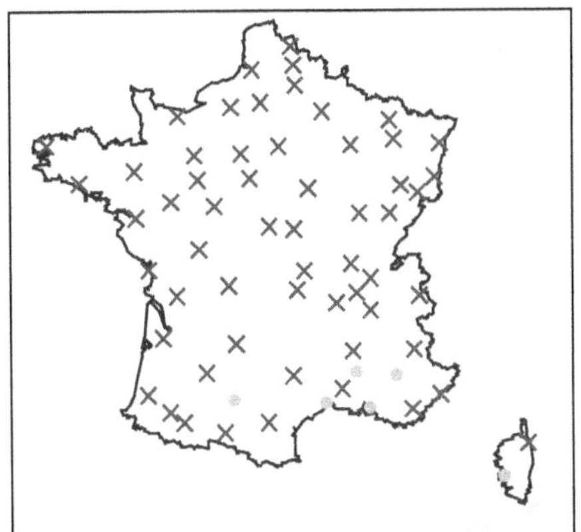

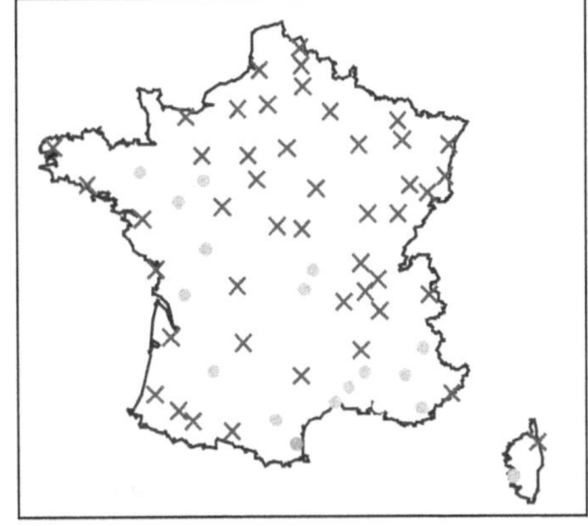

Figure 12-26. Ecoclimatic indices simulated with Climex for *Biscognauxia mediterranea* under current climate (1961-1990) (left) and for the 2070-2099 period, using the ARPEGE-Climat scenario (right).

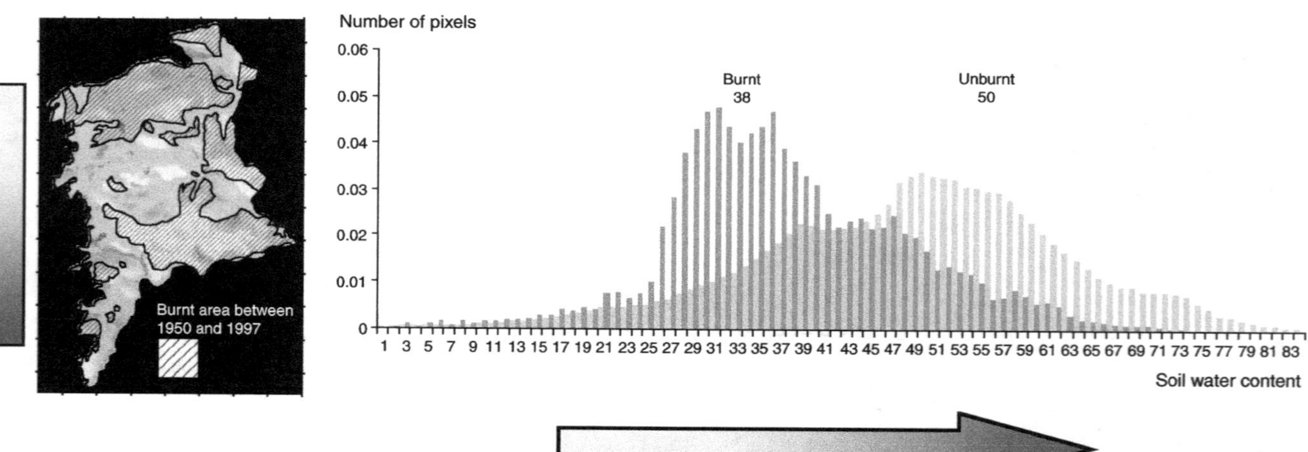

Figure 13-5. Water stress index and fire occurrence at the landscape level in Corsica.

The CLIMEX model

General description

CLIMEX is a dynamic simulation model developed by the Commonwealth Scientific and Industrial Research Organization (CSIRO) (Sutherst *et al.*, 1999) to predict the potential geographical distribution, or bioclimatic envelope, of a given species, that is, the locations were the target species can develop according to its climatic requirements. The specific values for parameters that describe the organism's response to temperature and moisture are inferred from the known geographical distribution and seasonal phenology of the species; laboratory data can also be used to fit some parameter values.

The main model output is an ecoclimatic index (EI), which gives an overall measure of suitability of locations for permanent occupation of the target species. This index is calculated from the combination of two indices: the annual growth index (GI) and the annual stress index (SI). Annual indices are the sum of weekly indices combining temperature and moisture indices as follows:

$$EI = GI \times SI \tag{2}$$

$$GI = 100 \times \Sigma GI_w / 52 \tag{3}$$

$$GI_w = TI_w \times MI_w \tag{4}$$

where TI_w is the weekly temperature index, scaled from 0 to 1, using four parameters: DV0 (lower temperature threshold); DV1 (lower optimum temperature); DV2 (upper optimum temperature); and DV3 (upper temperature threshold). MI_w is the weekly moisture index, scaled from 0 to 1, using four parameters: SM0, SM1, SM2 and SM3, defined as above.

The SI is also calculated from weekly indices for each of four primary stresses: heat stress, cold stress, wet stress and dry stress, and their interactions. A threshold value and an accumulation rate define all stresses. In the case of cold stress, two types of stress are defined: a lethal stress at very low temperatures and a chronic stress, when thermal accumulation is too low to maintain metabolism. In all cases, it is assumed that the accumulation of stress is exponential during the year.

Application

For all selected species, a thorough analysis of the literature dealing with the response to climate (including seasonal variation of susceptibility to disease) and geographical distribution was performed (Table 12.1). Geographical distributions derived from the DSF database were used for France.

The parameter values relating to species responses to temperature were defined from a laboratory experiment. For each species, three isolates (from the north-east, centre and south-west of France, in most cases) were used to establish mycelial growth-temperature curves. Isolates were sub-cultured on malt-agar or PDA media with five Petri dishes per isolate and temperature. In the case of oak powdery mildew, an obligate biotroph, isolates were inoculated on young excised oak leaves. After an initial growth period at 22°C (first growth assessment), Petri dishes were distributed in growth chambers at 2°C, 6°C, 10°C, 14°C, 18°C, 22°C, 26°C, 30°C, 34°C and 38°C. The second growth

Table 12.1.   Main references used for selecting initial values of parameters in the CLIMEX model.

Species	Map*	References
Melampsora pinitorqua	389	Desprez (1983)
		Kurkela (1969, 1973a, 1973b)
		Siwecki (1974)
Dothistroma pini	419 (1997)	Gibson (1972, 1974)
		Peterson (1973)
Sphaeropsis sapinea	459 (1992)	Blodgett *et al.* (1997)
		Keen and Smits (1989)
		Stanosz *et al.* (2001)
Phytophthora cinnamomi	302 (1991)	Benson (1982)
		Grant and Byrt (1984)
		Marçais *et al.* (1996)
		Phillips and Weste (1985)
		Shew and Benson (1983)
		Zentmyer (1981)
		Zentmyer *et al.* (1976, 1979)
Cryphonectria parasitica	66 (1994)	Anagnostakis and Aylor (1984)
		Guérin and Robin (2003)
		Guérin *et al.* (2001)
		Robin and Heiniger (2001)
Melampsora larici-populina	569 (1986)	
M. allii-populina		Somda and Pinon (1981)
M. medusae	547 (1991)	
Biscognauxia mediterranea		Vannini and Valentini (1994)

*Commonwealth Mycological Institute, then CABI Publishing: Distribution Maps of Plant Diseases.

assessment was performed when cultures had reached approximately half of the diameter of the dishes. Daily linear growth rate was calculated for each temperature. The four temperature parameter values used in CLIMEX were defined as follows: DV0 = lowest temperature with growth = 10% of maximal growth; DV1 = like DV0 with 70%; DV2 = highest temperature with growth = 70% of maximal growth, DV3 = like DV2 with 10%. Results are presented in Figure 12.1. Most species have a wide temperature range allowing growth, covering 25°C to 30°C, which reflects the great plasticity of fungal species and differs notably from the parameter templates provided with CLIMEX, mainly for animal and plant species. After the analyses of variance and comparisons of means, *B. mediterranea, S. sapinea, P. cinnamomi, C. parasitica, A. tabescens* could be assigned to a thermophilic group, *C. sororia* and *C. batschiana* to a psychrophilic group, the other species being intermediate. No general effect of geographical origin of isolates (north-centre-south of France) could be evidenced. Mean values were then used for each species.

258

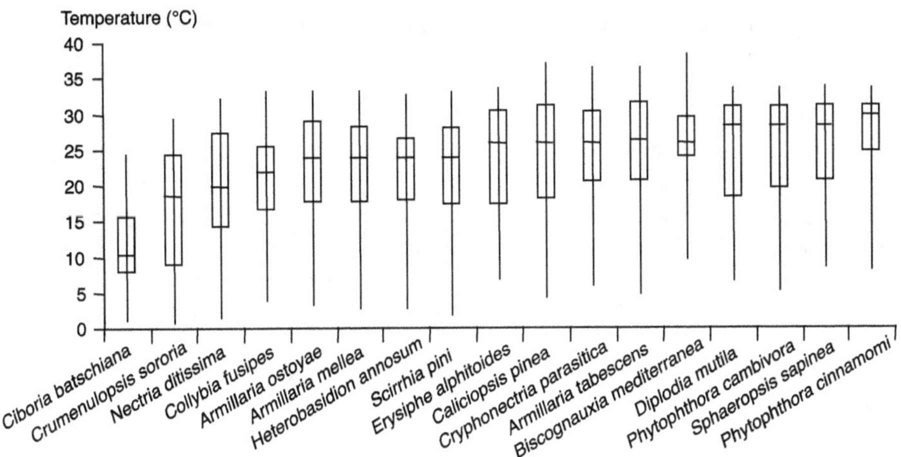

Figure 12.1. Response to temperature (*in vitro* growth) of several species of forest pathogenic fungi; the four values (lower ends of line and block, upper ends of blocks and lines) corresponds to 10%, 70%, 70% and 10% of maximal growth of the species (indicated by an horizontal line in the block).

For all other parameters, initial values were chosen from the templates provided with CLIMEX (i.e. temperate, Mediterranean and wet tropical) according to knowledge of species origin and distribution. A lethal stress at cold temperatures was only used for *P. cinnamomi*. In all cases, final values were obtained by an iterative parameter-fitting process until a good fit to geographical distribution and seasonal variation data was obtained.

Climatic data

Past climatic data for the 1961–90 period (or 1968–98 in the case of *P. cinnamomi*) were obtained from Météo-France. This past reference period was compared with the 2070–99 period by using simulated data extracted from the ARPEGE climate model. No intermediary period was considered in order to focus on the signal due to radiative forcing, which is more important than the natural fluctuations in climate beyond 2060. The comparison of observed and simulated data for the 1961–90 period clearly showed good agreement for means but the limiting ability of the climatic model to reproduce natural variability. To cope with this difficulty, we used the anomaly method, which enables the natural variability of climate to be retained while taking into account climate change at regional and temporal scales. Monthly mean anomalies between the final period and the initial reference period were calculated from simulated data for all climatic variables at all ARPEGE grid-points. These anomalies were added to the observed daily data for the 1961–90 period in order to generate comparable data for the 2070–99 period. In the case of the *P. cinnamomi* risk mapping, anomalies were interpolated to the 503 points corresponding to meteorological stations (Bergot *et al.*, 2004).

Distributions, simulation results and discussion

Melampsora pinitorqua

Most records of pine twisting rust in the DSF database are related to *Pinus pinaster* Ait. (62%) or *P. sylvestris* L. (33%). Reports are widely distributed within the geographical range of the two host species, except in the south-eastern part of the country (Figure 12.2). The absence of the disease in the Mediterranean area might be explained by the low occurrence or even absence of alternate hosts, mainly aspens, in this zone, preventing the fungus from completing its biological cycle. For both pine species, a trend of positive association between disease (standardized disease rates, SDR) and aspen abundance (% of occurrence in NFI plots) in departments with records of disease, was observed (r = 0.44 and 0.44, probability = 0.05 and 0.11 for *P. pinaster* and *P. sylvestris*, respectively). This relationship may also partly explain the lower SDR for *P. pinaster* in the Landes forest, included in the natural range of the species, as compared with more northern and eastern zones, where *P. pinaster* has more recently been used in plantations.

The relationship between aspen occurrence and pine disease may also be a result of common climatic requirements between the two species *M. pinitorqua* and *Populus tremula* L. *M. pinitorqua* is assumed to be adapted to a cool and humid temperate climate (Kurkela, 1973a, 1973b). The climatic envelope simulated for *M. pinitorqua* with CLIMEX 1961–90 climatic data is in good agreement with observations, especially for its absence in the Mediterranean area (Figure 12.3, Plate 22). The simulation with climatic data for the 2070–99 period indicates a general decrease in climatic suitability for the development of the fungus in the whole country as compared with the reference period (Figure 12.3). Nearly all the southern part of the country would become unfavourable for the disease.

Dothistroma pini

Dothistroma pini (= *D. septosporum*, teleomorph: *Mycosphaerella pini* = *Scirrhia pini*), the causal agent of red band disease of pines, was first described in France in 1967 (Morelet, 1967). Disease impact was relatively low in subsequent years (Lanier *et al.*, 1976), but has shown an increasing trend in the recent period, both in France (de Villebonne and Maugard, 1999) and other European countries (Brown *et al.*, 2002), and in North America (Woods *et al.*, 2005). *D. pini* is a major pathogen in intensively managed plantations of exotic pines, especially *Pinus radiata* D. Don, in the southern hemisphere (Gibson, 1972, 1974; Harrington and Wingfield, 1998). The fungus is assumed to be endemic in Central America (Evans, 1984) and in the Himalayas (Ivory, 1994).

In the DSF database, more than 92% *D. pini* reports occurred on *P. laricio* Poir, which confirms the high susceptibility of this species within pines. The highest standardized disease rates are observed in the western part of the country, except in the south-Atlantic zone (Figure 12.4). Two-thirds of reports were made between 1998 and 2002, corresponding to rainy years, especially in spring. These spatial and temporal patterns may be explained by climatic conditions required for successful infections, that is, mean daily temperatures close to 20°C and high moisture (Gibson, 1972; Gadgil, 1984). The low SDR in the eastern part of the country and in the Landes

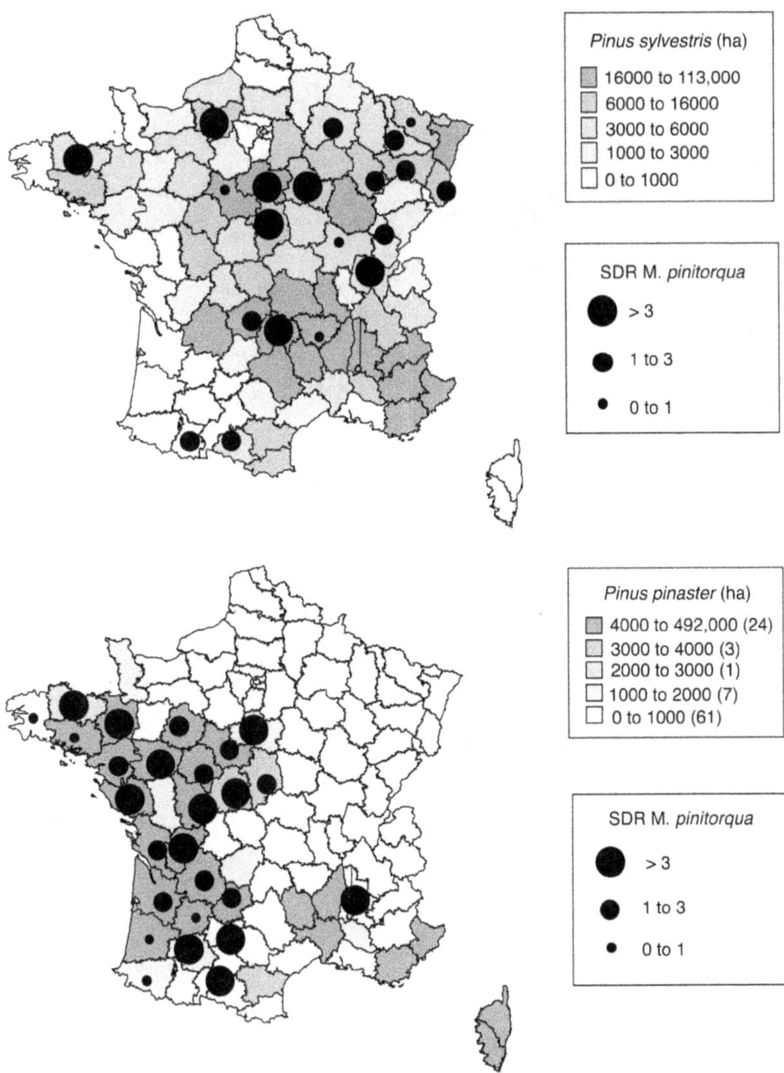

Figure 12.2. Distribution of *Melampsora pinitorqua* reports on *Pinus sylvestris* and *P. pinaster* (ratio of standardized disease rates (SDR) to the mean species rate).

area where *P. pinaster* Ait. is the main species, might be explained by the absence of susceptible hosts. CLIMEX simulations for the 1961–90 period suggest that most areas should be climatically suitable for *D. pini* development (Figure 12.5, Plate 22). Slightly less favourable conditions may result from climate change in the 2070–99 period in relation to decrease in precipitation during the growing season (mean anomaly in EI = − 5%, TI = + 5%, MI = − 12%) (Figure 12.5).

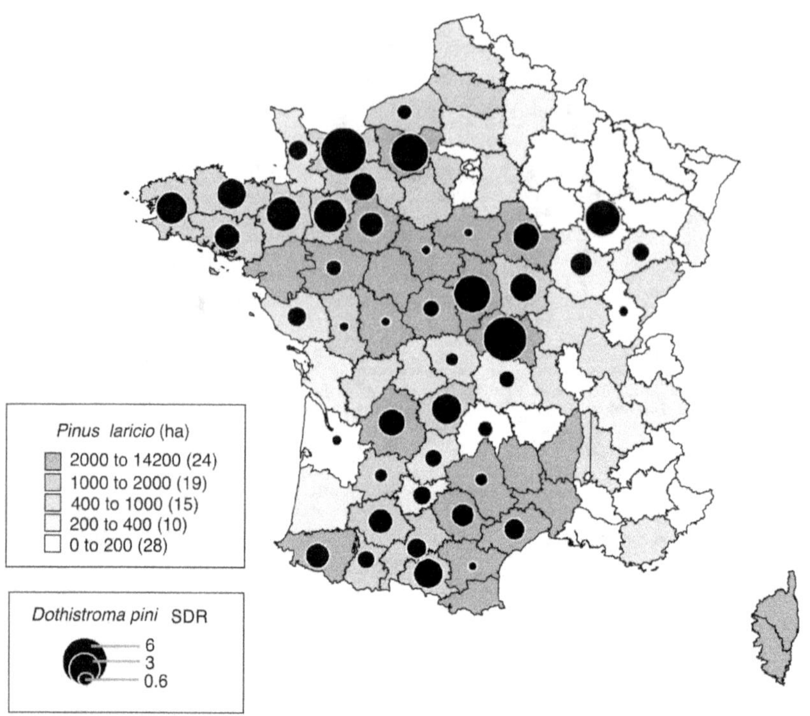

Figure 12.4. Distribution of *Dothistroma pini* reports (ratio of standardized disease rates (SDR) to the mean species rate).

Sphaeropsis sapinea

Sphaeropsis sapinea (anamorph: *Diplodia pinea*) was not considered to be a major pathogen in France in the 1970s (Lanier *et al.*, 1976), yet this parasite has been one of the most commonly reported in the DSF database during the last 15 years. Approximately 85% of reports are related to four *Pinus* species: *P. laricio* 46.1%, *P. nigra* Arnold 21.7%, *P. sylvestris* 12.2% and *P. pinaster* 4.7%. The high susceptibility of black pines had already reported in Europe and North America (Piou *et al.*, 1991). Most reports come from *P. laricio* in south-western and north-central France and *P. sylvestris* in the north-eastern part of the country (Figure 12.6, Plate 23).

Drought and hail have often been reported as predisposing factors frequently associated with *S. sapinea* damage (Swart and Wingfield, 1991; Smith *et al.*, 2002). In accordance with these reports, the highest SDR were observed during the 1990–92 period (Figure 12.7), following the severe 1989 drought. Moreover, 20% of hail reports in the DSF database were associated with *S. sapinea* damage and the zones affected by frequent hail episodes partly correspond to zones of high *S. sapinea* damage (Figure 12.8). Our laboratory experiment confirmed *S. sapinea* thermophily, with a growth optimum at around 30°C (Keen and Smits, 1989). The warm conditions prevailing in the last decade of the 20th century might explain the positive evolution of the disease.

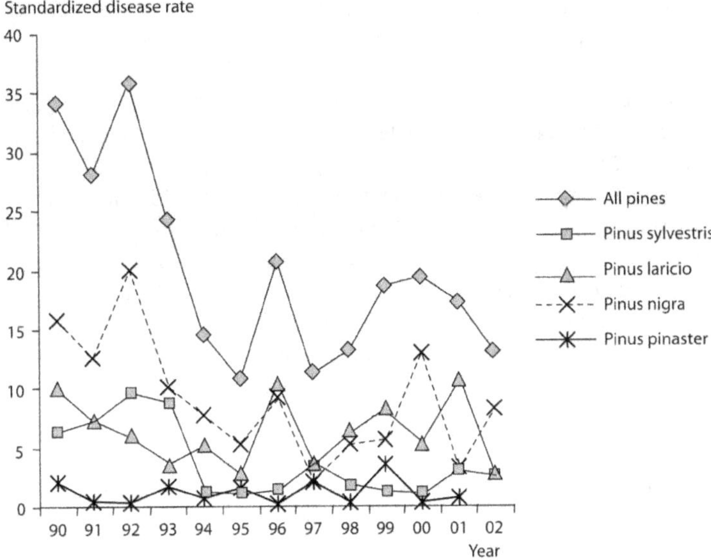

Figure 12.7. Temporal variation in *Sphaeropsis sapinea* reports in the DSF database.

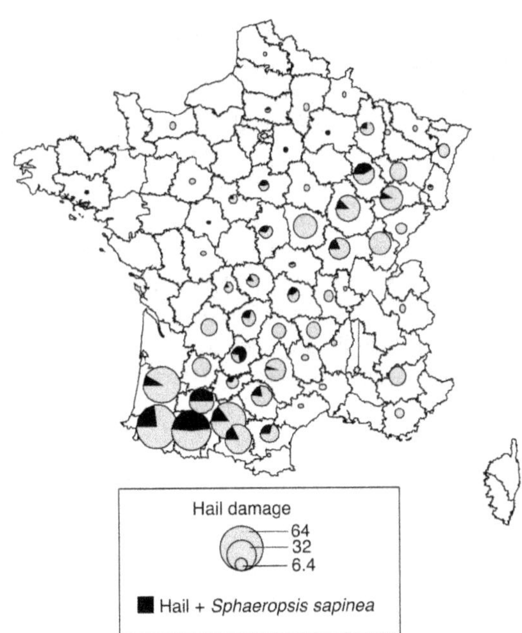

Figure 12.8. Distribution of hail damage reports in the DSF database from 1989 to 2002.

Simulations of *S. sapinea* climatic envelope with CLIMEX using available knowledge on the parasite ecophysiology and past climatic data showed a potential development of the disease all over the country, in accordance with observations. However, a higher potential was observed in the south in simulations, whereas high SDR were also observed in north-eastern France (Figure 12.9, Plate 23). A general increase in parasite activity was simulated for the 2070–99 period under the ARPEGE climate scenario.

The discrepancy between the climatic envelope simulated for the current period and the actual disease reports might be explained by the biology of the parasite. *S. sapinea* is an endophyte fungus, which turns pathogenic producing visible symptoms only after host trees are affected by stress factors, especially drought (Blodgett *et al.*, 1997; Stanosz *et al.*, 2001). The fungus might be present in a latent form in wide areas under various climates, but becoming visible only occasionally and/or locally in response to favouring factors (drought and hail).

Erysiphe alphitoïdes

Oak powdery mildew, caused by *Erysiphe (= Microsphaera) alphitoïdes*, suddenly appeared in France and other European countries in 1907 and has become one of the most common diseases on oaks since then (Desprez-Loustau, 2002). Reports of oak powdery mildew are distributed over the whole range of *Quercus robur* L., *Q. petraea* Liebl. and *Q. pubescens* Willd., its main hosts (Figure 12.10, Plate 24). A relatively high number of reports have been associated with a previous insect defoliation or frost, especially in the north-east (38% of all reports). When all other reports on mature trees were considered, the highest SDR were observed in the south-west (Pyrenean piedmont and the south-west edge of the Massif Central). An important annual variation was observed with three epidemic peaks in 1990, 1997 and 2001 within the period (Figure 12.11). These years were also characterized by relatively early disease reports (10% in May, instead of 2% as in other years), as a result of an early development of the disease, which unusually affected the first flush. This early development of infection may explain the higher disease severity at the end of the season due to a higher potential number of infection cycles.

The statistical analysis of SDR in the south-west in relation to several climatic variables showed that mean winter temperature was the main explaining variable of annual variation. A temperature threshold of 8.2°C was defined, above which all epidemic years occurred (Figure 12.12). The relationship between winter temperatures and disease severity, estimated by SDR, suggests that phenological synchrony between oaks and the parasite might be important in explaining the early initiation and final severity of oak powdery mildew epidemics in the south-west. Data from the European Forest Health network, available since 1997, are in good agreement, with two epidemic peaks for the reporting period associated with early budburst in oaks.

If the 8.2°C value is assumed to be the mean winter temperature threshold associated with the probability of occurrence of an epidemic in a given year, the frequency of years with a mean winter temperature above this threshold can be used as a risk index in a given location. Since this threshold was defined for the south-west region, the analysis was performed for the same area, using two meteorological stations for the 1968–98 and 2070–99 periods. A dramatic increase in the epidemic risk index was observed between the two periods, from approximately 10% in the initial reference period to 48% and 71%

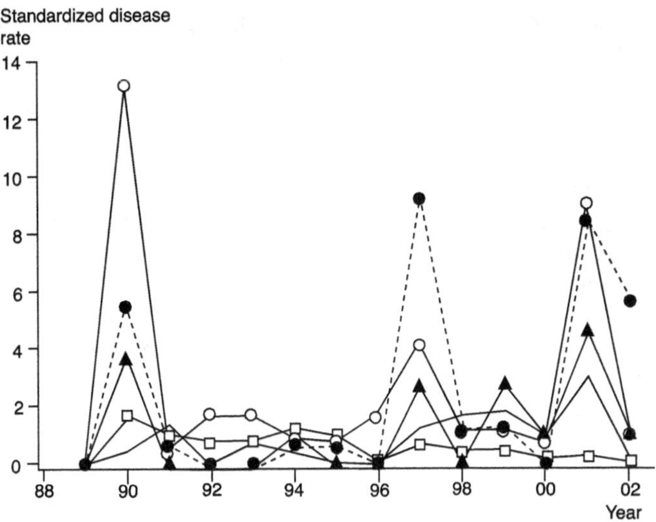

Figure 12.11. Annual variability of oak mildew *Erysiphe alphitoïdes* reports on mature *Quercus petraea, Q. pubescens* and *Q. robur,* with no previous defoliation by insects or frost damage (DSF database). Emply circles: Pyrenean piedmont; dots: north of Aquitaine basin; filled triangles: western plains; line without any symbol: central and northern plains; empty squares: north-eastern plain.

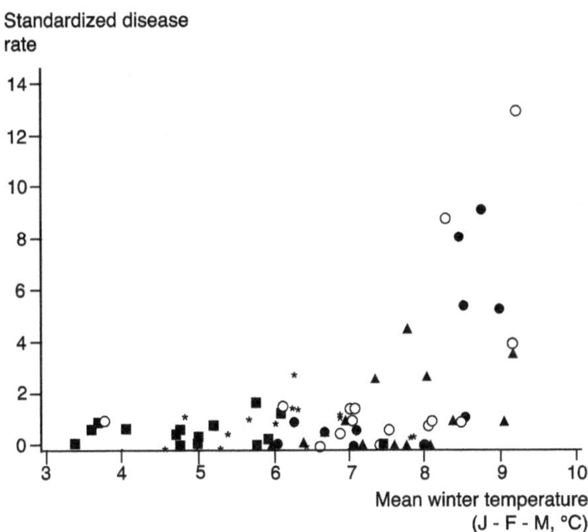

Figure 12.12. Relationship between oak mildew *Erysiphe alphitoïdes* severity (expressed as standardized disease rates (SDR) from the DSF database) and mean air temperature in January, February and March. O, Pyrenean piedmont; ●, north of Aquitaine basin; ▲, western plains; ★, central and northern plains; ■, north-eastern plain.

for the two meteorological stations in the future period. It is noteworthy that the risk index calculated for the 1968–98 period is already lower than the frequency of epidemic peaks observed during the last 15 years (approximately 20%). This may result from the warmer years, especially in the winters, during this latter period as compared with 1968–98.

Phytophthora cinnamomi

Phytophthora cinnamomi (Marçais *et al.*, 2004; Bergot *et al.*, 2004) is an exotic parasite in France, first reported in 1860 as the causal agent of ink disease of chestnuts, with primary symptoms of root necrosis leading to tree decline. Most chestnut grove areas were affected within 40 years after this first record of the disease (Grente, 1961). The first report of *P. cinnamomi* on oaks (red oak, *Quercus rubra* du Roi and pedunculate oak, *Q. robur*), causing a bleeding trunk canker, also called "ink disease", is much more recent (Moreau and Moreau, 1952). The current distribution of the parasite covers a large south-western part of the country, with a wider distribution on chestnuts (groves and forests) than on oaks (Figure 12.13, Plate 25). Chestnut is affected in most of its geographical range, except in the area around Paris and in Alsace. In this latter region, ink disease was shown to be caused by *P. cambivora* (Fleish, 2002). Ink disease on oaks is very restricted in its geographical range as compared with the host range. In Europe, *P. cinnamomi* has been reported as a chestnut or oak pathogen in Spain, Portugal, Italy, Switzerland, UK, Slovakia and Romania (Vannini and Vettraino, 2001). The distribution of *P. cinnamomi* in nurseries is much wider than in forest stands, both in France (Vegh and Bourgeois, 1975) and in Europe (Themann *et al.*, 2002).

The comparison of risk maps obtained with the *P. cinnamomi* specific model for the two study periods shows a great extension of zones at risk for disease whatever the oak species (see Figure 12.14, Plate 26 for pedunculate oak). Under current climate, high risk areas are restricted to the Mediterranean zone, the south-west and Brittany, including locations where disease was reported. In the future, the simulation results in a great extension of high risk areas ($F_{0.5}$ < 10%) for oak species, approximately 100 km eastward from the Atlantic coast and northward from the Mediterranean zone into the Rhone valley. Locations with a moderate disease risk (10% < $F_{0.05}$ (1968–98) < 40%) may become even more favourable to disease and, more remarkably, many locations previously excluded from disease risk (40% < $F_{0.5}$ (1968–98) < 75%) are expected to become at risk ($F_{0.5}$ (2070–99) < 40%). As a whole, the distribution of $F_{0.5}$ for the 503 stations in the three risk classes (high – moderate – null) changes from 20–25–55% to 36–37–27% for red oak (data not shown), and from 26–25–49% to 40–44–16% for pedunculate oak between 1968–98 and 2070–99, respectively.

Results of CLIMEX simulations were in very good agreement with the risk mapping for oaks (Figure 12.15, Plate 27). The different values of cold stress parameters used when *P. cinnamomi* was considered either as a root pathogen or as a canker agent resulted in a much wider range for the root disease (Figure 12.16, Plate 27). The simulation for the current period is consistent with disease reports. The evolution predicted for the root disease between the two periods is less important than for oaks, with a stable mean index over the country. Easternmost locations, previously unfavourable to the disease due to the temperature constraint, may become at risk; however, these positive evolutions are counterbalanced by a decrease in parasite development potential in many locations, resulting from a decreased moisture-growth index due to higher water constraints.

Cryphonectria (= Endothia) parasitica

Cryphonectria parasitica is also an exotic parasite, originating from China. It was first reported in France in 1956 from Ardèche (Rhone valley) and the disease is still spreading. The first systematic survey completed in 1996–97 showed that the disease was almost exclusively found in the southern part of the country (De Villebonne, 1997, 1998). Since then, new disease records have been made in Brittany, Île-de-France and Alsace (Figure 12.17, Plate 28). The main explaining variables for *C. parasitica* prevalence derived from the logistic analysis were the summer minimum temperature, spring precipitations and summer water deficit. The model showed a satisfactory level of external validation (correlation predicted-observed values in the validation data set = 0.6; df = 152). The odds ratio, 1.3 for each 1°C increase, 1.2 for each 10 mm increase in spring precipitations and 1.15 for each 10 mm increase in summer water deficit, indicate that *C. parasitica* occurrence is favoured by warmer and drier summer climate and increased spring precipitations. The latter effect may result from favourable conditions for ascospore dispersal (Guérin *et al.*, 2001). The map produced from this model with current climatic data fitted well with observations, especially for high disease incidence in the south (Figure 12.18, Plate 28). The model indicated low suitability of climate in the northern part of the country except in a few areas (e.g. Brittany and Alsace), where the disease had not been detected in the first survey but has been reported since then. EIs produced with CLIMEX under the same climate had the highest values in the south, in agreement with the observed distribution and that simulated with the logistic model, but also along the Atlantic coast, up to Brittany. Moderate values of EIs were observed in a large part of the country (Figure 12.19, Plate 29). It is noteworthy that EIs in France (from 17 to 52, mean value = 28) were much lower than in the region of origin of the fungus in China and in the other area of introduction, North America (up to > 80 in both areas). The slow extension of disease range in France within the last 50 years, as compared with the very rapid extension observed in the early 20th century in north-east America (Anagnostakis, 1987), might then be partly explained by a less favourable climate, in addition to the lower susceptibility of European chestnuts, the occurrence of hypovirulence and the application of quarantine regulations. For these reasons, the potential distribution of *C. parasitica* in France has probably not yet been realized. This would explain the discrepancy in northern areas between the logistic model outputs (based on observed distribution in France) and CLIMEX simulations (with parameter values fitted with distributional data in Asia and America).

Under the climatic scenario simulated by ARPEGE for the end of this century, an overall increase in PIT was predicted by the statistical model in most locations and stable EIs by Climex. The stability of EIs (mean values of 19 and 18 for the reference and future periods, respectively) was shown to be the result of increased temperature-growth indices (mean TI values of 43 and 53, respectively, with only positive anomalies) counterbalanced by decreased moisture-growth indices (mean MI values of 87 and 76, respectively, with only negative anomalies). Overall, the final maps of the logistic and CLIMEX models are in better agreement than the initial ones; the southern part of the country would be at high risk of disease, while the northern part at moderate risk. The already observed and predicted northwards expansion of *C. parasitica* distribution in France is likely to result not only from the effects of warmer climate but also from the progressive colonization of the potential climatic envelope.

Melampsora spp. on poplars

Spatial distribution

Poplar is a widely distributed species in France, however, cultivated poplars are mainly found in the north and in the Loire valley, and wild poplars mainly in the south (Figure 12.20, Plate 29). Three fungi causing poplar rusts were studied. The first two species, *Melampsora allii-populina* and *M. larici-populina*, are indigenous to Europe and are widely distributed within their host range, with a higher relative abundance of *M. allii-populina* in the south and west-central area (Figure 12.21, Plate 30) (Pinon, 1991).

The main explaining variables for *M. allii-populina* frequency, derived from the logistic analysis, were the summer temperatures (odds ratio of 1.6 for each 1°C increase) and water index (odds ratio of 1.2 for each 1 cm decrease). The spatial distribution simulated by this model with current climatic data is in good agreement with observed data from the SPV survey, with higher frequencies in the west and in the Rhone valley (no data for the Mediterranean area) (Figure 12.22, Plate 30). Under the climatic scenario for the period 2070–99, *M. allii-populina* would be highly favoured in most parts of the country. Only restricted zones in mountainous areas would keep low values. The situation in the north in 2100 would be similar to that in the present-day south-west.

CLIMEX simulations, using available knowledge on the ecophysiology of *M. larici-populina* and *M. allii-populina* (Somda and Pinon, 1981), could not explain the differences in distribution between the two species, especially the relative greater abundance of *M. larici-populina* in northern France and Europe (results not shown). Indeed, *M. allii-populina* is favoured at higher temperatures (above 15°C), but the two species have a very close response to low temperatures, *M. larici-populina* being favoured only after very severe frosts (Somda and Pinon, 1981). The relative frequencies of these two species with a similar niche might result from complex competitive interactions including factors other than temperature (Pinon and Berthelot, 2004).

Contrary to the two indigenous rust species, *M. medusae* distribution, as revealed by the SPV survey, was restricted to a small area in the south-west. The first observations of the species in Europe were made in Spain in 1925 (Fragoso, 1925). It was first described in France in 1943 (Dupias, 1943). *M. medusae* originates from North America, where it is widely distributed from Texas to Canada. It was also introduced into Australia and New Zealand, where it is mostly encountered in warmer areas (Spiers, 1998). The logistic regression of the frequency of occurrence of *M. medusae* in relation to climatic variables produced a highly significant model where frequency was explained by the mean minimal winter temperatures (odds ratio of 2.8 for each 1°C increase). The distribution map produced by this model with climatic data for the 1961–90 period is consistent with observations (Figure 12.23, Plate 31). Coastal zones along the Mediterranean Sea and Brittany appear as potentially favourable zones where the absence of detection of *M. medusae* may be explained by the poor presence of poplars. Using the climatic data simulated for 2070–99 with the same model led to an extension of favourable zones inward from the coasts. However, the simulated frequency would keep at low levels in most areas.

The determining positive effect of minimal temperatures strongly suggests that winter survival of the parasite may be a limiting factor for the development of disease.

Although teliospores, the sexual, thick-walled resting spores of most rust species, were observed in the affected area, their functional significance might be low due to a lack of infection of alternate hosts (larch occurs infrequently in these areas). According to this hypothesis, between-season survival would rely on urediniospores, the asexual spores, which are more sensitive to frost than teliospores (Pinon, 1986). Several observations are consistent with this hypothesis: (i) a population genetic analysis with molecular markers (Bourassa *et al.*, 2005) showed that most French isolates were clonal (unpublished results); (ii) in a fine targeted survey, *M. medusae* was detected in 1988 and 1999 in the west-central and east of the country but could not be found the following years in the same locations. Moreover, the overwintering of *M. medusae* as urediniospores has been demonstrated in the south of the USA (Sinclair *et al.*, 1987). Conversely, in the northern part of the natural range in Canada, larch infection by basidiospores (produced from teliospores) is functional, enabling the parasite to complete its biological cycle and to survive in cold areas.

Temporal variations

Variation in the time of infection of larch by *M. larici-populina* in spring has been studied for the last 15 years in Nancy (north-east). A significant negative correlation between the day of year for first infections and the mean daily temperature in March and April was demonstrated (r^2 adj $= 0.3$; $P = 0.016$) (Figure 12.24). The same type of relationship was demonstrated for larch budburst (Theurillat and Schlüssel, 2000; Worrall, 1993), suggesting phenological synchrony between host and parasite. Higher temperatures in March–April are, therefore, associated with earlier larch infection, which may result in earlier poplar infection. Due to the polycyclic nature of poplar rust, an earlier initial infection will potentially lead to a higher number of infection cycles and ultimately a higher severity at the end of the growing season. Based on the relationship which was found, a 1°C increase in temperature would result in an earlier larch infection by 11 days. Extending the poplar infection season by 11 days would then lead potentially to a 30% increase in the proportion of diseased tissues, using the classical equations of Van der Plank (1968) (Renaud *et al.*, unpublished results). The observed higher impact of poplar rust in the last two decades might be partly explained by such phenomena; spring temperature increased by 1.1°C between the 1923–88 period and the last 15 years (Figure 12.25).

Biscognauxia mediterranea

Biscognauxia mediterranea (= *Hypoxylon mediterraneum*, anamorph: *Nodulisporium sp.*, Collado *et al.*, 2001; or *Periconiella*) is the causal agent of charcoal disease of oaks and has been described so far only in the Mediterranean area, principally in Spain, Portugal, North Africa and in the southern USA. Only two reports are present in the DSF database. *B. mediterranea* is a highly thermophilic species (Figure 12.1). As *S. sapinea*, *B. mediterranea* is an endophyte turning pathogenic, that is, it produces symptoms after severe drought periods (Vannini and Valentini, 1994). CLIMEX simulations showed a very restricted bioclimatic envelope and low ecoclimatic indices in France under current climate (Figure 12.26, Plate 31). Under the scenario used for the 2070–99 period, a large extension from the Mediterranean region to the Atlantic coast is anticipated but ecoclimatic indices remain low.

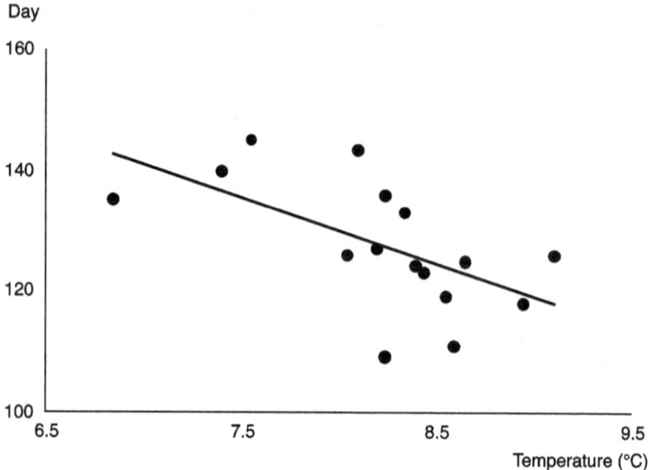

Figure 12.24. Relationship between the day of year for first infections due to *Melampsora larici-populina* on larch and mean daily temperature in March and April in Nancy.

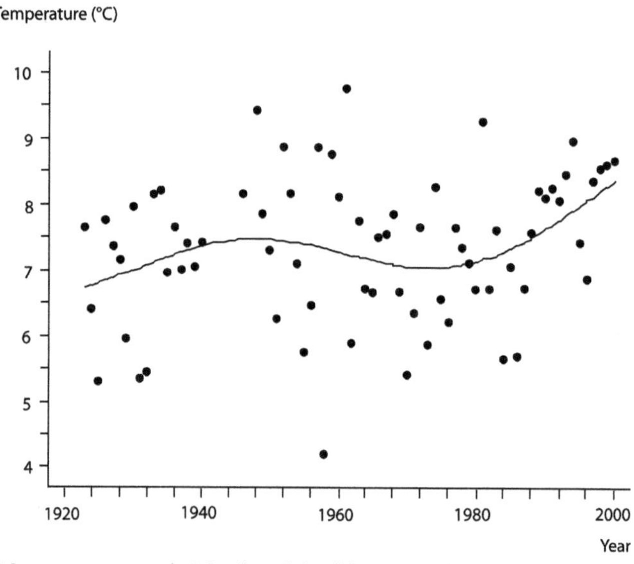

Figure 12.25. Mean temperature in March and April between 1923 and 2003 in Nancy.

General discussion and conclusion – future prospects

This study synthesizes available data on the geographical distribution of several diseases and parasites among the most currently reported in France, and focuses on climatic variables as explaining factors of spatial and temporal variations of occurrence.

Foliar and cortical pathogens were mainly selected since direct effects of climate are more likely than for root pathogens. For most studied pathosystems a significant effect of one or a few climatic variables could be demonstrated. The sensitivity of parasite ranges and/or impact to climate change could then be studied using different models. This quantitative approach adds to previous studies on potential effects of climate change, which were mostly based on general knowledge about the biology of the pathogens (e.g. Lonsdale and Gibbs, 2002).

As expected from the different ecological requirements of the parasite species, no general unidirectional trend in the evolution of diseases, both in geographical extension and severity, could be predicted. An important regression of potential range was simulated for *M. pinitorqua*, the only species typical of cool humid temperate climate included in the study. Conversely, increased ranges were simulated for more thermophilic species due to the release of temperature constraints. This is particularly the case for species for which winter survival is a critical stage limited by low temperatures in their introduced areas, such as *P. cinnamomi* and *M. medusae*. Some observations are consistent with such predictions (Fleisch, 2002). However, the climate change scenario did not always lead to range extensions or enhanced activity, even for rather thermophilic species. In the case of *C. parasitica*, for example, positive effects of warming might be counterbalanced by negative effects of water constraints. The same trend was observed for *D. pini* and *P. cinnamomi*. Only *B. mediterranea* and *S. sapinea*, which are clearly favoured by water stress, showed higher moisture-growth indices in CLIMEX simulations under future simulated climate. As a whole, the predicted evolution of temperatures, characterized by warming, should be favourable for most species whereas the evolution of precipitation regime, with lower precipitations during the growing season, will affect species very differently according to their biology. The response of parasites to water factors, therefore, is likely to be a determining process in their evolution with climate change (Woods *et al.*, 2005). However, water effects on parasites and the diseases they cause are much more complex, affected by local trends, and less documented than temperature effects, which dramatically limit our ability to predict evolutions linked to them.

More generally, the lack of knowledge of the ecophysiology of fungi is a clear limitation for the development of models allowing reliable simulations of climate effects on parasite population dynamics and disease. An additional difficulty is the specificities of the biology of these organisms. Pleomorphy, that is, the occurrence of different forms of a same organism, is a common trait of most fungi, which usually have both a mycelium and several spore forms (Burnett, 2003). These different forms correspond to different functional stages in the parasitic cycle (e.g. long-range or short-range dispersal, survival, etc.) and exhibit different ecological requirements. The release and germination of spores, which are the dispersal and infection propagules, often require free water whereas mycelium is able to grow in host tissue at relatively low water potentials. *B. mediterranea* offers an illustrative example of a parasite favoured by water stress for host colonization but requiring rain for ascospore dispersal (Vannini *et al.*, 1996). Different spore types usually have different ecological requirements, as exemplified by basidiospores, eciospores, urediniospores and teliospores in rusts (Kurkela, 1973a, 1973b). Only the *P. cinnamomi* model used in this study is a process-based model where disease severity is mainly explained by quantity of inoculum, resulting itself from winter survival as a function of temperature (Marçais *et al.*, 2004). This model needs further improvement to take into account the whole epidemiological cycle, including the potential effect of temperatures during the growing season and the effects of moisture. However, to our knowledge, this is the only mechanistic model including climate effects currently available for a forest parasite. In the absence of specific models, the use of generic models such as CLIMEX offers a potentially valuable alternative. The consistent results obtained between CLIMEX simulations and the mechanistic model for *P. cinnamomi* further confirmed its interest. However, parameter fitting in CLIMEX is mainly based on distribution data of organisms, which are not always available in sufficient detail. Moreover, a main limitation of this approach occurs if the potential distribution of the species has not been realized because of a non-equilibrium situation, for example (Thuiller *et al.*, 2005), as suggested for *C. parasitica* in France.

Another important consideration is that parasitism is the result of species interaction. Climatic effects, and especially water effects, should therefore be considered not only for their direct effects on the parasite but also through host physiology and its response to disease (Desprez-Loustau *et al.*, 2006). The positive effect of drought on the incidence of *S. sapinea* and *B. mediterranea* was mainly explained by water stress effects on host susceptibility (Vannini and Valentini, 1994; Blodgett *et al.*, 1997; Stanosz *et al.*, 2001). Effects of climate on host-parasite interaction may also include impact on phenological synchrony, in the case of a susceptibility window for the host and a limited period of infection for the parasite (Desprez-Loustau and Dupuis, 1994; Desprez-Loustau *et al.*, 1998).

The biogeographical approach used in this study only gives an indication of the potential range of parasites based on their climatic envelope on a long-term trend. The realized range will depend on species migration capacities. In addition, fungi, as other organisms, might also respond to climate change by adaptation (Davis and Shaw, 2001; Jump and Peñuelas, 2005). A further limitation of bioclimatic models in the case of parasites stems from the filter effect of host distribution, which is all the more important in that parasites are more specific. This was particularly illustrated in our study by heteroecious rusts, biotroph parasites alternating between two hosts species, such as *M. pinitorqua* and *M. medusae. D. pini* provides another example of the determining effect of susceptible host distribution on parasite distribution. The absence of reports of the fungus in some regions might be explained primarily by the low occurrence of susceptible hosts (*Pinus laricio)* more than climatic constraints. Conversely, for generalist parasites with wide host ranges, such as *P. cinnamomi*, the observed distribution on many different species may be more directly related to climatic effects. Human activities, by promoting parasite transport as well as affecting host distribution and susceptibility, have been and will be major factors potentially favouring diseases. Many emergent diseases of plants have been related to the introduction of exotic parasites or farming practices (Anderson *et al.*, 2004). The devastating epidemics caused by the introduction of *P. cinnamomi* in Australia and *C. parasitica* in North America are well documented examples (Weste and Marks, 1987; Anagnostakis, 1987). In the introduced areas, parasites were put in contact with very susceptible species, *Eucalyptus marginata* and many herbaceous species in Australia, *Castanea dentata* in North America. Conversely, in endemic zones, a long co-evolution between parasites and hosts has led to a dynamic equilibrium between virulence and resistance (e.g. Asian chestnuts resistant to the *Cryphonectria* canker) (Graves, 1950). The evolution of pine twisting rust damage is an example illustrating the effects of silviculture. Before the 1950s, the disease was almost exclusively reported on *P. sylvestris* (Hariot, 1908; Gaudineau, 1949; Dupias, 1965), whereas *P. pinaster* is now the most frequently reported host in the DSF database (0.073 and 0.122 reports per 1000 ha for *P. sylvestris* and *P. pinaster*, respectively, with similar surface areas for both species; SDR = 0.04 and 0.08). The extension of maritime pine plantations in areas favourable to the disease (due to climatic factors and alternate host abundance), associated with silvicultural practices promoting pine growth and resulting in higher susceptibility, may explain such patterns (Desprez-Loustau and Wagner, 1997).

In spite of the limitations addressed above, this study gives support for some recommendations for management. The availability of indicators is a prerequisite for any kind of decision taking. In particular, quantified and spatialized incidence data are required in order to assess spatial and temporal evolution of diseases. The DSF database used in this study proved to be very valuable; improvement could come from the combination of two types of data collection: (i) "spontaneous" reports as in most of the current DSF database; and (ii) monitoring of systematic networks, at European level, for example, which has been initiated for some diseases. This would allow both the detection of emerging diseases and quantitative assessments for established diseases required for the development of epidemiological models. Even more than for crops, disease management in forest ecosystems has to rely on an anticipatory and preventive approach, based on risk analysis. The first step is the identification of main risks, that is, pathogen species with possible higher impact in the future. The main result of our study is the demonstration of the high sensitivity of most pathosystems to climate, which strongly suggests an important evolution of the future "forest phytosanitary landscape" in relation to climate change. In most south-western Europe regions, climate change will be characterized by warming and decreased precipitations during the vegetative season. This should favour pathogens such as *S. sapinea* or *B. mediterranea*, which are typical opportunistic parasites affecting stressed hosts, especially after drought. These organisms are commonly found as endophytes, a latent form, which favours their anthropogenic dissemination in infected but symptomless plants or seeds (Burgess and Wingfield, 2002). Progress has recently been made in knowledge of such fungi (Smith and Stanosz, 1995; deWet *et al.*, 2002, 2003; Blodgett and Bonello, 2003), but the diversity of endophytes, both at inter- and intra-specific levels, and their significance in tree physiology (from mutualism to parasitism) clearly requires further research. Climate-matching could also be used to provide some indications of future risks. This method may allow identifying zones presently having the climate that is predicted in the zone of interest, and the pathogen species able to develop in these conditions. Finally, simulated geographical ranges are only potential envelopes in which the parasites may establish once they are present, therefore depending on their dispersal ability. Dispersal of parasites is greatly favoured by anthropic activities (e.g. long-distance transport). This points to the general need for strict hygiene measures based on the most probable dissemination pathways of organisms (i.e. in seeds, wood and plants), in order to avoid dissemination of parasites in climatically favourable zones where they could find naive host populations with potentially high susceptibility.

References

Anagnostakis S.L., 1987. Chestnut blight: the classical problem of an introduced pathogen. *Mycologia*, 79 (1), 23–27.

Anagnostakis S., Aylor D., 1984. The effect of temperature on growth of *Endothia (Cryphonectria) parasitica in vitro* and *in vivo*. *Mycologia*, 76 (3), 387–397.

Anderson P.K., Cunningham A.A., Patel N.G., Morales F.J., Epstein P.R., Daszak P., 2004. Emerging infectious diseases of plants: pathogen pollution, climate change and agrotechnology drivers. *Trends in Ecology and Evolution*, 19 (10), 535–544.

Benson D.M., 1982. Cold inactivation of *Phytophthora cinnamomi*. *Phytopathology*, 72, 560–563.

Bergot M., Cloppet E., Pérarnaud V., Déqué M., Marçais B., Desprez-Loustau M.L., 2004. Simulation of potential range expansion of oak disease caused by *Phytophthora cinnamomi* under climate change. *Global Change Biology*, 10, 1–14.

Blodgett J.T., Bonello P., 2003. The aggressiveness of *Sphaeropsis sapinea* on Austrian pine varies with isolate group and site of infection. *Forest Pathology*, 33, 15–19.

Blodgett J.T., Kruger E.L., Stanosz G.R., 1997. *Sphaeropsis sapinea* and water stress in a red pine plantation in central Wisconsin. *Phytopathology*, 87 (4), 429–434.

Bourassa M., Bernier L., Hamelin R.C., 2005. Direct genotyping of the poplar leaf rust fungus, *Melampsora medusae* f. sp. *deltoidae*, using codominant PCR-SSCP markers. *Forest Pathology*, 35, 245–261.

Brown A., Rose D., Weber J. 2002. *Red Band Needle Blight of Pine. Information Note*, No. 49. Forestry Commission, Edinburgh, 6 p.

Burgess T., Wingfield M.J., 2002. Quarantine is important in restricting the spread of exotic seed-borne tree pathogens in the southern hemisphere. *International Forestry Review*, 4 (1), 56–65.

Burnett J., 2003. *Fungal Population and Species*. Oxford University Press, New York.

Coakley S.M., 1988. Variation in climate and prediction of disease in plants. *Annual Review of Phytopathology*, 26, 163–181.

Coakley S.M., Scherm H., Chakraborty S., 1999. Climate change and plant disease management. *Annual Review of Phytopathology*, 37, 399–426.

Collado J., Platas G., Pelaez F., 2001. Identification of an endophytic *Nodulisporium* sp. from *Quercus ilex* in central Spain as the anamorph of *Biscognauxia mediterranea* by rDNA sequence analysis and effect of different ecological factors on distribution of the fungus. *Mycologia*, 93 (5), 875–886.

Davis M.B., Shaw R.G., 2001. Range shifts and adaptive responses to Quaternary climate change. *Science*, 292 (5517), 673–679.

Delatour C., 1986. Problem of *Phytophthora cinnamomi* on red oak (*Quercus rubra*). *Bulletin OEPP*, 16(3), 499–504.

Desprez M.L., 1983. Étude des sources de variabilité et des mécanismes de résistance du pin maritime (*Pinus pinaster* Ait.) à la rouille courbeuse (*Melampsora pinitorqua* Rostr.). Thèse de doctorat, Institut national agronomique Paris-Grignon, Paris, 153 p.

Desprez-Loustau M.L., 2002. L'oïdium des chênes, une maladie fréquente mais mal connue. *Les Cahiers du DSF 1-2002. La Santé des Forêts [France] en 2000 et 2001.* Ministry of Agriculture, and Fisheries (DERF), Paris, pp. 95–99.

Desprez-Loustau M.L., Capron G., Dupuis F., Wagner K., 1998. Phenological synchronization in the twisting rust-maritime pine pathosystem. *In: Proceedings of the First IUFRO Rusts of Forest Trees WP Conference, 2–7 August 1998. Research Papers,* No. 712. Finnish Forest Research Institute, Saariselkä, pp. 149–155.

Desprez-Loustau M.L., Dupuis F., 1994. Variation in the phenology of shoot elongation between geographic provenances of maritime pine (*Pinus pinaster*) – Implications for the synchrony with the phenology of the twisting rust *Melampsora pinitorqua. Annales des Sciences Forestières,* 51, 553–568.

Desprez-Loustau M.L., Marçais B., Bergot M., Perarnaud V., Levy A., 2002. Zonage des risques d'encre sur chêne et implications pour la gestion. *Les Cahiers du DSF 1-2002.* La Santé des Forêts [France] en 2000 et 2001. Ministry of Agriculture and Fisheries (DERF), Paris, pp. 115–118.

Desprez-Loustau M.L., Marcais B., Nageleisen L.M., Piou D., Vannini A., 2006. Interactive effects of drought and pathogens in forest trees. *Annals of Forest Science,* 63 (6), 597–612.

Desprez-Loustau M.L., Wagner K., 1997. Influence of silvicultural practices on twisting rust infection and damage in maritime pine, as related to growth. *Forest Ecology and Management,* 98, 135–147.

De Villebonne D., 1997. Le chancre du châtaignier: les résultats préliminaires de l'enquête nationale suggèrent une nette progression du pathogène vers le Nord. *Les Cahiers du DSF,* 1, 33–34.

De Villebonne D., 1998. Le chancre du châtaignier en forêt. Situation en France. Résultats de l'enquête 1996–1997. *Les Cahiers du DSF,* 4, 26 p.

De Villebonne D., Maugard F., 1999. *Scirrhia pini:* un pathogène du feuillage en pleine expansion sur le pin Laricio en France. *Les Cahiers du DSF,* 1, 30–32.

De Wet J., Burgess T., Slippers B., Preisig O., Wingfield B.D., Wingfield M.J., 2003. Multiple gene genealogies and microsatellite markers reflect relationships between morphotypes of *Sphaeropsis sapinea* and distinguish a new species of *Diplodia. Mycological Research,* 107 (5), 557–566.

De Wet J., Wingfield M.J., Coutinho T., Wingfiel B.D., 2002. Characterisation of the "C" morphotype of the pine pathogen *Sphaeropsis sapinea. Forest Ecology and Management,* 161, 181–188.

Dupias G., 1943. Contribution à l'étude des Urédinées de la Haute-Garonne. *Bulletin de la Société d'histoire naturelle de Toulouse,* 78, 32–52.

Dupias G. 1965. Les rouilles des peupliers dans les Pyrénées et le bassin sous-pyrénéen. *Bulletin de la Société Mycologique Française,* 81 (2), 188–196.

Etterson J.R., Shaw R.G., 2001. Constraint to adaptive evolution in response to global warming. *Science,* 294, 151–154.

Evans H.C., 1984. The genus *Mycosphaerella* and its anamorphs *Cercoseptoria, Dothistroma* and *Lecanosticta* on pines. *Mycological Paper* No.153, 102 p.

Fleisch M.R., 2002. Vers une recrudescence de la maladie de l'encre du châtaignier en forêt ? *Les Cahiers du DSF 1-2002.* La Santé des forêts [France] en 2000 et 2001. Ministry of Agriculture and Fisheries (DERF), Paris, pp. 63–66.

Fragoso R.G., 1925. *Flora Iberica II. Uredales.* Museo Nacional de Ciencias Naturales, Madrid, 424 p.

Gadgil P.D., 1984. Dothistroma needle blight. *New Zealand Forest Service, Forest Pathology in New-Zealand,* No. 5 (revised 2001).

Gaudineau M., 1949. Les champignons parasites du pin maritime. *Cahier des Ingénieurs Agronomes,* 4e trimestre, pp. 3–11.

Gibson I.A.S., 1972. *Dothistroma* blight of *Pinus radiata. Annual Review of Phytopathology,* 10, 51–72.

Gibson I.A.S., 1974. Impact and control of *dothistroma* blight of pines. *European Journal of Forest Pathology,* 4 (2), 89–100.

Goudriaan J., Zadocks J.C., 1995. Global climate change: modelling the potential responses of agrosystems with special reference to crop protection. *Environmental Pollution,* 87, 215–224.

Goussard F., Saintonge F.-X., Géri C., Auger-Rozenberg M.-A., Pasquier-Barre F., Rousselet J., 1999. Accroissement des risques de dégâts de la processionnaire du pin, *Thaumetopoea pityocampa* Denis and Schiff *(Lepidoptera: Thaumetopoeidae)* en région Centre dû au réchauffement climatique. *Annales de la Société Entomologique Française (N.S.),* 35 (suppl.), 341–343.

Grant B.R., Byrt P., 1984. Root temperature effects on the growth of *Phytophthora cinnamomi* in the roots of *Eucalyptus marginata* and *E. calophylla. Phytopathology,* 74 (2), 179–184.

Graves, A.H., 1950. Relative blight resistance in species and hybrids of *Castanea. Phytopathology,* 40, 1125–1131.

Grente J., 1961. La maladie de l'encre du châtaignier. *Annales des Épiphyties,* 12, 25–59.

Guérin L., Froidefond G., Xu X.M., 2001. Seasonal patterns of dispersal of ascospores of *Cryphonectria parasitica. Plant Pathology,* 50, 717–724.

Guérin L., Robin C., 2003. Seasonal effect on infection and development of lesions caused by *Cryphonectria parasitica* (Murr.) in *Castanea sativa* (Mill.). *Forest Pathology,* 33, 223–235.

Hariot M., 1908. *Les Urédinées.* O. Doin, Paris, pp. 256–264.

Harrington, R., Fleming R.A., Woiwod I.P., 2001. Climate change impacts on insect management and conservation in temperate regions: can they be predicted? *Agricultural and Forest Entomology,* 3, 233–240.

Harrington T.C., Wingfield M.J., 1998. Diseases and the ecology of indigenous and exotic pines. *In: Ecology and Biogeography of Pinus* (D.M. Richardson, ed.). Cambridge University Press, Cambridge.

Harvell C.D., Mitchell C.E., Ward J.R., Altizer S., Dobson A.P., Ostfeld R.S., Samuel M.D., 2002. Climate warming and disease risks for terrestrial and marine biota. *Science,* 296 (5576), 2158–2162.

Houston D.R., Valentine H.T., 1988. Beech bark disease: the temporal pattern of cankering in aftermath forests of Maine. *Canadian Journal of Forest Research,* 18 (1), 38–42.

Ivory M.H., 1994. Records of foliage pathogens of *Pinus* species in tropical countries. *Plant Pathology,* 43, 511–518.

Jahn M., Kluge E., Enzian S., 1996. Influence of climate diversity on fungal diseases on field crops – evaluation of long-term monitoring data. *Aspects of Applied Biology,* 45, 247–252.

Jump A.S., Peñuelas J., 2005. Running to stand still: adaptation and the response of plants to rapid climate change. *Ecology Letters,* 8, 1010–1020.

Keen A., Smits T.F.C., 1989 Application of a mathematical function for a temperature optimum curve to establish differences in growth between isolates of a fungus. *Netherlands Journal of Plant Pathology,* 95 (1), 37–49.

Kurkela T., 1969. Leaf rust on aspen in Finnish Lapland. *Folia forestalia,* 64, 1–4.

Kurkela T., 1973a. Release and germination of basidiospores of *Melampsora pinitorqua* (Braun) Rostr. and *M. larici-tremulae* at various temperatures. *Communicationes Instituti Forestalis Fenniae,* 78 (5), 1–22.

Kurkela T., 1973b. Epiphytology of *Melampsora* rusts of Scots Pine (*Pinus sylvestris* L.) and aspen (*Populus tremula* L.). *Communicationes Instituti Forestalis Fenniae,* 79 (4), 1–68.

Lanier L., Joly P., Bondoux P., Bellemère A., 1976. *Mycologie et pathologie forestières. Tome II, Pathologie forestière.* Masson, Paris, 478 p.

Lawson A.B., 2001. *Statistical Methods in Spatial Epidemiology.* Wiley, New York.

Lonsdale D., Gibbs J.N., 2002. Effects of climate change on fungal diseases of trees, *In: Climate Change: Impacts on UK Forests* (M.S.J. Broadmeadow, ed.). *Forestry Commission Bulletin* No. 125, Forestry Commission, Edinburgh, pp. 83–97.

Marçais B., Bergot M., Perarnaud V., Levy A., Desprez-Loustau M.L., 2004. Prediction and mapping of the impact of winter temperatures on the development of *Phytophthora cinnamomi* induced cankers on red and pedunculate oak. *Phytopathology,* 94, 826–831.

Marçais B., Dupuis F., Desprez-Loustau M.L., 1996. Modelling the influence of winter frosts on the development of the ink disease of oak, caused by *Phytophthora cinnamomi. Annals of Forest Science,* 53, 369–382.

McDonald B.A., Linde C., 2002. Pathogen population genetics, evolutionary potential, and durable resistance. *Annual Review of Phytopathology,* 40, 349–379.

Moreau C., Moreau M., 1952. Mycological study of ink disease of oak. *Revue de Pathologie Végétale,* 31, 201–231.

Morelet M., 1967. Une maladie des pins, nouvelle pour la France due à *Scirrhia pini* Funk et Parker et à son stade conidien : *Dothistroma pini* Hulbary. *Bulletin de la Société Linnéenne de Lyon,* 8, 361–367.

Parmesan C., Ryrholm N., Stefanescu C., Hill J.K., Thomas C.D., Descimon H., Huntley B., Kaila L., Kullberg J., Tammaru T., Tennent W.J., Thomas J.A., Warren M., 1999. Poleward shifts in geographical ranges of butterfly species associated with regional warming. *Nature,* 399, 579–583.

Parmesan C., Yohe G., 2003. A globally coherent fingerprint of climate change impacts across natural systems. *Nature*, 421, 37–42.

Peterson G.W., 1973. Infection of Austrian and *ponderosa* pines by *Dothistroma pini* in Eastern Nebraska. *Phytopathology*, 63, 1060–1063.

Phillips D., Weste G., 1985. Growth rates of four Australian isolates of *Phytophthora cinnamomi* in relation to temperature. *Transactions of the British Mycological society*, 183–185.

Pinon J., 1986. Situation de *Melampsora medusae* en Europe. *Bulletin OEPP*, 16 (3), 543–546.

Pinon J., 1991. Éléments de répartition des rouilles des peupliers cultivés en France. *Comptes-rendus de l'Académie d'Agriculture de France*, 77 (2), 109–115.

Pinon J., Berthelot A., 2004. La prise en compte des problèmes sanitaires par le GIS Peuplier. *Les Cahiers du DSF*, 1-2003/2004 (la santé des forêts [en France] en 2002). Ministry of Agriculture and Fisheries (DERF), Paris, pp. 84–87.

Piou D., Chandelier P., Morelet M., 1991. *Sphaeropsis sapinea*, nouveau problème sanitaire des pins en France ? *Revue Forestière Française*, 42 (3), 203–213.

Robin C., Heiniger U., 2001. Chestnut blight in Europe: Diversity of *Cryphonectria parasitica*, hypovirulence and biocontrol. *Forest Snow and Landscape Research*, 76 (3), 361–367.

Saxe H., Cannell M.G.R., Johnsen O., Ryan M.G., Vourlitis G., 2001. Tree and forest functioning in response to global warming. *New Phytologist*, 149 (3), 369–400.

Shew H.D., Benson D.M., 1983. Influence of soil temperature and inoculum density of *Phytophthora cinnamomi* on root rot of Fraser Fir. *Plant Disease*, 67 (5), 522–524.

Sinclair W.A., Lyon H.H., Johnson W.T., 1987. *Diseases of Trees and Shrubs*. Cornell University Press, Ithaca.

Siwecki R., 1974. A review of studies on the occurrence of *Melampsora pinitorqua* in Central and Eastern Europe. *European Journal of Plant Pathology*, 148–154.

Smith D.R., Stanosz G.R., 1995. Confirmation of two distinct populations of *Sphaeropsis sapinea* in the North Central United States using RAPDs. *Phytopathology*, 85, 699–704.

Smith H., Wingfied M.J., Coutinho T.A., 2002. The role of latent *Sphaeropsis sapinea* infections in post-hail associated die-back of *Pinus patula*. *Forest Ecology and Management*, 164 (1–3), 177–184.

Somda B., Pinon J., 1981. Ecophysiologie du stade urédien de *Melampsora larici-populina* Kleb. et de *M. allii-populina* Kleb. *European Journal of Plant Pathology*, 11, 243–254.

Spiers A.G., 1998. *Melampsora* and *Marssonina* pathogens of poplars and willows in New Zealand. International symposium on analysing pathogen and pest populations in poplar and willow, IACR-Long Ashton Research Station, UK, 23–25 September 1996. *European Journal of Forest Pathology*, 28 (4), 233–240.

Stanosz G.R., Blodgett J.T, Smith D.R., Kruger E.L., 2001. Water stress and *Sphaeropsis sapinea* as a latent pathogen of red pine seedlings. *New Phytologist*, 149 (3), 531–538.

Sutherst R.W., Maywald G.F., Yonow T., Stevens P.M., 1999. *Climex – Predicting the Effects of Climate on Plants and Animals*. CSIRO Publishing, Canberra.

Swart W.J., Wingfield M.J., 1991 Seasonal response of *Pinus radiata* in South Africa to artificial inoculations with *Sphaeropsis sapinea*. *Plant Disease*, 75, 1031–1033.

Themann K., Werres S., Lüttmann R., Diener HA., 2002. Observations of *Phytophthora* spp. in water recirculation systems in comercial hardy ornamental nursery stock. *European Journal of Plant Pathology*, 108, 337–343.

Theurillat J.-P., Sclhüssel A., 2000. Phenology and distribution strategy of key plant species within the subalpine-alpine ecocline in the Valaisan Alps (Switzerland). *Phytocoenologia*, 30 (3–4), 439–456.

Thuiller W., Richardson D.M., Pysek P., Midgley G. F., Hughes G.O., Rouget M., 2005. Niche-based modelling as a tool for predicting the risk of alien plant invasions at a global scale. *Global Change Biology*, 11 (12), 2234–2250.

Van der Plank J.E., 1968. *Disease Resistance in Plants*. Academic Press, New York, 201 p.

Vannini A., Paganini R., Anselmi N., 1996. Factors affecting discharge and germination of ascospores of *Hypoxylon mediterraneum* (De Not.) Mill. *European Journal of Forest Pathology*, 26, 12–24.

Vannini A., Valentini R., 1994. Influence of water relations in *Quercus cerris-Hypoxylon mediterraneum* interaction, a model of drought induced susceptibility to a weakness parasite. *Tree Physiology*, 14, 129–139.

Vannini A., Vettraino A.M., 2001. Ink disease in chestnuts: impact on the European chestnut. *Forest Snow and Landscape Research*, 76 (3), 345–350.

Vegh I., Bourgeois M., 1975. Preliminary observations on the aetiology of decline in ornamental conifers in French nurseries. Role of *Phytophthora cinnamomi* Rands. *Revue Horticole*, 153, 38–49.

Vettraino A.M., Morel O., Perlerou C., Robin C., Diamandis S., Vannini A., 2005. Occurrence and distribution of *Phytophthora* species in European chestnut stands, and their association with ink disease and crown decline. *European Journal of Plant Pathology*, 111 (2), 169–180.

Walther G.R., Post E., Convey P., Menzel A., Parmesan C., Beebee T.J.C., Fromentin J.M., Hoegh-Guldberg O., Bairlein F., 2002. Ecological responses to recent climate change. *Nature*, 416 (6879), 389–395.

Weste G., Marks G.C., 1987. The biology of *Phytophthora cinnamomi* in Australasian forests. *Annual Review of Phytopathology*, 25, 207–229.

Woods A., Coates K.D., Hamann A., 2005. Is an unprecedented *Dothistroma* needle blight epidemic related to climate change? *Bioscience*, 55 (9), 761–769.

Worrall J., 1993. Temperature effects on bud-burst and leaf-fall in subalpine larch. *Journal of Sustainable Forestry*, 1 (2), 1–18.

Zentmyer G.A., 1980. *Phytophthora cinnamomi* and the diseases it causes. *Monograph*, No. 10. American Phytopathological Society, St Paul, 96 p.

Zentmyer G.A., 1981. The effect of temperature on growth and pathogenesis of *Phytophthora cinnamomi* and on growth of its avocado host. *Phytopathology*, 71, 925–928.

Zentmyer G.A., Klure L.J., Pond E.C., 1979. The influence of temperature and nutrition on formation of sexual structures by *Phytophthora cinnamomi*. *Mycologia*, 71, 55–67.

Zentmyer G.A., Leary J.V., Klure L.J., Grantham G.L., 1976. Variability in growth of *Phytophthora cinnamomi* in relation to temperature. *Phytopathology*, 66, 982–986.

Forests, Carbon Cycle and Climate Change

Chapter 13

Mediterranean forests, fire and carbon budget: the threats of global change

FLORENT MOUILLOT, SERGE RAMBAL, ROLAND JOFFRE, JEAN-PIERRE RATTE

Introduction

Climate affects terrestrial ecosystems by directly driving their net productivity and their intrinsic vegetation structure and dynamic. The disturbance regime, particularly fires, can play a major role, probably more than any other process. A shift in fire regime can lead to permanent changes in the composition of vegetation communities, and the subsequent ecosystem functioning as carbon, nutrient or water cycles. Fire is by far the most frequent and widespread disturbance to vegetation. Overpeck *et al.* (1990) wrote that "in northern mid-latitude forests, the chief form of non-anthropogenic catastrophic disturbance is fire". This is particularly true in Mediterranean-type ecosystems where large areas are burnt each year and fire frequency is high (Moreno *et al.*, 1998). Besides the human influence on fire settings, the occurrence of fires depends on drought that drastically increases flammability during summer, the temperature reached during this period and the amount of fuel biomass (Trabaud, 1976; Swetnam and Betancourt, 1990; Whelan, 1995). Climate-induced changes in the overall flammability of plant material and the production of available fuel may in turn alter fire frequency and intensity, which influence the species composition of the vegetation. For example, changes in the available fuel biomass can result either from changes in total biomass, redistribution between over- and understorey layers, changes in floristic composition, or from an abundant above-ground dead or partly dead biomass resulting directly or indirectly (e.g. insect dieback) from exceptional droughts. Thus, with climate-mediated disturbances such as fire, the indirect effects may result in very complex responses to climate change (Swetnam, 1993) (Figure 13.1). The carbon cycle associated with these changes may result in different sources and sinks among the mixed fire-prone Mediterranean landscapes.

This chapter presents the main characteristics of fire in the Mediterranean region (see "Fire characteristics in the Mediterranean basin"), their effects on the carbon budget (see "Carbon budget associated with fires in Mediterranean-type ecosystems") and reviews the possible impacts of global change (see "The threat of global change").

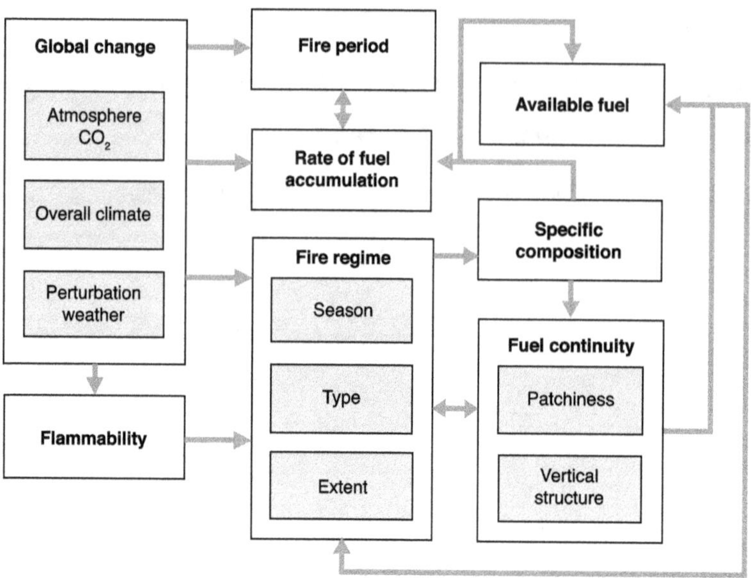

Figure 13.1. The main interactions between global change and the main components of fire risk in terrestrial ecosystems.

Fire characteristics in the Mediterranean basin

Statistics and history

About 0.62 million ha burn annually in the Mediterranean basin, representing 90% of the burnt areas in Europe and making this region the most fire-prone ecosystem in Europe. Based on the restricted area covered by Mediterranean-type ecosystems globally that fraction represents only 0.11% of the global burnt area (Mouillot and Field, 2005). Conversely, due to a long history of permanent agriculture, a high population and land-use density, and rainfall evenly distributed throughout the year, fires have never been important so far in northern Europe. The area burned was 0.02 million ha a year in 1900 and fell gradually to the current pattern of a quasi-total fire exclusion. Southern Europe (Albania, Italy, France, Greece, Portugal, Spain and the former Yugoslavia) has a Mediterranean climate with a prolonged summer drought, high temperatures and strong local winds, creating an extremely high fire risk. As a consequence, the moisture content of fine fuel decreases below 10%, greatly increasing ignition probability. Prevailing winds cause a further decrease in atmospheric humidity and also contribute to faster fire spread, crowning and spotting.

Despite the increase in fire prevention and suppression efforts during the last three decades, the surface area burnt has continued to grow markedly from 0.2 million ha per year in the 1960s to more than 0.6 million ha in the 1980s (Figure 13.2). The same trend is observed in various parts of the European continent (Pausas, 2004). It is unlikely that climate was fully responsible for this change as the southern part of the Mediterranean basin has not followed the same trend, despite being in the similar climatic area. Other factors are also important. The continuous urban-countryside interface and fuel management favours fire risk whereas increased prevention and suppression efforts have tended to reduce it. The sudden increase in area burnt cannot be explained by climatic parameters alone, and socio-economic causes also need to be considered. Much of the increase is a result of the relatively recent establishment of forest and shrubs on abandoned land, reflecting changes in agricultural policy (Moreira *et al.*, 2001; Mouillot *et al.*, 2003) and the decrease in rural population, which in Mediterranean countries decreased from 22.5 million in 1980, to 12 million in 1996 (Figure 13.3). In Spain, during the 1960s, industrial development involved depopulation of rural areas, decreased grazing pressure and wood gathering, and increased urbanization of rural areas (Le Houérou, 1992). The changes in traditional land use and lifestyle have implied the abandonment of large areas and farmland. Less exploitation, reduced grazing and the abandonment of wood harvest for fuel has led to an increase in accumulated fuel species in large parts of the Mediterranean landscape with large connectivity between fuel beds. This increase in large and continuous fuel beds together with the growth of population in the urban-countryside interface and the changing climate may explain the sudden rise in burned areas during the 1970s. The Mediterranean vegetation itself developing on abandoned lands has inherent chemical, physical and physiological properties that increase its flammability with age.

Factors controlling the spatial and temporal fire pattern

Besides the temporal trend observed in recent statistics, the surface area affected by fires particularly follows a strong inter-annual variability. Among the candidate mechanisms suspected of controlling a fire regime, temperature is the most considered. Even if we

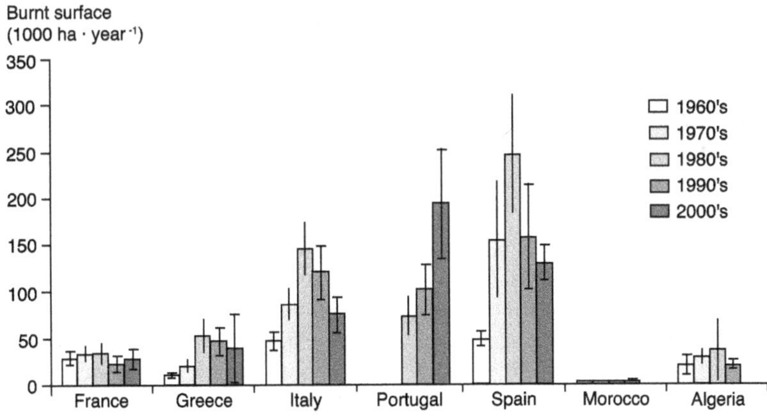

Figure 13.2. Burnt areas in the Mediterranean.

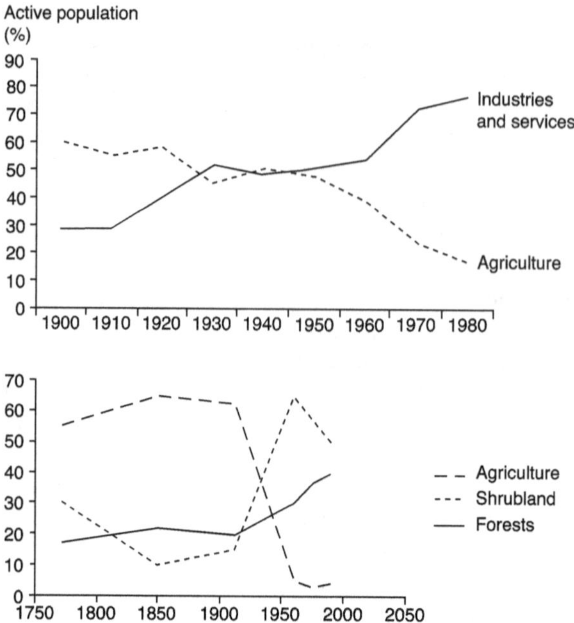

Figure 13.3. Land-use change in the Mediterranean from rural population (Pausas, 2004) and land cover statistics (e.g. in Corsica, Mouillot *et al.*, 2005).

consider that rainfall had not changed in the last five decades, the changes in temperature (i.e. 0.35°C per decade) (Pinol *et al.*, 1998; Lebourgeois, 2001) suggest an increase in ecosystem evapotranspiration and so drier conditions for plants, and in turn, fire hazard, which may partly explain the increase in the number of fires and area burnt. For example, the exceptional heat wave in 2003 significantly altered European carbon fluxes (Ciais *et al.*, 2005), and was coupled with a major fire season in Mediterranean Europe.

On the other hand, precipitation has tended to decrease over the last 50 years (5.2 mm·decade^{-1}). However, the high annual variability makes this possible trend imperceptible with the same observations in Greece (Amanatidis *et al.*, 1993), France (Lebourgeois *et al.*, 2001) and Italy (Piervitali and Colacino, 2003). If the climatic conditions became drier, fuel loads may also decrease and have significant feedbacks (Mouillot *et al.*, 2002), but summer rainfall remains an important factor in determining the area burned. During the last three decades, fires occurring in dry years tended to affect larger areas than fires in wet years (Mouillot *et al.*, 2003; Pausas, 2004). Rainfall is also cross-correlated with area burned with some delay (2 years), suggesting that high rainfall may increase fuel loads that burn 2 years later (Swetnam and Betancourt, 1998; Kitzberger *et al.*, 2001; Pausas, 2004). This is particularly true for herbaceous and shrubby vegetation (as savannas) rather than dense forests. Indeed, a positive relationship between fire and high humidity in prior years has also been reported for some ecosystems. For the Mediterranean, different drought indices based on rainfall amount and distribution during the season correlate with burnt areas (Figure 13.4). Also, at the landscape level, drier sites have often been observed as the most fire-prone such as, for example, southern aspects

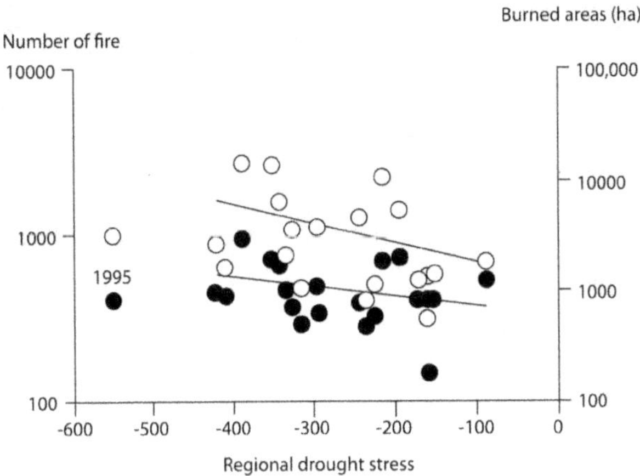

Figure 13.4. Number of fires (filled circles) and burned areas (empty circles) in the Languedoc-Roussillon region over the 1984–2005 period against the yearly integral of a drought stress index. More negative values mean increasing drought intensities. The trends between drought and both number of fires and burned areas have also been plotted (black lines). The 1995 data were excluded because the drought began in early spring during a cool period. Fire statistics: http://www.promethee.com/

and ridge tops with shallow soils (Figure 13.5, Plate 32), limiting water availability. Within a vegetation type, it has been clearly stated that fire risk is a function of stand age as a result of both the biomass accumulated and also the community composition reached at different stages of succession. The early stages of successions are more fire-prone (Mouillot *et al.*, 2002). The main reasons are their earlier desiccation due to shallower rooting depth and pioneer species accumulate fine fuel and dead biomass. In turn, the fire return interval in shrubland is around 20 years while it averages 60 years in forests and, as a consequence, 67% of area burnt in the Mediterranean affects shrubland only. This means that the relationship between fire and climate is not simple, and fire-spread processes are also driven by physical processes where large fires spread across landscapes irrespective of whether the vegetation is water-stressed or not.

Overall, the general hypothesis is that although fire ignitions may be determined by human factors, some of the inter-annual variability in the annual burnt area is explained by climatic parameters.

Fire ecology of Mediterranean species

Seeders versus resprouters

In a large part of the Mediterranean basin, two tree species coexist: a pine that is not able to resprout (Aleppo pine *Pinus halepensis* Mill.) and a resprouting oak (*Quercus ilex* L.) (Figure 13.6). Besides different management practices in the past, sites with high fire recurrence are losing their populations of pine but not oak. This is because the fire return interval in these places is shorter than the time that Aleppo pines need to produce viable seeds (15–20 years) or to accumulate a critical seed store for self-replacement. If

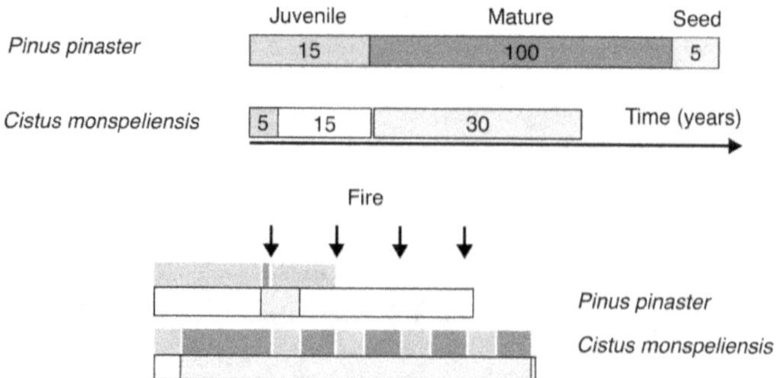

Figure 13.6. Concept of vital attributes for post-fire succession (Noble and Slatyer, 1980).

two consecutive fires occur in less than 20 years, the pine population may be eliminated. Although high fire recurrence or intensity may reduce the capacity to resprout from fires, most resprouting species will remain at the site after consecutive fires. For example, the large expansion of garrigues in the Mediterranean basin is often located on high fire recurrence areas where the shrub oak (*Q. coccifera* L.) resprouts vigorously after recurrent fires whereas pines are eliminated. In contrast other seeders such as *Cistus* spp. and *Ulex* sp. can be abundant if their age to maturity is less than an inter-fire period, or if their seed bank has not been exhausted by previous fires. Community composition in fire-prone ecosystems is not only determined by their regeneration strategies but also their life history traits in the case of repeated fires. Indeed, vegetation dynamics can be blocked at a given vegetation type when the seed bank is exhausted or has never been present due to long-standing agricultural activity.

Landscape processes

Following disturbances (e.g. land-use changes, natural disasters, etc.), shrubland dynamic is usually rapid. An efficient dispersal strategy, greater germination ability, better seedling survival and rapid growth to maturity have been widely identified as the traits favouring the large-scale invasion of shrubs at the expense of trees. However, shrubland persistence might largely depend on the lack of tree seed source than the result of internal dynamics. Besides ecological conditions, one major reason why a species is not present on a site is that no seed was able to reach that site (Pickett and White, 1985). Actually, the rate of forest area expansion is about 1% annually, with a slower dynamic on regularly burnt sites compared with unburnt sites. Some forest species have been shown to be weak invaders due to the shorter distance dispersal compared with anemochoric seeders and high seedling mortality due to soil summer drought in open areas. Indeed, the landscape pattern created by land uses or fires significantly affects forest colonization by seeders. Colonization occurs as a diffusion process from pre-disturbance forest patches and is more or less rapid according to dispersal abilities. Indeed, biomass regrowth also depends on the spatial structure of seed sources and the dispersal abilities of species. Therefore, at the landscape level, the accumulation of carbon stored in biomass can be slower than at the tree level. It includes two successive processes, that is, shrubland regrowth within 5 to 10 years, followed by a slower tree biomass accumulation.

Carbon budget associated with fires in Mediterranean-type ecosystems

General concepts

Until recently, studies on global fires aimed to quantify present trace gases by fuel combustion (Crutzen and Andreae, 1990; Hoelzemann *et al.*, 2004; Ito and Penner, 2004; Tansey *et al.*, 2004), but analyses of the global carbon budget generally ignored fires (Houghton, 2003). The general assumption was that fires have no net effect on a steady-state world, where losses due to combustion in some places were compensated for by gains due to biomass accumulation in others (Crutzen and Andreae, 1990). This steady-state view is relevant only if total carbon release from fires is constant over time. For fire-prone ecosystems, fire acts as a faster decomposing process, where all the debris and biomass is converted to CO_2 within a few days. When the disturbance regime is regular, the ecosystem can reach equilibrium on a long-term basis as far as vegetation is resilient enough to recreate the same vegetation type. When changes in the fire regime occur, fires can generate a carbon source or sink. Increasing fire frequency generates a net source of carbon, while decreasing fire frequency produces a net sink (Field and Fung, 1999; Chen *et al.*, 2000; Tilman *et al.*, 2000). The strength of the disequilibrium is determined by the carbon fluxes induced during the fire event. The amount of carbon emitted (E) directly to the atmosphere by a given fire is generally given by:

$$E = S \times \Sigma C_i \times B_i \tag{1}$$

where S is the burned area (m^2), C is the combustion efficiency (gC emitted·gC combusted), and B is the amount of available biomass ($gC \cdot m^{-2}$) of fuel types i (wood w, leaves l or litter lit). This simple equation is particularly true for homogenous areas with an even vegetation type and environmental conditions. However, for some vegetation types, a large proportion of fires affect only the ground layer, leaving trees alive and unburned. For these biomes, we usually introduce a tree mortality parameter M (i.e. percentage of burned area in which trees are burned) resulting in the following emission equation:

$$E = S \times C_l \times B_l + S \times C_{lit} + S \times C_w \times B_w \times M \tag{2}$$

This equation assumes that the whole leaf and litter carbon stocks are consumed whereas only M (%) of the wood is submitted to combustion processes. To account for fire effects on soil carbon, we include combustion of the soil carbon pools in the forested biomes (French *et al.*, 2002; Guild *et al.*, 2004; Soja *et al.*, 2004). Equation (1) has been adapted for more complex ecosystems at the $1° \times 1°$ resolution where global estimates are usually performed (Lavoue *et al.*, 2000). As calculations are mostly performed at a scale where vegetation types can be mixed within a reference unit (Lavoue, 2000), there is a growing need to know what do fires burn to estimate the completeness of the combustion (Cumming, 2001). The reason for such a need is that, in most ecosystems fires do not evenly affect the different fuel types available. For example, crown fires, affecting the barks and twigs, represent only 1% in savannas (Seiler and Crutzen, 1980; Hochberg *et al.*, 1994), 20% to 30% of fires in Russia (Shvidenko and Nilsson, 1996; Conard and Ivanova, 1997), but more than 90% in Canada (Kasischke *et al.*, 1995; French *et al.*, 2000). When they affect the crown, the completeness of the combustion is similar. To refine estimates of carbon emission due to fires, recent studies consider each biome as a mosaic of different vegetation types with different burning properties (forest, grassland and shrubland). Also each of these vegetation types has different fuel layers (i.e. canopy, surface and ground layers), which may burn differently. Consequently, for each grid cell, the burnt biomass M_c is given by:

$$M_c = \sum_{k=1}^{m} \left[(S_k \times A) \times \sum_{i=1}^{n} \left(\beta_{k,i} \times B_{k,i} \right) \right] \qquad (3)$$

where k is the number of vegetation types ($m = 4$ maximum), i is the number of fuel layers ($n = 3$ maximum), A is the area affected by fires (m²), S_k is the percentage of box area covered by the vegetation # k, $B_{k,I}$ is the biomass load (in g.m⁻²) of the fuel layer # i for the vegetation # k and $\beta_{k,i}$ is the combustion efficiency (in %) of the fuel # i for the vegetation # k. In such cases, regional or landscape studies link the vegetation maps with precise fire maps. Such improvements should be extended to the entire field burning. Indeed, landscape studies in fire ecology for mixed ecosystems such as the Mediterranean have recently emerged (Moreira *et al.*, 2001; Mouillot *et al.*, 2003).

In addition to the direct emissions during combustion, fires are also intrinsically coupled with indirect emissions from the decomposition of post-fire debris and significantly contribute to overall emissions related to fires. This process takes years or decades, for example, time for killed trunks to decompose. The amount of decomposing material (D) is then estimated as:

$$D = B \times S \times (1 - C) \qquad (4)$$

However, the three main parameters in Equation (1) for emissions are far from easy to collect and are subject to major uncertainties at the global or landscape level. These estimates based on the fraction of area and the biomass annually burnt were given with a factor of uncertainty of at least 3. Also, uncertainty in the assumptions of biomass density and combustion efficiency is still great and around $\pm$ 50% (Liousse *et al.*, 2004).

Burnt area (S)

Recent works illustrate the need to know what fires actually burn and, in turn, the importance of accurately delineating the area burnt where fires exactly occur. The knowledge of what fires burn is important for the amount of biomass consumed and how this biomass reconstitutes after fire depending on the vegetation type and species. Indeed, general statistics can be misleading regarding the carbon budget and vegetation dynamic. For example, when studying the spatial distribution of fires at the landscape level in the Mediterranean basin, statistics shift from 1 fire every 40 years, to 50% of the landscape subject to recurrent fires (5 to 12 years) and 50% never affected. This explains two contrasting patterns: (i) the development of forests in areas never burnt; and (ii) persistent shrublands at an early successional stage in areas subject to repeated fires. Mediterranean landscapes have then to be seen as more complex than a simple source of carbon during fire events followed by forest regrowth and carbon sequestration, as both mechanisms co-occur at the landscape level and early successional stages are more fire-prone. As a consequence, fires affect more or less biomass than the average biomass of the region, depending on what vegetation is affected.

Estimating burnt areas has been a growing concern in the last few decades for the developed countries for reasons of forest protection, expenditure and wealth lost estimations, and more recently, the impact on air pollution and global atmospheric circulation. Conversely, it was neglected for low populated regions such as tropical forest or regions where fires are a common aspect of the ecosystem processes as savannas, so that statistics are not evenly available across continents. For recent fires, we have the capacity to observe global fire activity from space at a range of spatial resolutions (Eva and Lambin, 1998; Dwyer *et al.*, 2000; Arino and Simon, 2001; Giglio *et al.*, 2003; Kasischke *et al.*, 2003; Van der Werf *et al.*, 2003; Simon *et al.*, 2004). Several sensors have been used for rapid and efficient detection of fires from the landscape to the global level (Levine, 2000). However, the satellite observations were mainly used for the detection of seasonal fire occurrence (Dwyer *et al.*, 2000), and not for estimates of the area burnt. This was due mainly to the spectral and geometric characteristics of the imagery used for these inventories, which were developed for meteorological applications and not for land surface applications. Other sensors use a variety of methods to estimate burned areas, including interpretation of smoke, unusually hot surface temperatures and post-fire scars on vegetation (Eva and Lambin, 1998), and even recent global studies based on remote sensing provide a range of estimates varying in their amount and spatial distribution (Boschetti *et al.*, 2004). The general output is a burnt area for a given region or country, or at best, a spatial representation of burnt areas at various spatial scales.

In the Mediterranean basin, there is a growing interest in estimating burnt surfaces, increasingly affecting the natural ecosystems since the 1970s, and we can rely on statistics for most of the countries at the national level (FAO, 2001). Initial statistics are delivered at the country level and can produce fine scaled statistics (Prométhee for France) (Vazquez and Moreno, 1995), but have not been compiled in a common database. In addition, they are established by foresters or fire fighters with different goals (that can lead to an over- or underestimation of the actual area burnt) and different perceptions of fires depending on the vegetation type affected or the minimum area burnt. There are also many fire types, not always registered, such as prescribed fires or agricultural fires. Finally, some countries reference only forest fires, occulting bush or grassland fires, the definition of which can vary from one country to another depending on the minimum tree cover considered to be a forest. In turn, we can observe significant differences within and between statistics available for countries.

Besides these observations in the field, ecological studies tend to use aerial photography for a better estimation of fire contours, but the 5 to 10 years time lag between successive campaigns is often too long to identify fire scars given the high fire frequency and the fast regrowth of vegetation. More recently, remote sensing tools tend to use classification methods to delineate fire boundaries and produce less biased fire statistics, but different algorithms or thresholds in the analysis can lead to overall accurate estimations of surfaces but different contour lines.

Combustion efficiencies (β)

Fire intensity is spatially highly variable and determines the combustion efficiency of biomass and soils, as well as post-fire regeneration of species. Fire intensity is the result of the amount of fuel biomass and its water status. Air temperature, wind speed and topography significantly modify fire spread, fire speed and, in turn, local fire intensity. The model mostly used is based on calorimetric measures and time of residence of fire. *Ir* (reaction intensity) is a function of biomass quantity, quality and structure. The rate of spread is then a function of *Ir* and local modifiers such as vegetation connectivity, slope, wind speed and direction, and vegetation water status. The actual intensity is $I = Ir \times R \times t$, where R is fire rate of spread and t is time. *Ir* determines combustion efficiency. The influence of wind and topography can be different from the initial model BEHAVE (Burgan and Rothermel, 1984) according to case studies (Weise and Biging, 1997).

Biomass

Uncertainty in the assumptions on combustion efficiency is still great and around ± 50%. Fuel composition influences combustion efficiency, which may control the emission factors. Trees are proportionally less combustable than shrubs due to the thinner structure of branches compared with the trunk, for example. A bibliography on combustion efficiency illustrates that trunks have a combustion efficiency of 20% to 30% in the boreal and up to 50% in the temperate forest, but the combustion efficiency of leaves and grass is almost 100%. For landscape-scale studies, estimations of fire intensity can be derived from field observations as bark or branch alterations (Perez and Moreno, 1998). For landscape-scale studies, remote sensing observation of normalized differential vegetation index (NDVI) before and after fires can provide a spatial representation of fire intensity within a fire contour line (Salvador *et al.*, 2000; Chafer *et al.*, 2004).

In grasslands particularly, fuel compaction and seasonal fuel moisture significantly affect combustion efficiency by a factor of 2 (Hoffa *et al.*, 1999). In the boreal forest, fuel moisture influences fire risk, but also drives fire spread and the subsequent combustion efficiencies of the litter and bark (Kasischke *et al.*, 2000). Dry vegetation is more flammable and combustion more complete. However, when fire spread is too fast, fires are less intense and allow some species to survive. Indeed, fire intensity is very variable and depends on intrinsic factors, but also spatial processes of fire spread such as wind speed and topography. Fire spread models integrate the major parameters, and there is growing interest in estimating seasonal vegetation water status from space.

Effects of fires on soil

A higher variability, however, happens on soils. The extent and duration of the effects of fire on soil properties depend firstly upon fire severity. The latter is in turn controlled by environmental factors that affect the combustion processes, such as the amount and nature of the above-ground biomass, the water content of live and dead fuel, air temperature and humidity, wind speed and the topography of the site. Fire severity consists of two components, that is, intensity and duration. Intensity is the rate at which a fire produces thermal energy. In some case, instantaneous values up to 700°C may be reached at the top soil surface. Temperature declines within the soil profile. The temperature profile largely depends on the soil physical properties, particularly the water content. Wet soils facilitate heat propagation. Heat is transferred faster and deeper. In such cases, the latent heat of vaporization prevents the temperature from exceeding 100°C. The most expected change soils experience during burning is the loss of organic matter. First, fire causes reduction or total destruction of the forest litter. Depending on fire severity, the impacts on the soil organic matter itself are very complex. Lowest temperatures (< 150°C) are practically ineffective. Highest temperatures (> 500°C) lead to complete oxidation. The main effects at intermediate temperatures have been detailed in González-Pérez *et al.* (2004) who considered two separate compartments in the soil organic carbon: soil free organic fractions and soil colloidal or humus fractions that comprise humic and fulvic acids. Oxygen-containing humic macromolecules are labile to heat. Part of the humic acids is transformed into humin. Fulvic acids are transformed in humic-like macromolecules and added to the newly formed aromatic and highly polymerized, or humic-like, compounds derived from the lignin and other brown products yielded by carbohydrate dehydration or dry distillation of plant residues. Concurrently, we observed the decrease in chain length of free lipids (i.e. fatty acids, resins and waxes) that contribute to the water repellence of the soil after wildfire. Other complex chemical reactions have been inventoried. Finally, there is also the production of a refractory and unalterable component, the so-called black carbon. All the above mentioned processes increase the stability of carbon substances to chemical and biological degradation. Hence, fire could be considered as a carbon stabilization factor with implications for the global geochemical cycle. Statistical analysis of results published in 48 studies by Johnson and Curtis (2001) yielded rather counter-intuitive results regarding the loss of organic matter. Their meta-analysis displayed a positive long-term effect of forest fires on the content of soil organic carbon. A significant (+ 8%) carbon increase in the A soil horizon was associated with fires that occurred more than 10 years earlier. Three reasons have been postulated: (i) the incorporation into the soil of charred residues protected against decomposition; (ii) the rearrangement of macromolecules of the soil organic material into more recalcitrant forms; and (iii) the increase of the occurrence of nitrogen-fixing understorey plants that enhance soil carbon sequestration. In another compilation summarizing results obtained for six Mediterranean forest ecosystems, except for a *Pinus halepensis* stand where a sharp decrease in soil carbon has been measured, Gonzalez-Perez *et al.* (2004) observed an increase of soil carbon suggesting a substantial incorporation of dead biomass. There is also the issue of so-called "black carbon". Black carbon is used as a synonym for soot, pyrogenic carbon and charcoal. It can be seen as a continuum from charred or incompletely burned plant material to graphite particles condensed from the combustion gas phase. This refractory material has been suspected to play a substantial part in the missing carbon of the global carbon budget. In the Andalusia region, under a semi-arid to sub-humid Mediterranean-type climate with a large occurrence of forest fires, the yearly amount of refractory material produced has been estimated to be more than 31,000 M g. Most of this black carbon remained as fire residues and will be further incorporated into the soil. About 850 M g has been emitted as soot into the atmosphere and, after a rather variable residence time, further precipitated into the Atlantic Ocean or the Mediterranean Sea (González *et al.*, 2002).

Available biomass before fires (*B*)

The development of vegetation and associated carbon fluxes after disturbances or land-use changes is now becoming of interest as a potential carbon sink in the biosphere. Many studies focus particularly on the replacement of previously cultivated or grazed areas by forest or shrubby vegetation, but a change in fire regimes can have the same effects. Due to the long time needed for carbon stock reconstructions, temporal series are rare and we use the diachronical analysis where stands are described at different ages. This method introduces some variability but allows studies on time series from decades to centuries. Based on these observations, biogeochemical models also simulate the dynamics of carbon stocks after disturbances.

The Mediterranean region is confronted by the concomitant processes of land abandonment and increase in burnt areas. Long-term vegetation dynamics since land abandonment give interesting results on the processes driving this long-term dynamic, such as border expansion for forest spread and fast invasion by shrubs. Post-fire vegetation dynamic has long been described as fast and giving rise to identical communities as before the fire (Trabaud and Lepart, 1981). However, multiple successive fires and fire intensity can mitigate this result. At the landscape level, short-term remote sensing studies can focus on post-fire vegetation response within a few months to a few years (Viedma *et al.*, 1997; Diaz-Delgado and Pons, 2001). These results suggest that fire intensity is not only important for the calculation of emission factors but also for species survival and, in turn, post-fire community composition.

Estimating biomass at the stand level has long been a major concern for foresters and ecologists (see Chapters 3 and 4). Diameter at breast height (DBH) for trees and crown area for shrubs are used to evaluate individual above-ground biomass. Ecosystem estimations require individual densities that are not easy to obtain at landscape, regional or global levels. Indeed, landscape maps are usually classified according to the dominance of species and vegetation layer (i.e. herbaceous, shrubs or forests) to differentiate open and closed canopy forest, for example. Remote sensing data as NDVI can describe foliage density and new products give an estimation of the percentage tree cover at the global or regional scale (De Fries *et al.*, 2000).

The estimation of soil carbon stocks and litter needs extensive field work, but remains of upmost importance for investigations into global carbon fluxes and stocks. The overall conclusion when compiling biomass estimations is that the high spatial variability depends on climate of course, but also on soil properties. Based on local observations, biogeochemical models estimate available carbon stocks in living biomass, litter and soil using climate data, soil type and vegetation type, and can be extrapolated at the landscape, continental or global levels. However, the lack of accurate data and models introduces large potential errors into these estimates. Improved monitoring and understanding of the extent of the landscape and the severity of fires and their effects on carbon storage, air chemistry, vegetation dynamics and structure, and forest health and productivity are essential to provide inputs into global and regional models of carbon cycling and atmospheric chemistry. For example, above-ground biomass estimations for Amazonia (Houghton *et al.*, 2001), boreal forest (Banfield *et al.*, 2002; Conard *et al.*, 2002) or temperate forest (Turner *et al.*, 2004) are still very variable as are estimations of soil carbon stocks (Krinner *et al.*, 2005).

However, biogeochemical models are useful for first estimates of the global carbon cycle and they are now being used for the calculation of fire emissions, and assume the biomass at equilibrium to be an accurate estimate, as well as the temporal dynamic of post-disturbance carbon reconstitution. At the landscape level, studies focusing on the estimate of carbon stocks and fluxes have to deal with the topo-climate and soil variability, which are difficult to apprehend. However, keystone events such as large fires in the tropical or boreal forest can be studied, for example, El Niño year 1997 in Indonesia (Liew *et al.*, 1998) or the fires in eastern Russian in 1987 (Cahoon *et al.*, 1994).

In Mediterranean-type ecosystems, biomass estimates on complex and fire-prone landscapes are difficult to estimate. Above-ground biomass is not directly linked with remotely sensed leaf area index (LAI) but depends on vegetation type and time since the last disturbance. Post-fire vegetation dynamic is rapid (within 4–5 years) and coupled with species replacement, so that vegetation maps are very dynamic and have to describe both the canopy level and the shrub/grass layer. Usual models do not deal much with response to disturbance, soil and climate and offer an estimate of biomass at the different layers of vegetation. Forestry indices commonly used in gap models (e.g. DBH) are generally of no use for resprouting species and shrubs, the most common vegetation type in Mediterranean type ecosystems. Some specific models now deal with biomass estimates and water status, as well as life traits. For example, SIERRA (Mouillot *et al.*, 2001; 2002) is a spatially explicit model, which combines processes determining vegetation dynamics and mechanisms affecting ecosystem scale flows. It takes into account the influence of space on the interaction between fluxes, ecosystems function and community dynamic. Space is considered by simulating the changes in soil and plant water status and plant cover at each grid-point of the landscape with implicit parameters such as soil and topographical properties and microclimate, and the spatial processes relating to adjacent cells such as runoff, seed dispersal and fire spread. Species are aggregated into cohorts and compete with each other for water and solar radiation. This model provides a basis for comparing the effects of climate on the coupled vegetation and fire dynamics. Carbon in litter and soils does not represent a large fraction of Mediterranean soils, but is usually considered according to the CENTURY model process. Based on these models, vegetation maps and remote sensing data, available biomass can be estimated in a spatially explicit manner and between the different layers from soil to canopy.

The threat of global change

Climate scenarios for the Mediterranean

Close to the Mediterranean Sea, early trends in temperature and rainfall changes are already observed as, for example, long periods of consecutive hot days without cool outbreaks during the dry season and increased water vapour into the warmer atmosphere associated with a reduction in stability leading to increased convection and variability in rainfall (Gordon *et al.*, 1992). Simulations of the future climate with an increased greenhouse effect give varying results depending on the general circulation model (GCM) used, although some general features appear in many of these simulations and provide the bases for "what-if" scenarios (see Chapter 6). The model of the Hadley Centre for Climate Prediction and Research of the UK Meteorological Office is a fully coupled atmosphere-deep ocean model with horizontal spatial resolution of 2.5° longitude × 3.75° latitude. ARPEGE is the Météo-France GCM at 50 km resolution. In both models, two runs are usually conducted: a control (CO_2 concentration at 320 ppm) and a perturbed model where greenhouse gas concentration is increased by 1% annually, which is close to the "business-as-usual" emission scenario. For grid-points close to the northern Mediterranean, we analysed the linear trends in deviations between both runs following Makridakis and Wheelwright (1978). Monthly warming of up to + 4°C for summer months and +1°C in winter are found. Daily potential evapotranspiration consequently follows an increase of 5% per °C according to McKenny and Rosenberg (1993). Under enhanced greenhouse conditions, GCMs have reported an increase in the frequency of convective rainfall and a possible decrease in less intense non-convective or large-scale rains. The predicted changes in rainfall intensity are more statistically significant and more spatially and seasonally uniform than changes in total amount (Pittock *et al.*, 1991). Actually, Gregory and Mitchell (1995) showed that, in the latitudes of Mediterranean-type climates of Europe, the average decrease in total annual rainfall is associated with a decrease in the number of rainy days. Precipitation is distributed over fewer days, but each with relatively larger amounts. As GCMs are not able to simulate the more intense rainfall events of the Mediterranean-type climate, the enhanced daily amounts are usually calculated from the observed ones assuming that the wet/dry sequences are not changed. As a consequence, the general pattern is a doubling in the frequency of occurrence of the 50 mm·day^{-1}, no change for the 20 mm class and a decrease in the amount of large-scale low intensity rainfall (Rambal and Debussche, 1990; Rambal and Hoff, 1998).

Consequences on vegetation functioning

The composition in dominant species under a changing climate is not drastically affected in the wettest northern part of the Mediterranean Sea. However, in extreme situations, some species may not develop leaves in the current year (Penuelas *et al.*, 2001), which may affect their flammability given that old leaves have lower water content and more energy content. Furthermore, prolonged drought can cause total or partial death of individuals with the consequent addition of dead material. A positive feedback would be the adaptation of tree density to water availability and limit tree death or water stress. Recent studies show some changes in biomass distribution between the main species but water fluxes seem to be more affected. For example, pine forest could experience a 15 mm decrease in transpiration and a corresponding increase in soil evaporation (Mouillot *et al.*, 2002). As a result the stress factor was increased by on average 1.53 equivalent week of stress per year. However, this increase can vary from 0 to 8 weeks, leading to more frequent high drought events rather than a systematic increase in light drought. We also observe a significant feedback of vegetation on the water stress index. There is an exponential decrease between summer rainfall and the water stress index, similar to the relationship between burnt areas and summer rainfall. Under a changing climate, this relationship follows the same exponential pattern but, for a same rainfall event, the water stress index is usually higher than current climate in the case of high summer rainfall and lower in the case of low summer rainfall. This illustrates that the water stress index under a changing climate does not follow the summer rainfall amount according to the model built under current climate. A negative feedback might occur in the response of the water stress index. A higher water demand under current climate due to a higher LAI increases water stress. That highlights the hydro-ecological balance that exists between rainfall amount and LAI in any given climate (Eagleson, 1982). The coupling in the change in rainfall pattern with temperature change leads to a new equilibrium with a higher drought stress.

Overall, there is little change in total annual rainfall between the climate scenarios, but monthly variations in the soil water balance reflect seasonal changes in rainfall patterns. The future climate expected in the Mediterranean basin will likely have little effect on the already poor productivity but will increase the length of the stress period as well as its frequency and, therefore, the risk of fire. This result supports the finding regarding the importance of inter-annual variability in determining fire events. It explains how water fluxes are sensitive to rainfall pattern and its inter-annual variability, but can be modified in the longer term due to changes in vegetation functioning. The observed increase is actually due to more frequent drought events than to the average increase in actual drought severity.

Impact on community composition

Indeed, the temporal pattern of fire events may be the major factor determining community composition. The species with shallower rooting depths such as shrubs present a higher stress index than trees but, might in turn, be less affected than trees adapted to greater water availability. Under changing climate the fire return interval might decrease from 72 to 62 years for forests and from 20 years to 16 years for shrublands. In turn, the enhanced fire frequency reduces the occurrence of pine forests as two successive fires close to each other prevents regeneration of pines and suppresses the community. This can lead to an expansion of the fast growing *Cistus* spp. At the landscape level, we observe nowadays a mixture of 28% of forest, 33% low maquis and 5% high maquis. Under a changing climate, forest would only account for 12%, whereas low maquis would cover almost 45% and high maquis 16% (Figures 13.7 and 13.8). The change in values caused by the different type of vegetation will probably involve less effort, both with regard to prevention and to alertness, which could accelerate the fire cycle (Trabaud and

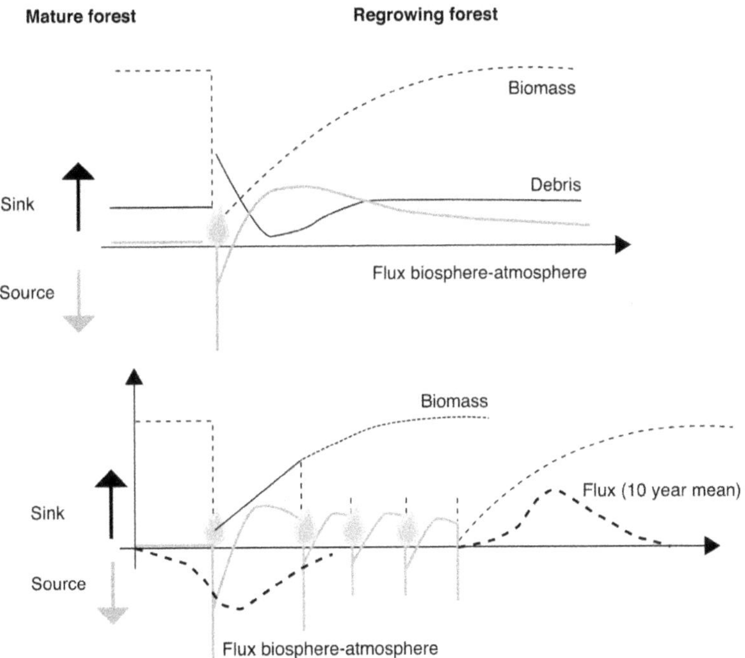

Figure 13.7. Time course of the stocks and fluxes of carbon associated with fires along successions compared between two contrasted fire regimes: low recurrence (upper); high recurrence (lower).

Galtié, 1996). Theoretically, the occurrence of fire is limited by fuel, so we could expect this process to decrease fire incidence, and higher fire fighting efficiencies would add to this process. However, fire appears to be mainly controlled by weather conditions, and in this case, many areas would be subjected to fires even in the early stages of succession, with the consequent risk of loss of soil fertility. Finally, it is difficult to forecast how situations caused by climate change will affect the people deliberately causing the fires. The persistence of high-risk situations will provide more opportunities for intentional fires. With regard to accidental fires, the higher climate-related danger could raise the chances of ignition sources ending in fire. To counteract this possibility, there might be a gradual improvement in information and training for the population, as well as greater awareness of the fire problem.

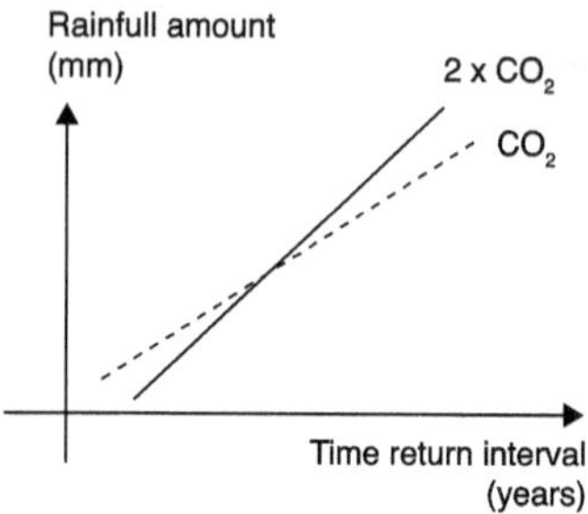

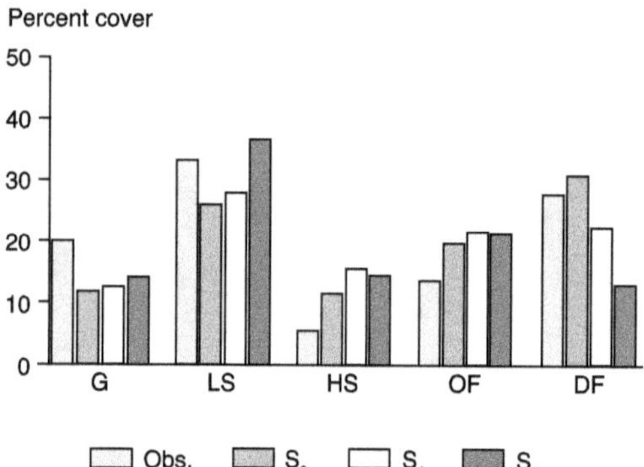

Figure 13.8. Effects of climate change on fire occurrence as time return interval of fire (above) and feedbacks on landscape composition (below) (Mouillot *et al.*, 2002). Climatic scenarios: S_1 involves a change in rainfall regime and S_2 changes in the rainfall regime associated with changes in temperature. S_0 refers to the current climate. Community type: G: grasslands and open shrublands; LS: low shrublands; HS: hight shrublands; OF: open forests; DF: dense forests.

References

Amanatidis G.T., Paliatsos A.G., Repapis C.C., Bartzis J.G., 1993. Decreasing precipitation trend in the Marathon area, Greece. *International Journal of Climatology*, 13 (2), 191–201.

Arino O., Simon M., 2001. The ERS-2 ATSR-2 world fire atlas and the ERS-2 ATSR-2 world burnt surface atlas project, *8th ISPRS Conference on Physical Measurements and Signatures in Remote Sensing, Aussois.*

Banfield G.E., Bhatti J.S., Jiang H., Apps M.J., 2002. Variability in regional scale estimates of carbon stocks in boreal forest ecosystems: results from West-Central Alberta. *Forest Ecology and Management*, 169 (1–2), 15–27.

Boschetti L., Eva H.D., Brivio P.A., Gregoire J.M., 2004. Lessons to be learned from the comparison of three satellite-derived biomass burning products. *Geophysical Research Letters*, 31 (21), DOI: 10.1029/2004GL021229.

Burgan R.E., Rothermel R.C., 1984. *BEHAVE: Fire Behaviour Prediction and Fuel Modelling System.* US Department of Agriculture, Forest Service, Ogden.

Cahoon D.R., Stocks B.J., Levine J.S., Cofer W.R., Pierson J.M., 1994. Satellite analysis of the severe 1987 forest-fires in northern China and southeastern Siberia. *Journal of Geophysical Research-Atmospheres*, 99 (D9), 18627–18638.

Chafer C.J., Noonan M., Macnaught E., 2004. The post-fire measurement of fire severity and intensity in the Christmas 2001 Sydney wildfires. *International Journal of Wildland Fire*, 13 (2), 227–240.

Chen J., Chen W., Liu J., Cihlar J., Gray S., 2000. Annual carbon balance of Canada's forests during 1895–1996. *Global Biochemical Cycles*, 14 (3), 839–849.

Ciais P., Reichstein M., Viovy N., Granier A., Ogee J., Allard V., Aubinet M., Buchmann N., Bernhofer C., Carrara A., Chevallier F., De Noblet N., Friend A.D., Friedlingstein P., Grunwald T., Heinesch B., Keronen P., Knohl A., Krinner G., Loustau D., Manca G., Matteucci G., Miglietta F., Ourcival J.M., Papale D., Pilegaard K., Rambal S., Seufert G., Soussana J.F., Sanz M.J., Schulze E.D., Vesala T., Valentini R., 2005. Europe-wide reduction in primary productivity caused by the heat and drought in 2003. *Nature*, 437 (7058), 529–533.

Conard S.G., Ivanova G.A., 1997. Wildfire in Russian boreal forests – potential impacts of fire regime characteristics on emissions and global carbon balance estimates. *Environmental Pollution*, 98 (3), 305–313.

Conard S.G., Sukhinin A.I., Stocks B.J., Cahoon D.R., Davidenko E.P., Ivanova G.A., 2002. Determining effects of area burned and fire severity on carbon cycling and emissions in Siberia. *Climatic Change*, 55 (1–2), 197–211.

Crutzen P.J., Andreae M.O., 1990. Biomass burning in the tropics: impact on atmospheric chemistry and biogeochemical cycles. *Science*, 250, 1669–1677.

Cumming S.G., 2001. Forest type and wildfire in the Alberta boreal mixed wood: what do fires burn? *Ecological Applications*, 11 (1), 97–110.

DeFries R.S., Hansen M.C., Townshend J.R.G., 2000. Global continuous fields of vegetation characteristics: a linear mixture model applied to multi-year 8 km AVHRR data. *International Journal of Remote Sensing*, 21 (6–7), 1389–1414.

Diaz-Delgado R., Pons X., 2001. Spatial pattern of forest fires in Catalonia NE Spain along the period 1975–1995: Analysis of vegetation recovery after fire. *Forest Ecology and Management*, 147, 67–74.

Dwyer E., Pinnock S., Gregoire J.-M., Pereira J.M.C., 2000. Global spatial and temporal distribution of vegetation fire as determined from satellite observations. *International Journal of Remote Sensing*, 21 (6–7), 1289–1302.

Eagleson P.S., 1982. Ecological optimality in water limited natural soil vegetation systems 1: theory and hypothesis. *Water Resource Research*, 18 (2), 325–340.

Eva H., Lambin E.F., 1998. Remote sensing of biomass burning in tropical regions: sampling issues and multisensor approach. *Remote Sensing of Environment*, 64, 292–315.

FAO, 2001. *Global Forest Fire Assessment, 1990–2000.* Food and Agriculture Organization of the United Nations, Rome.

Field C.B., Fung I.Y., 1999. The not-so-big US carbon sink. *Science*, 285, 544–545.

French N.H.F., Kasischke E.S., Stocks B.J., Mudd J.P., Martell D.L., Lee B.S., 2000. Carbon release from fires in the North American boreal forest. *Fire, Climate Change, and Carbon Cycling in the Boreal Forest*, 138, 377–388.

French N.H.F., Kasischke E.S., Williams D.G., 2002. Variability in the emission of carbon-based trace gases from wildfire in the Alaskan boreal forest. *Journal of Geophysical Research-Atmospheres*, 108, 8151, doi: 10.1029/2001JD000480.

Giglio L., Kendall J.D., Mack R., 2003. A multi-year active fire dataset for the tropics derived from the TRMM VIRS. *International Journal of Remote Sensing*, 24 (22), 4505–4525.

Gonzalez-Perez J.A., Gonzalez-Vila F.J., Almendros G., Knicker H., 2004. The effect of fire on soil organic matter – a review. *Environment International*, 30 (6), 855–870.

Gonzalez J.A., Gonzalez-Vila F.J., Polvillo O., Almendros G., Knicker H., Salas F., Costa J.C., 2002. Wildfire and black carbon in Andalusian Mediterranean forest. *In: Forest Fire Research and Wildland Fire Safety. Proceedings of IV International Conference on Forest Fire Research, 2002 Wildland Fire Safety Summit, Luso, Coimbra, Portugal, 18–23 November 2002*, p. 198.

Gordon H. B., Whetton P.H., Fowler A.B., Haylock M.R., 1992. Simulated changes in daily rainfall intensity due to the enhanced greenhouse effect: implications for extreme rainfall events. *Climate Dynamics*, 8, 83–102.

Gregory J.M., Mitchell J.F.B., 1995. Simulation of daily variability of surface temperature and precipitation over Europe in the current and $2 \times CO_2$ climates using UKMO climate model. *Quarterly Journal of the Royal Meteorological Society*, 121, 1451–1476.

Guild L.S., Kauffman J.B., Cohen W.B., Hlavka C.A., Ward D.E., 2004. Modeling biomass burning emissions for Amazon forest and pastures in Rondonia, Brazil. *Ecological Applications*, 14 (4), S232–S246.

Hochberg M.E., Menaut J.-C., Gignoux J., 1994. The influence of tree biology and fire in the spatial structure of the Western African savannah. *Journal of Ecology*, 82, 217–226.

Hoelzemann J.J., Schultz M.G., Brasseur G.P., Granier C., Simon M., 2004. Global Wildland Fire Emission Model (GWEM): Evaluating the use of global area burnt satellite data. *Journal of Geophysical Research-Atmospheres*, 109, D14S04, doi: 10.1029/2003JD003666.

Hoffa E.A., Ward D.E., Hao W.M., Susott R.A., Wakimoto R.H., 1999. Seasonality of carbon emissions from biomass burning in a Zambian savanna, *Journal of Geophysical Research-Atmospheres,* 104 (D11), 13841–13853.

Houghton R.A., 2003. Revised estimates of the annual net flux of carbon to the atmosphere from changes in land use and land management. *Tellus*, 55B, 378–390.

Houghton R.A., Lawrence K.T., Hackler J.L., Brown S., 2001. The spatial distribution of forest biomass in the Brazilian Amazon: a comparison of estimates. *Global Change Biology*, 7 (7), 731–746.

Ito A., Penner J.E., 2004. Global estimates of biomass burning emissions based on satellite imagery for the year 2000. *Journal of Geophysical Research-Atmospheres*, 109, D14S05, doi: 10.1029/2003JD004423.

Johnson D.W., Curtis P.S., 2001. Effects of forest management on soil C and N storage: meta analysis. *Forest Ecology and Management*, 140, 227–238.

Kasischke E.S., French N.H.F., Bourgeau-Chavez L.L., Christensen N.L., Jr, 1995. Estimating the release of carbon from 1990 and 1991 forest fires in Alaska. *Journal of Geophysical Research*, 100, 2941–2951.

Kasischke E.S., Hewson J.H., Stocks B., Randerson J., 2003. The use of ATSR active fire counts for estimating relative patterns of biomass burning – a study from the boreal forest region. *Geophysical Research Letters*, 30 (18), 1969.

Kasischke, E.S., Stocks B.J., Bourgeau-Chavez L.L., O'Neill K.P., French N.H.F., 2000. Direct effects of fire on the boreal forest carbon budget. *In: Biomass Burning and its Inter-relations with the Climate System* (J.L. Innes, M. Beniston, M.M. Verstraete, eds). Kluwer Academic Publishers, Dordrecht, pp. 51–68.

Kitzberger T., Swetnam T.W., Veblen T.T., 2001. Inter-hemispheric synchronic of forest fires and the El Nino Southern Oscillation. *Global Ecology and Biogeography*, 10 (3), 315–326.

Krinner G., Viovy N., de Noblet-Ducoudré N., Ogée J., Polcher J., Friedlingstein P., Ciais P., Sitch S., Prentice I.C., 2005. A dynamic global vegetation model for studies of the coupled atmosphere-biosphere system. *Global Biogeochemical Cycles*, 19, article 1015.

Lavoue D., 2000. Transport vers la région arctique de l'aérosol carbone émis par les feux de biomasse des régions boréales et tempérées. Thèse de doctorat, université Paris VII, 267 p.

Lavoue D., Liousse C., Cachier H., Stocks B.J., Goldammer J.G., 2000. Modeling of carbonaceous particles emitted by boreal and temperate wildfires at northern latitudes. *Journal of Geophysical Research-Atmospheres*, 105 (D22), 26871–26890.

Lebourgeois F., Granier A., Breda N., 2001. An analysis of regional climate change in France between 1956 and 1997. *Annals of Forest Science*, 58 (7), 733–754.

Le Houérou H.N., 1992. Vegetation and land use in the Mediterranean basin by the year 2050: a prospective study. *In: Climatic change and the Mediterranean: Environmental and societal impacts of climatic change and sea-level rise in the Mediterranean region* (L. Jeftic, J.D. Milliman, G. Sestini, eds). Hodder and Stoughton, London, pp 175–232.

Levine J.S., 2000. Global biomass burning: a case study of the gaseous and particulate emissions released to the atmosphere during the 1997 fires in Kalimantan and Sumatra, Indonesia. *In: Biomass burning and its inter-relationship with the climate system* (J.L. Innes, M. Beniston, M.M. Verstraete, eds). Kluwer, Dordrecht, pp 15–31.

Liew S.C., Lim O.K., Kwoh L.K., Lim H., 1998. A study of the 1997 fires in South East Asia using SPOT quicklook mosaics. *In: International geoscience and remote sensing symposium proceedings, Seattle. IEEE International 2, 6–10 July 1998.* IGARSS apos; 98, pp. 879–881.

Liousse C., Andreae M.O., Artaxo P., Barbosa P., Cachier H., Grégoire J.M., Hobbs P., Lavoué D., Mouillot F., Penner J., Scholes M., Schultz M.G., 2004. Deriving global quantitative estimates for spatial and temporal distributions of biomass burning emissions. *In: Emissions of Atmospheric Trace Compounds* (C. Granier, P. Artaxo, C.E. Reeves, eds). Kluwer, Dordrecht, pp. 71–114.

Makridakis S., Wheelwright S.C., 1978. *Interactive forecasting: Univariate and multivariate methods.* Holden-Day, Oakland.

McKenny M.S., Rosenberg N.J., 1993. Sensitivity of some potential evapotranspiration estimation methods to climate change. *Agricultural and Forest Meteorology*, 64, 81–110.

Moreira F., Rego F.C., Ferreira P.G., 2001. Temporal 1958–1995 pattern of change in a cultural landscape of Northwestern Portugal: Implications for fire. *Landscape Ecology*, 16 (6), 557–567.

Moreno J.M., Vazquez A., Llovet J., 1998. Recent history of forest fires in Spain. *In: Large forest fires* (J. M. Moreno, ed.). Backhuys, Leiden, pp. 159–186.

Mouillot F., Field C.B., 2005. Fire history and the global carbon budget: a 1 × 1 fire history reconstruction for the 20th century. *Global Change Biology*, 11, 398–420.

Mouillot F., Rambal S., Joffre R., 2002. Simulating climate change impacts on fire frequency and vegetation dynamics in a Mediterranean-type ecosystem. *Global Change Biology*, 8 (5), 423–437.

Mouillot F., Rambal S., Lavorel S., 2001. A generic process-based SImulator for meditERRanean landscApes (SIERRA): design and validation exercises. *Forest Ecology and Management*, 147 (1), 75–97.

Mouillot F., Ratte J.P., Joffre R., Moreno J.M., Rambal S., 2003. Some determinants of the spatio-temporal fire cycle in a Mediterranean landscape (Corsica, France). *Landscape Ecology*, 18 (7), 665–674.

Mouillot F., Ratte J.P., Joffre R., Mouillot D., Rambal S., 2005. Long-term forest dynamic after land abandonment in a fire-prone Mediterranean landscape (central Corsica, France). *Landscape Ecology*, 20 (1), 101–112.

Noble I. R., Slatyer R.O., 1980. The use of vital attributes to predict successional changes in plant communities subject to recurrent disturbances. *Vegetatio*, 43, 5–21.

Overpeck J.T., Rind D., Goldberg R., 1990. Climate induced changes in forest disturbance. *Nature*, 343, 51–53.

Pausas J.G., 2004. Changes in fire and climate in the eastern Iberian peninsula Mediterranean basin. *Climatic Change*, 63 (3), 337–350.

Penuelas J., Lloret F., Montoya R., 2001. Severe drought effects on Mediterranean woody flora in Spain. *Forest Science*, 47 (2), 214–218.

Perez B., Moreno J.M., 1998. Methods for quantifying fire severity in shrubland fires. *Plant Ecology*, 139, 91–101.

Pickett S.T.A., White P.S., 1985. Patch dynamics: a synthesis. *In: The Ecology of Natural Disturbance and Patch Dynamics* (S.T.A. Pickett, P.S. White, eds). Academic Press, New York, pp. 371–984.

Piervitali E., Colacino M., 2003. Precipitation scenarios in the Central Western Mediterranean basin. *In: Mediterranean climate variability and trends* (H.J. Bolle, ed.). Springer, Berlin, pp. 245–258.

Pinol J., Terradas J., Lloret F., 1998. Climate warming, wildfire hazard, and wildfire occurence in coastal eastern Spain. *Climatic Change*, 38, 345–357.

Pittock A. B., Fowler M.A., Whetton P.H., 1991. Probable changes in rainfall regimes due to enhanced greenhouse effect. *In: International hydrology and water resources symposium.* Institution of Engineers, Canberra, pp. 182–186.

Rambal S., Debussche G., 1990. Water balance of Mediterranean Ecosystems under a changing climate. *In: Global change and Mediterranean-type ecosystems* (J.M. Moreno, W.C. Oechel, eds). Springer-Verlag, New York, pp. 386–407.

Rambal S., Hoff C., 1998. Mediterranean ecosystems and fire: the threats of global change. *In: Large forest fires* (J.M. Moreno, ed.). Backhuys, Leiden, pp. 187–213.

Salvador R., Valeriano J., Pons X., Diaz-Delgado R., 2000. A semi-automatic methodology to detect fire scars in shrubs and evergreen forests with Landsat MSS time series. *International Journal of Remote Sensing*, 21 (4), 655–671.

Seiler W., Crutzen P.J., 1980. Estimates of gross and net fluxes of carbon between the biosphere and the atmosphere from biomass burning. *Climatic Change*, 2, 207–247.

Shvidenko A., Nilsson S., 1996. Fire and the carbon budget of Russian forests. *In: Fire, climate change and carbon cycling in the boreal forest* (E.S. Kasischke, B.J. Stocks, eds). Springer, New York.

Simon M., Plummer S., Fierens F., Hoelzemann J.J., Arino O., 2004. Burnt area detection at global scale using ATSR-2: The GLOBSCAR products and their qualification. *Journal of Geophysical Research-Atmospheres*, 109, D14S02, doi: 10.1029/2003JD003622.

Soja A.J., Cofer W.R., Shugart H.H., Sukhinin A.I., Stackhouse P.W., Mcrae D.J., Conard S.G., 2004. Estimating fire emissions and disparities in boreal Siberia 1998–2002. *Journal of Geophysical Research-Atmospheres*, 109, D14, pp. D14S06.1–D14S06.22 (2 p.1/2).

Swetnam T.W., 1993. Fire history and climate change in giant Sequoia Groves. *Science*, 262, 885–889.

Swetnam T.W., Betancourt J.L., 1990. Fire – Southern Oscillation relations in the Southwestern United States. *Science*, 249 (4972), 1017–1020.

Swetnam T. W., Betancourt J.L., 1998. Mesoscale disturbance and ecological response to decadal climatic variability in the American Southwest. *Journal of Climate*, 11, 3128–3147.

Tansey K., Gregoire J.M., Stroppiana D., Sousa A., Silva J.M.N., Pereira J.M.C., Boschetti L., Maggi M., Brivio P.A., Fraser R., Flasse S., Ershov D., Binaghi E., Graetz D., Peduzzi P., 2004. Vegetation burning in the year 2000: Global burned area estimates from SPOT VEGETATION data. *Journal of Geophysical Research-Atmospheres*, 109, D14S03, doi: 10.1029/2003JD003598.

Tilman D., Reich P., Phillips H., Menton M., Patel A., Vos E., Peterson D., Knops J., 2000. Fire suppression and ecosystem carbon storage. *Ecology*, 81 (10), 2680–2685.

Trabaud L., 1976. Inflammabilité et combustibilité des principales espèces des garrigues de la région méditerranéenne. *Oecologia Plantarum*, 11 (2), 117–136.

Trabaud L., Galtié J.F., 1996. Effects of fire frequency on plant communities and landscape pattern in the Massif des Aspres southern France. *Landscape Ecology*, 11 (4), 215–224.

Trabaud L., Lepart J., 1981. Changes in the floristic composition of a *Quercus coccifera* L. garrigue in relation to different fire regimes. *Vegetatio*, 46, 105–116.

Turner M.G., Tinker, D.B., Romme, W.H., Kashian, D.M., Litton, C.M., 2004. Landscape patterns of sapling density, leaf area, and aboveground net primary production in postfire lodgepole pine forests, Yellowstone National Park USA. *Ecosystems*, 7 (7), 751–775.

Van der Werf G.R., Randerson J.T., Collatz G.J., Giglio L., 2003. Carbon emissions from fires in tropical and subtropical ecosystems. *Global Change Biology*, 9 (4), 547–562.

Vazquez A., Moreno J.M., 1995. Patterns of fire occurrence across a climatic gradient and its relationship to meteorological variables in Spain. *In: Global change and Mediterranean-type ecosystem* (M.J. Moreno, W.C. Oechel, eds). Springer-Verlag, New York, pp. 408–434.

Viedma, O., Melia J., Segarra D., Garcia-Haro J., 1997. Modeling rates of ecosystem recovery after fires by using Landsat TM data. *Remote Sensing of Environment*, 61 (3), 383–398.

Weise D.R., Biging G.S., 1997. A qualitative comparison of fire spread models incorporating wind and slope effects. *Forest Science*, 43 (2), 170–180.

Whelan R.J., 1995. *The Ecology of Fire*. Cambridge University Press, Cambridge.

Abbreviations

ASR	annual survival rate
AUC	area under ROC curve
AWBG	above-ground woody biomass growth
BEF	biomass expansion factor
CEA	Agency for Atomic Energy
CIRAD	Centre for Agricultural Research for Developing Countries
CITEPA	Interprofessional Technical Centre for Studies on Atmospheric Pollution
CNRS	National Centre for Scientific Research
CSIRO	Commonwealth Scientific and Industrial Research Organization
CUE	carbon-use efficiency
DBH	diameter at breast height
DOY	day of the year
DSF	Département de Santé des Forêts
EI	ecoclimatic index
ETR	evapotranspiration
GCM	global circulation model
GI	annual growth index
GHG	greenhouse gases
GPP	gross primary production
HTOT	total height
IFN	National Forest Inventory

INRA	French National Institute for Agricultural Research
IPCC	Intergovernmental Panel on Climate Change
ISBA	Interactions Soil-Biosphere-Atmosphere
IUFRO	The Global Network for Forest Science Cooperation
LAI	leaf area index
MAE	mean absolute error
MAI	mean annual increment
MAT	mean annual temperature
MTR	mean time of residence
NBP	net biome production
NDVI	normalized differential vegetation index
NEE	net ecosystem exchange
NEP	net ecosystem production
NFI	National Forest Inventory
NPP	net primary production
PAR	photosynthetic active radiation
PAI	plant area index
PCA	principal component analysis
PET	potential evapotranspiration
PFT	plant functional type
PIT	percentage of infected trees
R	respiration
Reco	ecosystem respiration
R_a	autrotrophic respiration
Rh	heterotrophic respiration
REW	relative root extractable water
RMSE	root mean square error
ROC	receiver-operator characteristic curves
RPM	resistant plant material
SATMOS	Space Observation Meteorological Processing and Storage Service
SDR	standardized disease rates
SI	stress index
SOM	soil organic matter

SPV	Plant Health Service
SST	sea surface temperature
TOC	total organic carbon
VPD	vapour pressure saturation deficit
WH	water holding capacity
WUE	water-use efficiency

Contributors

Authors

Badeau Vincent, INRA, UMR INRA-UHP 1137 Écologie et écophysiologie forestières, équipe Phytoécologie, centre INRA de Nancy, F-54280 Champenoux, France. *badeau@nancy.inra.fr*

Bakker Mark, INRA, UR1263 EPHYSE, domaine de la Grande Ferrade, F-33140 Villenave-d'Ornon, France. *Bakker@bordeaux.inra.fr*

Balesdent Jérôme, INRA, UR1119 Géochimie des sols et des eaux, Europôle méditerranéen de l'Arbois, BP 80, F-13545 Aix-en-Provence Cedex 04, France. *Jerome.Balesdent@aix.inra.fr*

Belrose Valérie, DERF-DSF, 19 avenue du Maine, F-75732 Paris, France. *valerie.belrose@agriculture.gouv.fr*

Bergot Magali, Météo-France, direction de la Production, 42 avenue Gaspard Coriolis, F-31057 Toulouse Cedex, France. *magali.bergot@meteo.fr*

Bonal Damien, INRA, UMR 745 Écologie des forêts de Guyane, campus agronomique, BP 709, F-97387 Kourou Cedex, French Guiana. *bonal@kourou.inra.fr*

Bosc Alexandre, INRA, UR1263 Écologie fonctionnelle et physique de l'environnement (EPHYSE), centre de Bordeaux – Aquitaine, 71 avenue Édouard Bourlaux, F-33140 Villenave-d'Ornon, France. *Bosc@pierroton.inra.fr*

Capron Gilles, INRA, UMR 1202 BIOGECO, équipe Pathologie forestière, domaine de la Grande Ferrade, F-33140 Villenave-d'Ornon, France.

Caraglio Yves, CIRAD, UMR AMAP, TA 40/PS2, 2196 boulevard de la Lironde, F-34398 Montpellier Cedex 5, France. *yves.caraglio@cirad.fr*

Cazin Antoine, INRA, laboratoire d'étude des Ressources forêt-bois, UMR INRA/ENGREF 1092, 14 rue Girardet, F-54042 Nancy Cedex, France.

Chuine Isabelle, CNRS, UMR 5175 CEFE-CNRS, 1919 route de Mende, F-34293 Montpellier Cedex 5, France. *isabelle.chuine@cefe.cnrs.fr*

Ciais Philippe, laboratoire des Sciences du climat et de l'environnement, CEA-CNRS-UVSQ, centre d'études Orme des Merisiers, F-91191, Gif-sur-Yvette Cedex, France. *Philippe.Ciais@lsce.ipsl.fr*

Cloppet Emmanuel, Météo-France, direction de la Production, 42 avenue Gaspard Coriolis, F-31057 Toulouse Cedex, France. emmanuel.*cloppet@meteo.fr*

Cluzeau Catherine, Inventaire forestier national, 14 rue Girardet – CS 4216, F-54042 Nancy Cedex, France. *catherine.cluzeau@ifn.fr*

Colin Antoine, INRA, unité de recherche Écologie fonctionnelle et physique de l'environnement (EPHYSE), centre de Bordeaux – Aquitaine, 71 avenue Édouard Bourlaux, F-33140 Villenave-d'Ornon, France. *antoine.colin@ifn.fr*

Davi Hendrik, laboratoire Écologie systématique et évolution (ESE), UMR 8079, université Paris-Sud Orsay/CNRS/AgroParitech, département Écophysiologie végétale, université Paris-Sud XI, bât. 362, F-91405 Orsay Cedex, France. *davi@avignon.inra.fr*

Delage François, laboratoire des Sciences du climat et de l'environnement, CEA-CNRS-UVSQ, centre d'études Orme des Merisiers, F-91191, Gif-sur-Yvette Cedex, France. *francois.delage@lsce.ipsl.fr*

Delzon Sylvain, INRA, UR1263 Écologie fonctionnelle et physique de l'environnement (EPHYSE), centre de Bordeaux – Aquitaine, 71 avenue Édouard Bourlaux, F-33140 Villenave d'Ornon, France.

Present address: université de Bordeaux, UMR BIOGECO, avenue des Facultés F-33405 Talence. *s.delzon@ecologie.u-bordeaux1.fr*

Déqué Michel, Météo-France, CNRM, GAME, F-31057 Toulouse 1, France. *michel.deque@meteo.fr*

Desprez-Loustau Marie-Laure, INRA, UMR 1202 BIOGECO, équipe de Pathologie forestière, domaine de la Grande Ferrade, F-33140 Villenave d'Ornon, France. *Loustau@bordeaux.inra.fr*

Dhôte Jean-François, INRA, laboratoire d'étude des Ressources forêt-bois, UMR INRA/ENGREF 1092, 14 rue Girardet, F-54042 Nancy Cedex, France. *dhote@nancy.inra.fr*

Disnar Jean-Robert, Institut des sciences de la Terre d'Orléans (ISTO), université d'Orléans/CNRS/université François Rabelais-Tours (UMR 6113), campus Géosciences, 1A rue de la Férollerie, F-45072 Orléans Cedex, France. *jean-robert.disnar@univ-orleans.fr*

Drapier Jacques, Inventaire forestier national, 14 rue Girardet – CS 4216, F-54042 Nancy Cedex, France. *jacques.drapier@ifn.fr*

Dufrêne Eric, CNRS, laboratoire Écologie systématique et évolution (ESE), UMR 8079, université Paris-Sud Orsay/CNRS/AgroParitech, département Écophysiologie végétale, université Paris-Sud XI, bât. 362, F-91405 Orsay Cedex, France. *eric.dufrene@ese.u-psud.fr*

Dupouey Jean-Luc, INRA, UMR 1137 INRA-UHP Forest Ecology and Ecophysiology, INRA-Nancy, F-54280 Champenoux, France. *dupouey@nancy.inra.fr*

François Christophe, CNRS, laboratoire Écologie systématique et évolution (ESE), UMR 8079, université Paris-Sud Orsay/CNRS/AgroParitech, département Écophysiologie végétale, université Paris-Sud XI, bât. 362, F-91405 Orsay Cedex, France. *christophe.francois@ese.u-psud.fr*

Granier André, INRA, UMR INRA-UHP 1137 Écologie et ecophysiologie forestières, centre INRA de Nancy, F-54280 Champenoux, France. *agranier@nancy.inra.fr*

Guillet Bernard, Institut des sciences de la Terre d'Orléans (ISTO), université d'Orléans/ CNRS/université François Rabelais-Tours (UMR 6113), campus Géosciences, 1A rue de la Férollerie, F-45072 Orléans Cedex, France.

Hamza Nabila, IFN, direction de la Valorisation, BP 1001 Maurin, F-34972 Lattes Cedex, France. *nhamza@uep.ifn.fr*

Husson Claude, INRA, UMR1136 INRA/université Henri Poincaré, Interactions arbres-micro-organismes, F-54280 Champenoux, France. *Claude.Husson@nancy.inra.fr*

Joffre Richard, CEFE-CNRS, 1919 route de Mende, F-34293 Montpellier Cedex 05, France. *richard.joffre@cefe.cnrs.fr*

Kowalski Andrew S., INRA, UR1263 EPHYSE, 69 route d'Arcachon, F-33611.

Current address: Departamento de Física Aplicada, Universidad de Granada, Calle Fuente Nueva, S/N, 18071 Granada, Spain. *andyk@ugr.es*

Le Bas Christine, INRA, unité INFOSOL – INRA, centre de recherche d'Orléans, avenue de la Pomme de Pin, BP 20619 – Ardon, F-45166 Olivet Cedex, France. *Christine.Le-bas@orleans.inra.fr*

Le Maire Guerric, laboratoire Écologie systématique et évolution (ESE), UMR 8079, université Paris-Sud Orsay/CNRS/AgroParitech, département Écophysiologie végétale, université Paris-Sud XI, bât. 362, F-91405 Orsay Cedex, France. *guerric.le_maire@cirad.fr*

Lebourgeois François, INRA, UMR 1092, ENGREF, LERFOB, Écologie et écophysiologie forestières, F-54042 Nancy, France. *Lebourgeois@engref.fr*

Longdoz Bernard, INRA, UMR1137 INRA-UHP Écologie et écophysiologie forestières, centre INRA de Nancy, F-54280 Champenoux, France. *Longdoz@nancy.inra.fr*

Loustau Denis, INRA, UR1263 Écologie fonctionnelle et physique de l'environnement (EPHYSE), centre de Bordeaux – Aquitaine, 71 avenue Édouard Bourlaux, F-33140 Villenave-d'Ornon, France. *loustau@pierroton.inra.fr*

Marçais Benoît, INRA, UMR IAM, équipe de Pathologie forestière, centre de Nancy, F-54280 Champenoux, France. *marcais@nancy.inra.fr*

Meredieu Céline, INRA, UR1263 Écologie fonctionnelle et physique de l'environnement (EPHYSE), centre de Bordeaux – Aquitaine, 71 avenue Édouard Bourlaux, F-33140 Villenave-d'Ornon, France. *meredieu@pierroton.inra.fr*

Mouillot Florent, IRD-UR 060, CEFE-CNRS, 1919 route de Mende, F-34293 Montpellier Cedex 05, France. *florent.mouillot@cefe.cnrs.fr*

Nouvellon Yann, CIRAD, Persyst, UPR80, TA10/D, F-34398 Montpellier Cedex 5, France. *yann.nouvellon@cirad.fr*

Ogée Jérôme, INRA, UR1263 Écologie fonctionnelle et physique de l'environnement (EPHYSE), centre de Bordeaux – Aquitaine, 71 avenue Édouard Bourlaux, F-33140 Villenave-d'Ornon, France. *Ogee@bordeaux.inra.fr*

Pignard Gérôme, IFN, direction de la Valorisation, BP 1001 Maurin, F-34972 Lattes Cedex, France.

Present address: DDAF de l'Hérault, place Chaptal, F-34261 Montpellier Cedex 2, France. *gerome.pignard@agriculture.gouv.fr*

Piou Dominique, INRA, département Santé des forêts, domaine de l'Hermitage, Pierroton, F-33610 Cestas, France. Piou@pierroton.inra.fr

Porté Annabel, INRA, UR1263 Écologie fonctionnelle et physique de l'environnement (EPHYSE), centre de Bordeaux – Aquitaine, 71 avenue Édouard Bourlaux, F-33140 Villenave-d'Ornon, France. *porté@pierroton.inra.fr*

Rambal Serge, CEFE-CNRS, 1919 route de Mende, F-34293 Montpellier Cedex 05, France. *Serge.Rambal@cefe.cnrs.fr*

Ratte Jean-Pierre, CEFE-CNRS, 1919 route de Mende, F-34293 Montpellier Cedex 05, France. *jean-pierre.ratte@cefe.cnrs.fr*

Reynaud Grégory, INRA, UMR 1202 BIOGECO, équipe de Pathologie forestière, domaine de la Grande Ferrade, F-33140 Villenave-d'Ornon, France.

Robin Cécile, INRA, UMR 1202 BIOGECO, équipe de Pathologie forestière, domaine de la Grande Ferrade, F-33140 Villenave-d'Ornon, France. *Robin@bordeaux.inra.fr*

Saint-André Laurent, CIRAD, UPR80, Écosystèmes de plantation, F-34398 Montpellier, France. *standre@cirad.fr*

Saugier Bernard, Member of Académie d'agriculture de France, université Paris-Sud, Écologie, systématique et évolution, bât. 362, F-91405 Orsay Cedex, France. *Bernard.Saugier@u-psud.fr*

Shilong Piao, Department of Ecology, College of Urban and Environmental Sciences, and Key Laboratory for Earth Surface Processes of the Ministry of Education, Peking University, Beijing 100871, China. *slpiao@pku.edu.cn*

Ulrich Erwin, Office national des forêts, département Recherche, F-77300 Fontainebleau, France. *erwin.ulrich@onf.fr*

Vallet, Patrick, laboratoire d'étude des Ressources forêt-bois, UMR INRA/ENGREF 1092, 14 rue Girardet, F-54042 Nancy Cedex, France.

Present address: UR Écosystèmes forestiers, CEMAGREF, F-45290 Nogent-sur-Marne, France. *patrick.vallet@cemagref.fr*

Viovy Nicolas, laboratoire des Sciences du climat et de l'environnement, CEA-CNRS-UVSQ, centre d'études Orme des Merisiers, F-91191 Gif-sur-Yvette Cedex, France. *Nicolas.Viovy@cea.fr*

External reviewers

Augspurger Carol, University of Illinois, USA.

Broadmeadow Marc, Forest Research, Alice Holt, UK.

Dewar Roderick, INRA, Bordeaux, France.

Ellert Ben H., Agriculture Canada, Alberta, Canada.

Gibbs John, Forest Research, Alice Holt, UK.

Guldberg Annette, Danish Meteorological Institute, Copenhagen, Denmark.

Hänninen Heikki, University of Helsinki, Finland.

Hansen Everett, Oregon State University, Oregon, USA.

Kjellstrom Erik, Swedish Meteorological and Hydrological Institute, Norrköping, Sweden.

Lehtonen Alexis, Finnish Forest Research Institute, Vantaa, Finland.

Medlyn Belinda E., MacQuarie University, Australia.

Usoltsev Vladimir, Ural Branch of Russian Academy of Sciences, Ekaterinburg, Russia.

Zimmermann Niklaus, Swiss Federal Research Institute WSL, Birmensdorf, Switzerland.

Assistant editors

Bouchon Anne-Marie, INRA, UR1263 EPHYSE, France.

Hayes Stéphanie, INRA, UR1263 EPHYSE, France.

Reviron Marie-Pierre, INRA, UMR 1202 BIOGECO, France.

Imprimé pour vous par Books on Demand (Allemagne)